Advances in Plant Physiology

Advances in Plant Physiology

Editor
P.C. Trivedi
Head, Department of Botany and
Co-ordinator P.G. Course in Biotechnology
University of Rajasthan, Jaipur-302 004, India

I.K. International Publishing House Pvt. Ltd.
New Delhi • Mumbai • Bangalore

Published by
I.K. International Publishing House Pvt. Ltd.
S-25, Green Park Extension
Uphaar Cinema Market
New Delhi 110 016 (India)
E-mail: ik_in@vsnl.net

Branch Offices:
A-6, Royal Industrial Estate, Naigaum Cross Road
Wadala, Mumbai 400 031 (India)
E-mail: ik_mumbai@vsnl.net

G-4 "Embassy Centre", 11 Crescent Road
Kumara Park East, Bangalore 560 001 (India)
E-mail: ik_bang@vsnl.net

ISBN 81-88237-69-8

Published by Krishan Makhijani for I.K. International Publishing House Pvt. Ltd., S-25, Green Park Extension, Uphaar Cinema Market, New Delhi 110 016. Printed by Rekha Printers Pvt. Ltd., Okhla Industrial Area, Phase II, New Delhi 110 020.

Dedicated to renowned
Botanist & My Teacher
Prof. B. Tiagi

Preface

Plant Physiology is a "science concerned with processes and functions, the responses of plants to changes in environment and the growth and development that results from responses." A great deal of information about plant physiological processes has been gathered during the last century, largely because of the growth in these cognate sciences. The present era may be termed as the era of realistic approach for integrating the efforts of scientists working in different fields of specialisation to intensify their activities for sustainable development. Sustainable development embodies the sense of development in such a manner in which we could ensure the environmentally safe welfare activities. In the present scenario role of plant physiologists in environmental protection, in enhancing the agricultural and horticultural production in afforestation programme with reasonable to economic viability, in quality maintenance of horticultural and agricultural produce in elevating the export potential and also in meeting the requirements of certain industries relating to the production of agro chemicals etc is very important. Looking to the present need plant physiologists shall have to shed their habit of working in isolation. They cannot afford to be defamed as basic scientists. They will have to come forward with more vigour and enthusiasm to bridge the gap between traditional ecological knowledge, traditional farm practices and the modern agro-biotechnologies. In this age of very rapidly expanding information technology, plant physiologists also shall have to think globally, act globally as well as locally. They should make all round efforts to collect feedback from grass-root workers agriculturists, environmentalists and foresters. The holistic approach is the only solution to tackle problems relating to the protection of environment while maintaining the pace of development.

Plant physiologists shall have to be more active for finding constraints in low productivity for agricultural, horticultural to forest species especially in comparison with other countries like the U.S.A. If there is an interaction between genetics and environment responsible for such a situation, what role may be played by eco-physiologists and plant biologists to break the barrier. In the new millennium, we will have to produce more food and other agricultural commodities under the condition of diminishing per capita arable land and irrigation water resources and fast expanding biotic stresses. Enhancing the production of agricultural sector without taking care of environmental safety shall not be acceptable to society, as they will go against the concepts of sustainable development. For sustainable agricultural development, we should focus on productivity and our R & D priorities must be carefully formulated.

Advances in Plant Physiology contains 17 articles on different facets of Plant Physiology. The book embodies original and thought provoking review articles and research papers contributed by experienced and eminent specialists in their respective fields. Articles on diagnosis and management of physiological disorders; advances in physiology of vegetable crups; effect of sulphur dioxide on growth, photosynthesis and antioxidant enzyme activities; effect of herbicide on pollen germination; protected cultivation of vegetables in north Indian condition etc. have been reviewed covering latest information in the field. Topics on abiotic stresses and physiology of drought resistance; stress caused by pollutant and micronutrient stresses and their management have added to the value of the book. Articles on system of rice intensification—a physiological perspective; effect of different photoquality on germination and germination and early seedling growth; absolute growth and growth rate of guava of different maturities during the ontogeny provide information on role of different physiological processes. It also covers information on breeding efficient crops for dryland condition; physiology of groundnut; seed priming and potash nutrition as the management tools for drought investigation; interactive effect of ABA and cytokinin on mustard under water stress condition etc.

I am grateful to all the contributors for writing authoritative and informative articles for this volume. The publication of the present volume could not have been possible without the sincere co-operation and hard work of the contributors. I have tried to honour their ideas in the original shape, however, the onus of technical content rests with the contributors.

Advances in Plant Physiology is dedicated to prof. B. Tiagi in recognition of his contributions to Botany. His main contributions are in the field of plant embryology and plant nematology. Under his leadership Botany department of Rajasthan University received due recognition at national and international level. He has served for a long time on most of the University academic and administrative bodies. Prof. Tiagi is not only a true friend, guide and philosopher but above all, a wonderful human being. I am indebted to Prof. B. Tiagi for his able guidance and constant encouragement throughout my career. I wish him many more happy and fruitful years of service to the cause of botany and society.

I wish to thank my wife Kusum, daughter Priyanka and son Rohit for endurance during the compilation of the text.

The book will certainly be found useful by scientists, teachers, planners and administrators interested in disciplines like Botany, Agricultural Botany, Plant Breeding, Plant and Crop Physiology, Horticulture, Forestry and related industries.

I am confident that the book will be widely accepted by all students, teachers and researchers in the field of Plant Physiology.

P.C. Trivedi

Contents

Preface — *vii*

1. **Diagnosis and Management of Physiological Disorders in Subtropical Fruit Production** — **1**
   Gurcharan Singh and Harminder Kaur
2. **Breeding Efficient Crops and Varieties for Dryland Conditions** — **22**
   A. Henry
3. **Degree of Submergence Stress Tolerance of Lowland Rice at Varying Growth Stages** — **44**
   B. Haloi, P.C. Dey
4. **Effect of SO_2 on Growth, Photosynthesis and Antioxidant Enzyme Activities in Blackgram (*Vigna mungo* L. Hepper)** — **50**
   Sarvajeet Singh and Nafees A. Khan
5. **Stimulatory Effect of Herbicide (Atrataf 50W) on Pollen Germination and Tube Growth of Successive Flowers of Apocynaceae: A Critical Review** — **60**
   S.A. Salgare
6. **Abiotic Stresses and Physiology of Drought Resistance in Chickpea** — **70**
   S.C. Gupta
7. **Stress Caused by Nagaon Paper Mill Effluents on Growth, Development and Yield of Rice (*Oryza sativa* L. var. Mahsuri) Plant** — **95**
   Sarat Kumar Dutta

8. **Advances in Physiology of Vegetable Crops: A Brief Background** 116
S.K. Arora and Rakesh Mehra

9. **Protected Cultivation of Vegetables in North Indian Conditions** 139
S.K. Arora, A.K. Bhatia, S.P.S. Yadav and Rakesh Mehra

10. **System of Rice Intensification: A Physiological Perspective** 150
Anitha S. and Jose Mathew

11. **Study of the Interactive Effect of ABA and Cytokinin on Mustard under Soil Water Stress Condition** 163
Meera Srivastava

12. **Effect of Different Photoquality on Germination and Early Seedling Growth of three Varieties of *Vigna radiata* (L.) Wilczek** 170
Arvind Kumar

13. **Absolute Growth and Growth Rate of Winter Crop Guava cv. Sardar of Different Maturities during the Ontogeny** 174
K. Bhupinder, K. Nirmaljit, S. Tejinder, K. Harinder and K. Gursharan

14. **A Comparative Analysis of Vegetation and Soil Characteristics of Montane Broad-Leaved, Mixed Pine and Pine Forests of Northeast India** 185
B.P. Mishra and R.C. Laloo

15. **Macronutrient Stresses and Their Management in Crop Plants** 198
A.L. Singh

16. **Bambara Groundnut: Its Physiology and Introduction in India** 235
A.L. Singh and M.S. Basu

17. **Seed Priming and Potash Nutrition as the Management Tools for Drought Mitigation in Upland Summer Rice** 250
B. Haloi, R.K. Saud and P.C. Dey

Index *259*

CHAPTER 1

Diagnosis and Management of Physiological Disorders in Subtropical Fruit Production

Gurcharan Singh and Harminder Kaur

The quality of fresh fruits offered to consumers is constrained by the level of quality achieved at harvest and postharvest storage. To maintain the quality attributes of perishable fruits, it is necessary to check the preharvest and postharvest physiological disorders of the fruits. Physiological disorders may result from some adverse environmental conditions and may be associated with some varieties and not with others. Some disorders may affect only certain tissues while others affect the entire fruit. Some are visible at harvest while others occur only after a period of storage.

PREHARVEST PHYSIOLOGICAL DISORDERS

Preharvest physiological disorders result from adverse environmental conditions or from inadequate nutrition during growth and development of the fruit. Examples of adverse environments are frost, high temperature, limited or heavy irrigation. Plants require a balanced mineral intake for proper development of plant as a whole and fruits show various disorders due to mineral deficiencies. These disorders can be prevented by correct nutritional balance during growth or to some extent by postharvest application. Calcium has been associated more with deficiency disorders than other minerals e.g. water core or internal breakdown in apple, soft nose in mango and cork spot in pear. Calcium suppresses respiration and several other metabolic sequences in plant tissues leading to senescence. Calcium is also associated with pectic substances in the middle lamella and may prevent disorders by strengthening structural components of the cells.

*Address for correspondence: Department of Horticulture, PAU, Ludhiana

POSTHARVEST PHYSIOLOGICAL DISORDERS

Postharvest physiological disorders during storage result from postharvest handling and transportation stresses, chilling injury and high temperature injury.

Postharvest Handling and Transportation Stresses

Disorders due to stresses or bruises or injuries occurring in harvesting, handling, packaging and transportation of fruits from the orchard to the stores or markets lower quality and post-harvest life of fruits. Compresses or bruises result due to overfilling and staking of boxes one above the other. Higher the moisture content of the fruits, more susceptible they are to compression damage.

Impact bruising results either from the fruit drop or something hitting the fruits. Damage may be on the surface or internal. Similarly, vibration bruising occurs when the loosely packed fruits are transported.

Mechanical injuries occur in the epidermal oil cells due to rough handling and picking at advance stage of maturity.

Chilling Injury

Low temperature storage is the most effective tool used to maintain quality and extend shelf life of the fruits. However, many tropical and sub-tropical fruits develop chilling injury as a result of their exposure to low but nonfreezing temperature (generally 15°C down to 0°C). The symptoms most commonly seen are surface lesions such as pitting in pear, scald in citrus and papaya, and internal discoloration of avocado. Physiological changes that are associated with chilling injury are cell membrane alteration, changes in respiration rate and ethylene synthesis. These changes result in the loss of structural integrity and overall quality of the fruits. Increase in 1-aminocyclopropane-1-carboxylic acid (ACC) level or ethylene production have been shown to be a response of chilling sensitive tissue.

Chilling injuries may affect some species of the fruits while others are not influenced. Within a species, some genotypes are more susceptible than others. Within a genotype, some organs are more affected. Chilling injury generally becomes evident after the fruit is exposed to non chilling temperature.

Chilling injury is one of the most intensely studied postharvest disorder and affects many high value crops as citrus, banana, avocado, mango etc. The most obvious way to prevent chilling injury is to store the susceptible commodities above the temperature at which injury is incurred, though it may somewhat shorten the shelf life.

High Temperature Injury

High temperature injury can result from chronic exposure for days or weeks to temperature above typical room temperature (above 30 to 40°C) and acute exposure of short duration to high temperature. Much less is known about high temperature injury than chilling injury since the former can be prevented with normal handling practices. Short term injury can be induced during harvesting operations in warm climates and during storage at ambient temperature.

Banana and pear are among the crops shown to be susceptible to high temperature injuries. High temperature response has also been studied in apple and papaya and most extensively in tomato and pear which fail to ripen above 30°C. Since electrolyte leakage is observed above 40°C, it has been suggested that injury is mediated through membrane lipid.

A. PREHARVEST PHYSIOLOGICAL DISORDERS

(a) Climatic stress
(b) Nutritional disorders
(c) Interaction of plant growth factors, nutrition and growth regulators

B. POSTHARVEST PHYSIOLOGICAL DISORDERS

(a) Postharvest handling and transportation of fruits.
(b) Physiological storage disorders

C. MISCELLANEOUS PHYSIOLOGICAL DISORDERS

PREHARVEST PHYSIOLOGICAL DISORDERS

(a) Climatic stress

1. Spongy tissue in mango

In Alphonso variety, the incidence of this disorder is as high as 30 per cent.

Symptoms: A non-edible, yellowish, sour smelling sponge-like patch with or without air pockets develops in the mesocarp of fruit during ripening, predominantly at the distal end of fruit. Postharvest exposure of fruits to the sun during mid day causes maximum spongy tissue in the mid portion of the fruit. Affected tissue is visible only when the ripe fruit is cut.

Causes: Heat emission from soil as convective flux at fruit maturity keeps the starch unhydrolysed at the affected part due to biochemical disturbances in the pulp, consequently the pulp remains unripe.

Management: It is alleviated by mulching or sod culture that reduces the ground temperature as compared with clean cultivation/bare soil.

2. Black tip of mango/Black tip-Necrosis of guava

This is widely prevalent in Punjab, Uttar Pradesh, Bihar and West Bengal.

Symptoms: Development of a small etiolated area at the distal end of the fruit, which gradually spreads and turns nearly black and covers the tip completely.

Causes: Gases like CO_2, $SO_{2,}$ acetylene and fluorine emitted as fumes from adjacent brick kiln are the causal factors. Cultivars having dense lenticels on fruit surface are more susceptible. In guava, trees growing in the vicinity of factories producing S products prone to this disorder.

LVS show discolouration, necrosis of tips and margins, distal end of fruit turns black, resembling black tip of mango.

Management: (i) This disorder can be avoided by planting the orchard at a distance at least 1.6 km in east and west and 0.8 km in north and south of brick kiln. (ii) Chimney height of brick kiln 15 to 18 m minimises the damaging effects. (iii) Three sprays of Borax (0.6%) and Caustic soda (0.8%) each before and during flowering and at fruit set stage. These basic chemicals neutralise the acidic fumes of brick kilns.

3. Sun Scald and Sunburn (of leaves, bark and fruits in citrus and fig)

Symptoms: Sunburn injury occurs as whitish patches on leaves as in lemons; cracking and peeling off bark as in young fig and orange plants.

Causes: High temperature, desiccating winds, low relative humidity, long exposure to sun high transpiration and subsequently sunburn injuries. Alternating high temperature during day followed by low temperature after sunset causes sun scald injuries on exposed trunk bark of citrus trees. Heavy pruning in fig exposing the trunk and branches is responsible for sunburn.

Management: (i) Avoid heavy pruning in fig. (ii) Whitewash exposed parts. (iii) Planting of wind break. (iv) Do not plant lemons close to cemented walls in kitchen garden. (v) Light but frequent irrigation from April-June.

4. Purple/Reddish Brown spot/Wind Injury of Loquat

Symptoms: Purple/reddish brown spots develop in late ripening cultivars of Loquat during first fortnight of April and the fruits lose their succulence and become insipid in taste.

Causes: Abrupt shift in temperature from low to higher side coupled with low relative humidity and soil moisture and strong winds.

Management: (i) Plant early ripening cultivars. (ii) Light but frequent irrigation in April. (iii) Wind break/Inter planting in loquat. (v) Application of grey paper bags, black from inside with low light transmittance reduced the incidence.

5. Split pit and Gumming (Peach)

A serious quality defect affecting both cling and free stone peaches.

Symptoms: This disorder is serious where the downward translocation of food is interfered with so that top and twigs are high in carbohydrate. Cross diameter of the fruit is relatively greater in proportion than in normal. Embryo of split pit is aborted in a high per cent of cases but the development of flesh to maturity is normal. Fracture of the pit takes place mainly along the line corresponding to dorsal and ventral sides.

Abortion of embryo is associated with gumming at pit hardening stage either on the dorsal side opposite the ventral suture about one-third distance from distal end. Fruit falls after 7-10 days of gumming. No organism is associated with gumming.

Causes: Split pit mostly occurs at pit hardening stage. Favourable temperature and soil moisture that affect fruit growth contribute to potential split pit.

Management: Since pit size and fruit growth are genetic in nature as likely solution is through the genetic approach.

Fig. 1.1: Potassium deficiency in Plum (leaves and fruits)

(b) Nutritional Disorders

Nitrogen: Its excessive use or calcium deficiency can cause disorders in fruits.

1. Soft-nope of mango and peach/Tip pulp/Insidious fruit rot

Symptoms: Break down of flesh towards the apex of fruit before ripening is apparent through marked cell separation and cell wall disintegration in mesocarp cells. Cultivars differ in susceptibility to this disorder.

Fig. 1.2: Foliar magnesium deficiency in Peach

Causes: It increases with increase in N fertilization that reduces leaf calcium.

Management: This can be reduced by maintaining leaf Ca at 2.5% either by supplying Nitrogen as calcium nitrate or by adding lime stone or Gypsum.

2. *Bud-killing in grape*

Bud killing has been reported to be associated with poor productivity in Thompson Seedless.

Symptoms: About 50% of the differentiated buds are found shrivelled and brownish black in colour under microscopic studies. These fail to sprout.

Causes: Excessive nitrogen application may be the cause of this disorder.

Management: No direct association of fungus has been ascertained so far. Girdling and exogenous application of auxins have been found to reduce bud-killing in Thompson Seedless.

3. *Water berry in grape*

Symptoms: It refers to the watery, soft and flabby berries resulting from interrupted flow of sugars and other ripening metabolites into the berries.

Causes: Excessive application of nitrogen fertiliser and irrigation resulting over bearing are the causal factors.

Management: Judicious irrigation and fertilizers should be applied to manage this problem.

Phosphorus

1. *Bronzing and Marginal Scorching of Leaves (Peach, Plum and Pear)*

Common problem in low-chilling peach varieties under subtropical conditions.

Symptoms: The symptoms first appear in the rainy season; with leaves exhibiting burning purple margins. In acute cases the whole tree becomes dark purple causing leaf shedding.

Causes: High temperature of soil during rainy season ceases the uptake of phosphorus by the roots in soil of high pH resulting into its deficiency in leaves.

Management: Confirm the phosphorus deficiency by leaf and soil analysis and apply phosphate fertiliser according to age and vigour of the tree.

Potassium

1. *Fruit splitting/Cracking and Creasing (Lime, lemon and sweet orange)*

Symptoms: Fruit cracking takes place at both planes viz., longitudinally or transversely. Former splitting is more common.

Creasing is noticed in some varieties of oranges in albedo in the last stage of ripening. Flavedo sinks in and gives the fruit a creased appearance.

Causes: In addition to fluctuating soil moisture and atmospheric temperature, potassium nutrition is also associated with splitting of fruits via the general strengthening of the fruit rind. Similarly, creasing of the fruit is also mediated through potassium nutrition.

Management: (i) Apply light irrigation frequently (ii) Regular picking of fruits (iii) Check direct contact of fruits to soil. (IV) Fruit splitting and creasing can be checked by the application of K_2SO_4 [(6–8%) and KNO_3 (4–6%)] during May in lemon and oranges respectively.

2. *Scorching, Crinkling and Mottling of leaves (Peach, Plum, Pear, Grape and Mango)*

Symptoms: Marginal scorching and crinkling along the mid-rib of basal leaves, small leaves, upward curling and etched browning of the lower surfaces near the lateral margins, interveinal chlorosis in peach and plum on shallow light textured soils in summer months are the common symptoms.

Causes: K deficiency or chloride toxicity.

Management: (i) Follow the recommendations of K fertilisation as given for certain areas. (ii) Regulate moisture supply during summer season. (iii) In high pH soils, the application of Gypsum is advisable.

Calcium

Since Ca is not freely mobile, therefore, its deficiency first appears on shoot tips and at youngest leaves.

1. *Cork spot of pears*

Symptoms: It is a hard compressed tissue which is either more green or red than the surrounding appearing early in the development of fruit.

Causes: Low level of Ca in fruits has been related to this disorder. Calcium level in the normal pear fruit is 40–66 ppm while in the cork-spot fruit its range is 33–42 ppm.

Management: This storage disorder can be prevented by calcium salt sprays as chlorides nitrate or bi-phosphates on the trees or fruits. However, $CaCl_2$ consistently increased Ca concentration of fruits and reduced this disorders after storage.

2. *Blossom end rot of grapes*

Symptoms: Black sunken spots develop at the blossom end of the berry which later on spreads with water-soaked regions around it.

Causes: Defective Ca nutrition and assimilation appear to be the cause of this disorder.

Management: Spray of $CaNO_3$ @ 1% during mid fruit development period reduces this incidence.

3. *Stalk Necrosis (Grape)*

Symptoms: Dead areas appear on pedicels of flower or rachis of panicle that leads to the drying of cluster beyond dead areas.

Causes: This necrosis is associated with the low level of Ca in leaves.

Management: K is highly antagonistic to Ca uptake. Sprays of $CaCl_2$ or $CaNO_3$ during full bloom check this disorder.

Magnesium

1. *Islands/Herringbone Pattern of chlorosis/V-Pattern green Island at Base: (Peach, Plum, Grape, Mango, Citrus).*

Symptoms: V-pattern of green island, interveinal chlorosis followed by necrosis in citrus, peach, mango and plum leaves. In the basal leaves of grapes, reddish-yellowish leaf discolouration begins first from margins between the main veins, followed by 2/3 chlorosis of main veins from tip and margin side with a green pattern at base giving a herringbone appearance of chlorosis.

Fig. 1.3: Magnesium deficiency in grapes

:auses: Mg deficiency can occur in orchards in low Mg soil or soil with high Ca : Mg ratio or ιose with heavy K fertiliser.

Management: Mg deficiency can be corrected by the foliar spray of magnesium sulphate or xide or nitrate. In Florida, $Mg(NO_3)_2$ 1.26 kg/100 l of water as sprays are recommended for itrus orchards. Spray residue should remain on leaf surface for 72 hours or longer.

For proper control of Mg deficiency, excessive use of KCl should be avoided and N supply hould be adequate. Mg has synergetic effect on Zn and Mn content in leaves of citrus.

Aicronutrients

:inc

. *(a) Little leaf, Rosette; Mottle leaf, Frenched leaf/Frenching of Citrus, Peach, Plum, Pear and Grape*

(b) Bronzing in guava and litchi leaflets

ymptoms:

(a) Smalling of leaves, reduction in internodal distance, rosette formation at apical portion of twigs, interveinal chlorosis of newly developing small leaves and variegated colour of leaves are the common Zn deficiency symptoms in citrus.

Fig. 1.4: Zinc deficiency in citrus

These deficiency symptoms develop in deciduous fruits only on sandy, arid, and high pH soil or on sandy coastal plains. In grape the affected leaves exhibit widening of the basal lobes of blades from petiole, leading to a low fruit set that results in straggly clusters with shot berries. In citrus, the fruits with severe Zn deficiency are small, misshapen and lose their green colour prematurely.

(b) In guava, bronzed leaves show inter-veinal chlorosis of older leaves on a branch turning red to purplish-red. The terminal 1st or 2nd pair of leaves remains green. Incidence is severe during rainy season (up to 80% of trees show these symptoms) and very little during winter (up to 10% in December).

Causes:

(a) Zn deficiency in citrus, grapes and other fruits is responsible for "*Little leaf/Mottling/Rosette*" of leaves.

(b) Bronzing is a complex nutritional disorder due to P, K and Zn deficiencies. Water soluble P status of leaves is a better index in identifying bronzing due to nutrient deficiencies.

Management:

(a) To control Zn deficiency, give three sprays of $ZnSO_4$ each at 0.3–0.45% concentration depending upon the intensity of disorder during April-May, June and August-September.

(b) (i) To check bronzing apply phytosanitary measures in orchards (ii) Allahabad cv. is less susceptible than L-49. (iii) Apply recommended doses of balanced fertilizers and Zn as foliar sprays. (iv) Avoid planting guava in acidic soil or soil having high water table. (v) Avoid extremes of water stress.

Copper

1. *Die back, Multiple bud/Witch broom/Ammoniation, Exanthema, Red Rust, Gumming on rind, Tending of Cracking and Splitting (Citrus), Peach, Litchi*

Symptoms: Partial dying back of terminal shoots followed by buds growing below dead area result in bushy appearance of new growth. Leaves become chlorotic and small. Some species show brown/bronze areas and necrotic spots in acute cases. In citrus, gum pockets appear at nodes. Gum excretions on rind of fruits and a tendency of cracking/splitting can be seen in fruits.

Causes: Cu deficiency in affected trees is sure.

Management: Normally Bordeaux mixture, the usual formulation employed to control several fungal diseases in fruit plants is beneficial. It is very effective in correcting Cu deficiency in citrus. In case of orchards on mineral soils, a single application of Copper sulphate (0.2-0.5%) is enough. On organic soils, higher doses are recommended. Besides, copper chloride, nitrate and finely ground ores of copper can also be applied.

Boron

1. *Internal Fruit Necrosis (Aonla)*

Symptoms: Necrosis starts with the browning of the mesocarp in the fruits extending towards the upper surface and terminating into brownish-black areas on the fruit surface in 2nd and 3rd week of October. Finally pulp turns black, corky with gum pockets.

Causes: Problem is non-pathogenic and is due to B deficiency.

Fig. 1.5: Multiple branching of citrus due to copper deficiency

Management: Francis cultivar is highly susceptible (60-80%). Three sprays of Borax @ 0.6% starting from early September at 15 days interval are given to check this disorder.

2. Measels of Pear twigs

Symptoms: Irregular, superficial, dark or light coloured bark cankers slightly raised with uneven borders on young branches giving a somewhat flattened appearance to affected parts. A die-back of twigs follows and tree gradually dies unless B deficiency is corrected.

Fig. 1.6: Manganese deficiency in citrus (from R-L)

Causes: B deficiency is associated with measles development.

Management: Spray Borax @ 0.1-0.2% in early fall. It prevents B deficiency for 3 years. If soil pH is > 8.5 spray 236 g Borax/100 l of water/acre. Three sprays are suggested.

3. Blossom Blast (Pear)

Symptoms: Blossoms wilt, die/dry, fail to abscise and persist until the following year. Sometimes leaves may die and limbs may not put new buds within a few weeks after the blast. This disorder is random in distribution usually on heavy soil.

Causes: B-deficiency.

Management: Spray 236 g of Boric acid/100 l a week before full bloom.

4. External and Internal Cork or Drought Spot (Pear)

Symptoms

External Cork. It usually develops before the fruit is half-grown and soon after fruit set. Fruits take on a gnarled misshapen appearance caused by depressions, usually under laid by corky tissues which turn brown rapidly when exposed to air. In some cvs. cracking may occur if both Ca and B are low and may give a russetted look.

Internal Cork. It may develop any time from fruit set to harvest. Cork formation results from the large lesions with less defined margins in the large proportion of flesh. Texture of lesion is soft and spongy.

In acute case, the terminal buds fail to open and twig die-back occurs. Terminal leaves become malformed "boatlike". Fruit symptoms usually occur before vegetative symptoms are apparent.

Causes: Boron deficiency often associated with dry weather.

Management: B deficiency disorders can be corrected by the application of Borax, Boric acid, Sodium borate and colemanite. For spray application, Boric acid is preferred over Borax because of its greater availability of B. For soil application both are equally good.

To prevent early season B deficiency in pear spray Boric acid @ 600-1200 g/500 l

Iron

1. Chlorosis or Lime induced Chlorosis (Citrus, Peach, Grape, Plum, Pear)

Symptoms: Initially veins become darker green than interveinal area in young leaves. In more advanced stage the interveinal areas are yellow and eventually the entire leave may become of ivory colour. In severe cases, trees become partially defoliated, causing die-back and sometimes trees die.

In peach and plum tissue, breakdown occurs in patches on leave surface and patches of dead tissue drop away. The Fe deficiency increases the susceptibility of peach trees to *Bacterial canker and Short life.*

Causes: Light textured soils with high pH (> 7.0) containing high lime contents make the Fe fixed and unavailable to plant creating its deficiency in plants. High temperature, precipitation, fast vegetative growth in growing season and leaching loss of the element may also be the causes of this problem. In citrus, low temperature in winter may also result into Fe deficiency.

Management: Materials like Fe EDTA, Fe DDHA (chelates for soil application) have been used to correct Fe deficiency in citrus and grapes. Ferrous sulphate ($FeSO_4 7H_2O$) @ 0.3–0.5% is the commonly used measure to check chlorosis disorder. For rates higher than 0.3% add half quantity of unslaked lime. Sprays should be done only if Fe deficiency is observed in field or confirmed by leaf/soil test, since range between deficiency and toxicity of the element is very narrow.

Molybdenum

1. Yellow-spot/Orange-spot (Citrus)

Symptoms: These disorders develop in 3-4 months old leaves as translucent spots along their sides. These are irregular in contour and size and resemble another condition. Mesophyll collapses in mid to late summer. During fall affected areas impregnated with resinous gum which extrudes through lower surface of leaf and make reddish-brown spot forming a yellow halos around gum soaked areas.

Causes: MO deficiency in older leaves associated with interruption in NO_3 reduction to NO_2 in N assimilation. It is an indication of low soil pH and inadequate liming.

Management: Sodium molybdate/Ammonium molybdate @ 28 g/380 l within 2-4 weeks of deficiency are highly effective in correcting this disorder. Phosphoric fertilisers also contain MO in significant amount as impurity.

(b) Interaction of Plant Growth Factors, Nutrition and Growth Regulators

1. Granulation of citrus

It is a problem of citrus fruits harvested late in the season and during storage and affecting marketability of oranges particularly.

Analogy of Granulation: Selerocystosis, Dry-end in Florida, Crystallisation in California, Koa Saran in Thailand, Corkiness in West Indies.

Symptoms: It is a disorder of juice sacs of citrus fruits which are characterised by enlarged, hardened, and colourless or grayish colour of juice vesicles appearing first at the stem end. Cell walls in granulated vesicles become thicker and gelatinous thereby increasing fruit volume and decreasing seed number.

The alcohol insoluble fraction of juice vesicles increases with severity of granulation whereas sugars, organic acids and carotenoid decrease. The water is bound in gel and, therefore, not available as juice. Granulated fruits become tasteless.

Causes: Several factors influence its occurrence. Humid climate, high temperature in spring and low temperature in winter, high light intensity in grown up trees (10-12 years and above), older trees, declining young trees, vigorous trees, large fruit size, vigorous root stocks (rough-lemon karna khatta) and varietal susceptibility aggravating this disorder problem. This problem is reported to be serious in Hamlin, Mosambi, Blood Red, Dancy, Clementine, Thoronton and Pearl.

Management: (i) Picking of large size fruits early in the season is recommended. (ii) Decrease the amount and frequency of irrigation, it does not affect quality. (iii) Lime spray @ 18-20 kg/ 450 l of water; controls incidence. Do not spray during desiceating winds. (iv) NAA @ 100 ppm/2,4,-D @ 12 ppm control and delay the occurrence of agranulation. (v) Combination of Zn and Cu sprays are effective in checking the incidence. (vi) Avoid using rough lemon as rootstock.

2. Fruit Drop (Citrus, Mango) Non-pathogenic

Preharvest fruit drop is a common problem in sweet oranges. Mosambi and Blood Red are more prone to this drop. Drop may be pathological and physiological.

Symptoms (physiological drop): Fruit drop is due to the formation of an abscission layer at stem end. The dropped fruits look sound and without any incidence of pest or diseases.

Causes: (i) Formation of abscission layer at the stem point. (ii) Imbalance of growth regulators. (iii) Excess or deficiency of certain essential nutrients. (iv) Faulty cultural practices. (v) Unfavourable climate and soil high temperature, hails, lack of available soil moisture, high pH.

Management: (i) Regular and light irrigation during fruit setting and development. (ii) Application of growth regulators e.g. NAA @ 5-125 ppm or 2, 4-D at 10-20 ppm during pea stage of fruits in mango or 2 months before the harvesting of citrus fruits. (iii) 2,4, 5-T @ 5-30 ppm before the harvesting of citrus fruits. (iv) Application of $ZnSO_4$ @ 0.3 to 0.5% in September.

In mango, spray of NAA @ 40 ppm @ pre-bloom stage; CCC (Chlormequat) @ 200 ppm at pea stage and $AgNO_3$ @ 200 ppm at marble stage effectively checked the fruit drop.

3. *Fruit splitting/cracking (Lemon, Pomegranate, Loquat, Litchi, Fig)*

Symptoms: Splitting usually begins at the stylar end of the fruit, although it may develop mid way between ends. The peel/rind seemingly fails to enlarge as fast as the mesocarp/pulp/juice vesicles and develops one or more fissure.

Cracked fruits are also liable to be invaded by certain fungi and bacteria, thus become unfit for consumption.

Causes: (i) A combination of fluctuating environmental and soil factors e.g. high temperature, irregular irrigation, high soil moisture and low relative humidity thinness of peel at stylar end and varietal susceptibility (ii) Boron deficiency (iii) Sudden fluctuation in day and night temperature (Mrig Bahar crop of pomegranate).

Management: (i) Regular light irrigation (ii) Planting of wind break around northwest side of orchard. (iii) In pomegranate spray Borax @ 0.1%; GA_3 250 ppm in June. (iv) Pomegranate cultivars whose rind is thicker are less prone to cracking as Bedana, Jalore Seedless, Ganesh and Kandhari. In case of lemon, spray NAA @ 40 ppm and K_2SO_4 (6-8%) in May. (vi) In litchi, spray Boron @ 4% at pit hardening stage. Covering the fruit bunches in paper bags minimizes cracking by 4-18%. Calcutta and Hong Kong cultivars are less susceptible to cracking. (vii) Use mulching in orchards.

POSTHARVEST PHYSIOLOGICAL DISORDERS

Postharvest Handling and Transportation

Fruit picking, grading, packaging and transportation need careful handling since fruits on ripening become too soft to get easy injury that reduces eating and keeping quality of fruits. Some important fruits like guava, peach, plum and soft pears grown in subtropical climate are more tender than citrus, mango, grape, litchi, etc. which are likely to be bruised in packaging and transportation.

Compression bruising

In the packed fruits of guava, peach, plum and pears where the downward force on the fruits is above a threshold level can be bruised. Fruits lose their quality rapidly.

Causes: Compression bruising is the result of overfilling of boxes and then staking them so that the weight is supported by the fruits in lower boxes. It is commonly found in the lower layer of guava, mango, peaches and oranges in bulk stores.

Compression damage varies with the fruit species and cultivars and also related to the moisture content. Higher the moisture content more susceptible they are, which may be related to preharvest cultural practices.

Management: (i) To overcome this problem it may be necessary to redesign boxes and avoid overfilling. (ii) In fruits like mango, grape and orange certain level of compression bruising may be accepted to maximise the capacity of stores for economic reasons.

Impact bruising

This results either from the fruit being dropped or something hitting the fruits. Damage might be obvious on the surface or it might be internal. This damage can occur if the box is dropped or overfilled.

Symptoms: Internal blackening of mango and peach is common. Internal blackening is also temperature sensitive and occurs more frequently, when the pulp temperature is above 20°C. Impact bruising is commonly experienced during picking and sorting or collection of fruits in rooms/sheds for packaging. The fruits thus bruised suffer great losses during subsequent storage.

Management: (i) To avoid impact damage fruits must be handled with great care particularly where the fruits are harvested and handled mechanically. (ii) Chutes and cushioned pads should be used to collect the falling fruits. (iii) Distance of fall should be kept minimum.

Vibration bruising

This occurs when fruits are being transported especially in lorries. It is common where the fruits are packed loosely in boxes or in the lorry. This bruising is largely the result of fruits moving and impacting with each other or the walls of the box or lorry. It can result in an increase in the respiration rate of crop as well as surface bruising.

Management: To minimise vibration bruising the fruit crop needs to be packed tightly after grading. Fruits of each grade need to be packed separately in different boxes. Tight packaging reduces the movement effect.

Oleocellosis of Limes, Lemons and Oranges

Causes: Mechanical injury to the epidermal oil cells occurring during picking and handling procedures.

Symptoms: After mechanical injury, a blemish is left on the fruit surface, it degreens rapidly which turns to brown-black discolouration.

Management: (i) Handle the fruits gently. (ii) Fruits should not be picked while wet or damp. (iii) Overlapping of fruits in field boxes should be avoided.

Stylar-End Breakdown of Limes (SEB); Blossom end Clearing of Grape fruits

Causes: (i) Mechanical injury during rough handling of fruits. (ii) Picking at advanced stage of maturity. (iii) Picking in moist or humid weather.

Symptoms: Water soaked area close to stylar end which progresses until 1/3 – 1/2 of the rind is affected.

Management: (i) Avoid rough handling of fruits during picking, sorting and packaging. (ii) Picking at pre- and post-optimal stage of maturity should be checked. (iii) Avoid picking during moist or humid weather.

Brown Stem of Oranges

Causes: Postharvest water loss.

Symptoms: The brown blemish, depressed, hard and leathery touch around the button or stem-end of orange.

Management: Fruits in dry areas are resistant to Brown stem incidence than the groves located in high rainfall area. In fruit of dry areas cells in the peel are small, thick walled, with thick cuticle and sunken stomata and these have moisture conserving modifications.

Physiological Disorder in Storage

Chilling Injury

Chilling injury is different from freezing damage. It occurs at a temperature which is low but much above the freezing point of the tissue and in some cases damage may be observed even at a temperature as high as 14°C. This type of spoilage is more acute in subtropical fruits during storage or even in prolonged refrigerated shipment.

Symptoms: "Pitting" seems to occur in at least 60% of subtropical or tropical fruits. "Water Soaking" and failure to ripen properly more evident in thin skinned fruits e.g. papaya.

Table 1.1: Chilling injury symptoms of some subtropical fruits.

Fruits	*Approx. temp. at which chilling starts (°F)*	*Symptoms*
Citrus		
Grapefruit	Variable	Pitting of Flavedo, oil glands rarely elevated above the rest of depressed area, fairly uniform browning.
Lemon	50-53	Pitting of Flavedo, slow degree-ring, oil glands darker, pits reddish brown.
Lime	40-43	Pitting as brown sunken spots, on coalescing form irregular shaped patches.
Litchi	32-40	Dull skin, brown spots inner side of peel
Mango	40	Dull skin with brown patches, improper ripening, lack of sweetness.
Papaya	43	Skin pitting, water soaking of flesh, hydrolysis of sucrose to reducing sugars lacking, sensitivity to attack by *Alternaria* spp.
Guava	32-40	Dull skin, brown patches, lack of sweetness.
Pineapple and Valencia oranges	32-40	Pitting and excessive decay.

Source: Postharvest Physiology, Handling and Utilisation of Tropical and Subtropical Fruits and Vegetables. Ed. By EB. B. Pantastico (1975), Univ. of Philippines.

Control of chilling injury: There are no available means of minimising chilling injury of tropical and subtropical fruits other than (i) Temperature regulation/preconditioning (ii) Regulation of humidity – a humidity close to 100% ameliorated chilling injury but low humidity aggravated the symptoms. (iii) Waxing – waxed fruits of mandarins and grapefruit exhibit reduction in the degree of pitting. (iv) Controlled atmosphere (CA) storage. Modification of storage affects pitting. A concentration of 7% O_2 was optimal in preventing chilling injury. All these means comprise modest and readily available measures of reducing injury. A gradual reduction in temperature mitigates the effect rather than prompt cold storage. (v) Breeding for resistant lines of subtropical fruits can be a unique approach.

Chilling injury in mango can be prevented by storing fruits in CA (5% CO_2 + 10% O_2). This can be created in 4 kg. Carton lined; with a film – Micro-perforated polyethylene film (PE) or X tend film (XF). XF is most effective.

Individual seal packaging of lime, lemons, grapefruit and mandarin in thick film of high density polyethylene (HDPE) has been found to inhibit the development of chilling injury. It also heals small wounds before it turns to visible scar.

Bleaching of Grape-berries

Causes: SO_2 injury during fumigation to reduce the infection by molds.

Symptoms: Entry of too much SO_2 through lenticels damages the tissue resulting into the bleaching.

Management: Intensity and duration of SO_2 fumigation should be regulated.

Sweating of Cold stored fruits

Symptoms: Appearance of moisture on fruit surface after it is removed in cool room to ambient temperature directly.

Causes: Condensation of water vapours on the cold surface of fruits in the ambient environment.

Management: Sweating is difficult to overcome if produce directly comes from cool room at 0°C to ambient temperature. The extent of sweating can be reduced by placing produce at 7°C or 10°C room before shifting it to ambient temperature.

"Ethylene burn" of Temple oranges

Causes: Injury to peel by C_2H_4 related to the concentration and duration during degreening of fruits.

Symptoms: Appearance of burns on the peel after 2 days of the removal of fruits from degreening room.

Management: Prohibit degreening of Temple oranges.

Miscellaneous Physiological Disorders

Albedo Browning, Peteca, Membraneous stains, Endoxerosis of lemons

Symptoms: Albedo browning is darkening of soft tissues below the Flavedo. Peteca resembles pitting with round edges. Membrane stains is the browning of carpellary membrane between segments. Endoxerosis is internal breakdown and drying of tissues near the stylar end.

Causes: It is believed to be related to the moisture supply of tree and fruits.

Management: Proper irrigation and early picking of green lemons are the preventive measures.

Zebra sking of Tangerine

Symptoms: Dark sunken areas that follow the division between the fruit segments.

Causes: Related to the moisture regimes in the grove and degreening conditions. Incidence is high (a) In heavily irrigated orchards. (b) Degreening with high C_2H_4 level. (c) Mixing of coloured and green fruits in degreening. (d) Rough handling during polishing in packaging house.

Management: (i) Water regulation and regulation of degreening. (ii) Avoid mixing of green and coloured fruits during degreening and rough handling of fruits.

Puffiness of Mandarins

Affected fruits have loose, soft rind, may crack sometimes.

Causes: Loss of water from segments, resulting into shrinkage and separation from rind.

Management: Avoid prolonged dry storage and exposure to sun. Handle fruits carefully during packaging.

Sap-burn of Mango

Causes: Caustic latex exuding from the cut/broken pedicels during harvesting enters through lenticels of fruits on the skin causing enzymatic browning.

Management. Injury could be largely eliminated by removing stalks and allowing fruits to de sap in $Ca(OH)_2$ at 10% solution.

5. Internal Browning of Plum

Causes: Heat waves prior to harvest cause this disorder. Depletion of O_2 as a result of high respiration rate resulting into anaerobic respiration deep in the tissue leads to the beak-down, browning and death of tissue.

Fruit compression increases browning as tissue injury releases enzymes responsible for browning reactions.

Management: (i) Preharvest sprays of GA_3 and urea one month before harvest (ii) Picking the fruits still firm (iii) Preventing undue pressure and storing the fruit at low temperature at –1°C to 0°C.

Gum spot and stem end shrived in plum

Gum spots are induced by moisture stress in fruits and can be reduced by decreasing crop density.

Drought stress causes shrivelling of fruits at the stem side of fruit due to enhanced transpiration.

Black tip/Hard-end of Pear

This disorder appears in fruits on *Pyrus pyrifolia* rootstock.

Symptoms: Fruits approaching maturity become hard and black at the blossom end.

Causes: May be attributed to unfavourable water relations between fruits and tree limbs.

Management: Sweating is difficult to overcome if produce directly comes from cool room at 0°C to ambient temperature. The extent of sweating can be reduced by placing produce at 7°C or 10°C room before shifting it to ambient temperature.

REFERENCES

Bose T.K., S.K. Mitra and D. Sanyal (2002). Fruits, Tropical and Subtropical Vol.I & II Partha Sankar Basu. NAYA UDYOG Calcutta.

Chadha T.R. (2001). *Textbook of Temperate Fruits.* ICAR Publication. New Delhi.

Galletta G.J. and D.J. Himelrick (1990). *Small Fruit Crop Management.* Prentice Hall-Eaglewood Cliffs-New Jersey.

Mitra S.K., D.S. Rathore and T.K. Bose (1991). *Temperate Fruits.* Horticulture and Allied Publishers, Calcutta.

Pantastico B. (1975). *Postharvest Physiology, Handling and Utilization of Tropical and Subtropical Fruits and Vegetables.* Westport, Connecticut The AVI Publishing Co.

Sprague H.B. (1964). *Hunger Signs in Crops: A Symposium.* David Mckay Company Inc. New York.

Thompson A.K. (1996). *Postharvest Technology of Fruits and Vegetables.* Hartnolls Limited, Bodmin, Cornwell.

Westwood M.L. (1993). *Temperate- Zone Pomology Physiology and Culture.* Portland, Oregon Timber Press Inc.

CHAPTER 2

Breeding Efficient Crops and Varieties for Dryland Conditions

A. Henry

INTRODUCTION

Rainfed areas cover 75% of the total cropped area and account for 40-42% of the foodgrain output. So yield in drylands must increase. Even when the entire irrigation potential will be tapped, unirrigated area will continue to be very important. Almost all major crops have a significant area under rainfed conditions except sugarcane. The most important rainfed crops are cereals, millet, pulses, oilseed and cotton. It is found that 62% of the rice area and 30% of the wheat area are rainfed. More than 80% production of sorghum, pearlmillet small millets, pulses and oilseeds in India come from rainfed areas. Improvement in the productivity of these crops holds the key to the overall improvement of dryland agriculture.

Though in recent past technological developments have helped in increasing production in drylands through the choice of suitable crops and varieties, efficient cropping programmes, judicious use of fertilizers, water management and plant protection, the quantum jump, as witnessed under the irrigated conditions, is still awaited.

Evaluation and release of efficient and widely adapted varieties, particularly of cereals for the irrigated and intensive management conditions have played a pivotal role in revolutionising crop production. However, no concerted effort has so far been made to evolve high yielding and suitable varieties for rainfed areas. The varieties which are presently listed as dryland varieties were not specifically bred for such conditions. The general belief that the varieties which excel under intensive management conditions, would also do well under rainfed conditions, have been rather defied. The plant breeders face much stiffer challenge in evolving efficient varieties for dryland conditions because of extreme variability in the agro-climatic conditions of such stress areas.

*Address for correspondence: Central Arid Zone Research Institute, Jodhpur-342 003 (Raj.)

OBJECTIVES

The breeder's objectives would be to maximise and stabilse production/gallon water/time. The approach is to evolve genotypes, which make the most efficient use of the limited available water. In India, particularly in the northern and north-western plains, the *kharif* and *rabi* seasons pose distinct problems. In *kharif,* rainfall is often unevenly distributed and the moisture stress may coincide with the critical stages of the crop growth and cause maximum damage. In *rabi,* the precipitation is rather negligible and the farmer has to make use of the residual moisture in such a way that he harvests a respectable yield. Under such a situation the breeders must aim at evolving genotypes which respond favourably to congenial weather conditions and yield reasonably high even under stress conditions. In *rabi,* the farmers are also confronted with the problems of sowing early when soil moisture is adequate, but the temperature during the period is usually high. Such a combination of moisture and temperature, though ensures high germination percentage, invites a variety of problems of pests, diseases and thermo-sensitivity of the crops. Therefore, the breeder is confronted with the superimposed problem of breeding varieties which are as thermo-insensitive besides being highly resistant to diseases and pests.

From the above it is obvious that breeding crop varieties for dryland conditions is much more difficult than for conditions where inputs are controlled and the outputs are more or less assured. Therefore, the breeding objectives should be considered an integral part of the specific environmental conditions under which the improved varieties will be grown. In general, the objectives would be:

(i) Evolution of efficient plant type/types
(ii) Wide adaptation
(iii) Moderate level of fertiliser responsiveness
(iv) Tolerance, endurance, resistance, avoidance to water stress conditions.
(v) Superior quality of the economic product
(vi) Resistance to major diseases and pests
(vii) And tolerance to difficult soil conditions, e.g. salinity and alkalinity.

In view of the complexity of the environment a proper appraisal of the climatic, edaphic, biotic and cultural conditions of the production areas which may impose restriction of production should be made and the parameters defined before launching upon a breeding programme.

DROUGHT RESISTANCE

Drought resistance may be defined in a number of ways.

(i) **Drought resistance:** It is a total expression of plant ability to stay alive, grow and ultimately produce grain with at least part of its life cycle under water stress.
(ii) **Drought avoidance:** It is the ability of the plant to keep the water potential high either by absorbing water and conducting it to the sink, or retention of water losses by certain other physiomorphological adaptation.

(iii) **Drought tolerance:** The ability of cells surviving and metabolically functioning although the tissues are desiccated or are reduced with potentials.

BREEDING APPROACHES FOR DRYLAND CONDITIONS

The approaches may be classified in three groups:

(a) Physiological breeding: To increase efficiency of the plant population to make most efficient use of the solar and soil energy and the biotic factors so that total productivity as well as economic yields are maximised.

(b) Breeding for wide adaptability and stability.

(c) Removal of bottleneck genes: Breeding of varieties resistant to diseases and pests, improvement in nutritional and market quality of the ultimate products.

FEATURES OF VARIETIES SUITABLE FOR DRYLAND CONDITIONS

Plant Height and Early Vigour

Medium to medium tall varieties with high early seedling vigour are usually preferred so as to smother the weeds and to build up adequate biological yield. Extra dwarf varieties should be discouraged, varieties with moderately long, even droopy may have more plasticity in leaf rolling under stress conditions. It may be noted that this is in contrast to what constitutes an efficient plant type for intensive management conditions.

Branching Characteristics

Experiences have shown that in case of pulses and oilseed crops that moderately branched varieties consistently out yielded single stemmed varieties e.g. guar, til, etc.

Root/Shoot Ratio

Deep and branched root system is always preferred under dryland conditions. As a matter of fact positive correlations among root growth, stem height and seed weight have been revealed and since these traits are under genetic control, it would be possible to constitute genotype with above trait combinations suitable for rainfed conditions.

Earliness

Earliness confers escape from drought, however, it is negatively correlated with productivity. Efficient, early maturing genotypes with appreciable yielding ability which can match the rainfall pattern of the region are desirable. Early types are particularly useful for multiple/mixed cropping systems.

The plant type would depend on the climatic and edaphic environments under which a particularl crop is to be grown. To illustrate, if the rainy season is short and distribution of rainfall is uniform within the season, a short duration variety with low to moderate level of drought avoidance may likely escape drought and produce satisfactory yields. If rainfall distribution is abundant and uniform, any high yielding variety with medium late maturity and having low level of drought resistance might produce high yields when properly fertilised and weeded. But when the total rainfall is low and the distribution is rather erratic, an intermediate statured variety that is moderately branched, with deep and thick roots and other drought resistance properties, will provide reasonably good and stable yield under modest levels of management over the years. Therefore, the choice of an efficient plant type is more location specific for dryland conditions than for irrigated conditions. In all cases, however, the variety should have a high grain to straw ratio (harvest index) and a relatively high root to shoot ratio in order to yield efficiently.

Harvest Index

The proportion of the economic yield should be maximised per unit area/time /water.

Photo- and Thermo-insensitivity

For wide adaptability and multiple cropping a variety should be photo- and thermo-insensitive, the latter being most important for *rabi* crops, particularly those to be sown in October when temperature is still high.

Genotypes Suitable for Inter/Mixed Cropping

Selection of cultivator of different pulses and oilseed crops which are commonly grown as inter/mixed crops with Bajra, sorghum etc., the selection of both component genotypes should be done under actual conditions. Therefore, breeding for crop geometry would be another avenue for increasing water use efficiency and this increasing productivity.

Ear Characters

It has been shown that with adequate water supply the number of ears had the most potent influence on grain yield whereas under moisture stress, number of grains per ear and sometimes 1000 grain weight has as much effect as ear number. Since tillering capacity was not fully expressed under non-irrigated cultivation, and the main axis produce most of the ear population, its yield (grain number and grain weight) should have an important influence on total grain yield per unit area.

Leaf Growth

Under condition of moisture stress, it would be desirable to maintain leaf areas commensurate with minimal water loss and optimum photosynthetic output and growth.

Leaf Arrangement

Under non-irrigated conditions where a high plant population cannot be sustained, horizontally disposed leaves will intercept light better. It has been reported that horizontal leaves in wheat retain dew better and longer than vertically disposed leaves and, therefore, maintain a better water balance under non-irrigated conditions.

AN IDEOTYPE OF WHEAT FOR NON-IRRIGATED CONDITION

Asana (1969) suggested a model for wheat plant for barani cultivation, the main characteristics are described below.

Ear: It must have a large number of fertile spikelets, or a fertile branched ear, to provide for a large number of grains.

Peduncle: Its length should form a substantial proportion of the length of stem, because photosynthesis in the peduncle and flag leaf-sheath contributes to yield.

Leaves: About seven on main axis (with a large sized flag leaf), preferably horizontally disposed.

Root System: Must be substantial in the third and fourth foot of soil depth.

Time of earing: The ear must emerge at such a time that grains can develop for at least five weeks at a mean maximum temperature around 25°C. Asana and Saini (1962) also emphasized the importance" of seed number and 1000 grain weight under rainfed cultivation other attributes that will also be desirable under water stress are: long awns, hairy and light-green leaves, waxy stem etc.

PLANT TYPES FOR PULSES

As regard to pulses, they have been grown mostly in marginal lands characterised by moisture and fertility stress with the result that they have been selected for such primitive characteristics as a bushy, spreading, and in some cases indeterminate growth habit associated with prolonged maturity duration and non-synchronised development. Above all, they have a low harvest index. These characteristics are of considerable adaptive value in the dryland conditions but they also account for the fact that most of the energy and nutrients are diverted to the formation of vegetative parts rather than the grains. The total dry matter production in pulses like chickpea and pigeonpea are very high but only a small part of it is recovered in the form of grains. These traditional plant types/varieties are not able to take advantage of improved agronomic practices and they cannot be fitted in intensive cropping pattern.

The new strategy is, therefore, based on restructuring of pulses so as to evolve new plant type that can be fitted in a series of multiple and intercropping pattern with the cereal and other crops. The main effort in evolving these new plant types has been to reduce maturity duration to induce more synchronised development in terms of pod formation and to repartition the dry matter so that more of it will be diverted to the formation of seeds.

SCREENING VARIETIES FOR DROUGHT RESISTANCE

Many techniques have been developed and are modified in the past for measuring water stress in plants, however, no one method can be considered most reliable. For large scale screening it is suggested that under strictly rainfed conditions large population of germplasm or segregating materials may be raised at short intervals during the crop season, assuming that crop in each date of seeding will receive different amount of moisture. Genotypes which suffer least reduction in plant height, root number, root depth, dry matter production, harvest index and economic yield, would be considered drought tolerant/resistant. The threshold of the different cultivars can be studied by varying soil moisture in controlled fashion. Plant response which may indicate drought resistance, such as:

(i) Plasticity in leaf rolling and unfolding during stress
(ii) Death of lower leaves
(iii) Injury to upper leaf
(iv) Degree of stunted growth
(v) Delay in flowering and other, may serve as guidelines for selection.

Physiological parameters which supplement the screening are the following:

(i) Cuticular resistance
(ii) Resistance to potassium chlorate toxicity or water absorption power of germinating seeds in specific solutions.
(iii) The water retention capacity of excised plants
(iv) And the changes in the sugar content of the seedling.

Certain changes in amino acid metabolism and protein synthesis have been associated with resistance to moisture stress the concentration of soluble nitrogen compounds being increased. The increased accumulation of proline in leaf tissues has been reported under stress conditions in cereals. It may be emphasised that the screening based on physiological and biochemical parameters should be correlated with findinds from field experiments.

Accumulation of favourable genes adopting population improvement approaches coupled with multilocation testing, should prove effective in evolving desirable genotypes for dryland conditions.

RECENT ADVANCES IN THE BREEDING OF VARIETIES OF OILSEED SUITABLE FOR DRYLAND AREAS

Among nine oilseeds grown in the country groundnut and rapeseed-mustard together account for 62% of total oilseeds production. Total area under oilseeds was 25.26 million hectares with the annual production of 24.46 million tons (Anonymous 1997). The average productivity of oil seeds was 368 kg per ha. during 1951 while during 1995-96 it was 851 kg per hectare. The average productivity in India is very low as compared to the world average of 1262 kg per hectare.

The total oilseed production in Rajasthan is 3.1 million tons (Anonymous 1996-97) which is derived from groundnut, rapeseed-mustard, soybean, sesame, castor seed, niger seed and linseed (Table 2.1).

The major oilseed crops of the arid zone are sesamum and castor in *kharif* season and Rape-Mustard in *rabi* season. The other oilseed crops like sunflower and safflower also hold promise for the area.

Sesame: Sesame is grown in about 3 lac hectare area in western Rajasthan with 33.7 thousand metric tons of production. The average productivity is about 106 kg per ha. It is a major oilseed crop of western Rajasthan in *kharif* season.

Table 2.1: Area, production and productivity of oilseed crops in Rajasthan

Crops	Sesame		Castor		Groundnut		Rape-mustard		Linseed	
	1996-97	1997-98	1996-97	1997-98	1996-97	1997-98	1996-97	1997-98	1996-97	1997-98
Area ('000 ha.)	420	268	36	37	245	329	2616	2812	19	11
Production (000t)	88	53	48	47	273	369	2622	2041	11	6
Productivity kg/ha	209	197	1352	1287	1112	1122	1002	726	550	549

Castor: India is one of the world's principal producer of castor, contributing more than 40% of the total world production. Annually the crop is raised over an area of about 7.16 lakh hectares and its production in 8.9 lakh tons in India. With the availability of hybrid varieties, the productivity of castor sharply increased from 409 kg per hectare (1969-70) to 1243 kg per hectare (1997-98). In western Rajasthan, castor is grown to a limited extent only in some areas.

Sunflower: The commercial cultivation of sunflower in India began in 1972-73 with the introduction of open pollinated varieties of Russian origin. The area under the crop was 15,000-20,000 ha. in 1972-73 while in 1993-94 it touched 2.4 million hectares with a production of 1.6 million tons and a productivity of about 700 kg per hectare. The introduction of the crop in the arid zone was successful and promising. (Daulay, Singh and Henry, 1984).

Rapeseed Mustard: India is the largest mustard producing country, occupying an area of 6.47 million ha. and producing about 6.10 million tonnes which is about 32 per cent acreage of world under Rape/mustard. Rajasthan stands first in average and production i.e. 26.9 lakh; hectares and 23.68 lakh tons, respectively. (*Agriculture in Rajasthan: some facts* March, 1997 Directorate of Agriculture, Jaipur). The area under these crops was 13.68%.)

Safflower: In the country it is grown in about 7.5 lakh hectare area mostly in peninsular states. The crop proved successful during *rabi* season in arid areas (Henry and Daulay, 1990, Henry and Kumar 1991).

Breeding Objectives

The major breeding objectives for the improvement of the productivity of these crops may be grouped as under:

(i) Adaptation of the species to the environment
(ii) Increasing yield of oil per hectare
(iii) Improvement or change in oil quality
(iv) Improvement in protein quantity and quality of oilseeds.

1. **Adaptation of the species to the environment:** Most oilseed crops, upon their introduction to a new area, give a less than promising performance. Adjustment can be of two types:

   (a) Modification of the environment, (b) Modification of heredity.

   (a) Modification of the environment: It is identifying the best agricultural situation for a crop. Having identified the agricultural environment for the species, optimum production practices, in terms of dates of seeding, plant population, irrigation etc.

must be developed. In arid zone, castorbean has been introduced successfully and was found very promising. Similarly there is scope of introducing sunflower cultivation in *kharif* season. Sunflower holds promise during *rabi* season under limited irrigation conditions.

(b) **Modification of heredity:** The major adjustment on introducing a new crop to an area is in its heredity. Traditional breeding techniques are generally used, including evaluation of introduction, selection and hybridisation. Recently biotechnology is employed to develop high yielding varieties in mustard.

1. The first step was to identify the species with most promise
2. The identification of genotypes within these species adapted to the new environment. Selection is an obvious first step in the improvement of introduction (what to select?)
3. The introduced selection can be improved upon by hybridisation of different types to combine good qualities of 2 or more introduction. The technique and role of hybridisation in the improvement of oilseed crops is discussed by Fehr and Hadley (1980).

In most oilseed crops, efforts have been made to develop hybrid cultivars. Greatest success have been achieved with sunflower (Fick, 1978) where a (cytoplasmic male sterile) line, B. (maintainer) lines and R (restore) lines were used to produce single crosshybrid. Seetharam et al (1980) developed 8 hybrids by using four male sterile lines and two fertility restorer lines. Out of these BSH-1 and BSH-2 were identified as the most desirable hybrids. Compared to open pollinated and synthetic cultivars, hybrid cultivars are higher in yield and more uniform in maturity.

Efforts are being made in other oilseed crops to develop hybrid cultivars, thus for with little success. Botoson and Gairdner (1921) discovered the cytoplasmic male sterility (CMS) in flax but no hybrid varieties have become available because of low yield of CMS lines in crossing block. In India, CMS lines, maintainer and restorer lines are available in rape and mustard. Genetic male sterility is available in safflower (Heaten and Knowles, 1982), but the cost of removal of male fertile types in crossing blocks has made the price of seed high. Rac (1982) also indicated tremendous potential of hybrid in safflower for raising per hectare yields and oil. In castor, after introduction of exotic pistillate line from texas (USA) namely TSP-10R, India initiated systematic research efforts on hybrid castor. Six hybrids were released for commercial cultivation.

2. **Increasing yield of oil per hectare:** Increased yield of oil may be achieved in two ways (i) by increasing the seed yield per hectare (ii) and by increasing oil content of the seed. Increased seed yield can be achieved through combination of better cultivars and better production practices. Increased oil content can be achieved in 2 ways (a) by decreasing the proportion of the seed that is hull or seed coat area (b) by increasing the oil content of the embryo. For example, the first major step in the development of sunflower as an important oil crop was the development of thin hulled types in Russia. Seeds of such types had an oil content ranging from 40-45 per cent, whereas the older unimproved types had 25-35 per cent. Their introduction to other countries led to the development of many new breeding programmes on sunflower as an oil crop. A similar development has taken place in safflower (Rubis, 1967, Urie and Zimmer, 1970

and Urie, 1981) where reduction in hull has resulted in higher oil content. In rapeseed, yellow seeded cultivars have higher oil content because of thinner seed coats (Downey et al, 1975). Likewise in Flax, yellow seed cultivars have thinner seed coats and higher oil content than brown seeded cultivars.

Changes in the oil content of the embryo, have been achieved in soybeans, usually with opposite changes in protein content (Hartwig and Hinson, 1972). A similar situation sometimes prevails in cruciferous species (Downey et al, 1975).

3. **Improvement or change in oil quality:** Change or improvement in oil quality have been achieved in large part by modifying the fatty acid composition of the oil. Greatest success has been achieved in rapeseed species and in safflower. Canadian and European rapeseed breeders have been able to remove crucic and eicosonic acids from rapeseed oil (Downey et al, 1975). Thus providing a new oil crop with a fatty acid profile somewhat like that of soybean oil. The new oil was much more acceptable for use in edible products.

   In safflower modification involves changes in the proportion of stearic, oleic and linoleic acids (Knowles, 1969, 1972). In safflower and to a large extent in rapeseed, genetic control of fatty acid composition was expressed in the embryo. Both embryo and sporophytic control of fatty acid composition was found in sunflower (Fernandez-Mand) and rapeseed (Kondra and Stefansoon, 1970). In soybean, Brim *et al* (1968) found only sporophytic control. Breeding procedures will be governed to some extent by the location of genetic control.

4. **Improvement in protein quantity and quality of oilseeds:** After extraction of the oil from oilseeds, the residue, which is a cake or meal, is high in protein. Most of the cake and meal is used as livestock and poultry feeds.

(i) Changes in protein quantity

As long as hull and/or seed coat amount remain constant, there is an inverse relationship between oil and protein content. The reduction in hull of safflower has resulted in increase in oil content, there has been a parallel increase in protein content also. A similar situation exists in rapeseed where the seed coat has been reduced (Downey et al, 1975).

(ii) Changes in meal quality

Protein quality of an oilseed meal may be measured by its amino acid composition. Almost all oilseed meals contain products that are toxic or undesirable in some way. Examples are trypsin inhibitors in soybean. Aflotoxin in groundnut, chlorogenic acid in sunflower, gossypol in cotton seed, matairesinal monoglucoside and lignan glicoside in safflower, oxalates in the seed coat of sesame and ricin in castorbean. In some these toxic substances removal or reduction is not possible by genetic means. The toxic glucosinolates in the meal of rapeseed were reduced by Canadian workers to very low levels by genetic means (Downey *et al*, 1975) such that meal quality was almost equivalent to that of soybean. The improved rapeseeds, free of erucic acid and much reduced in levels of glucosinates, are referred to as "double zero" or CANOLA types (Anonymous, 1981).

The important emphasis for the improvement of the oilseeds viz; sesame, castor, sunflower, safflower and rapes mustard are as follows:

Sesame (*Sesamum indicum* L.)

Locally known as *till*, this is a major *kharif* oilseed crop grown essentially in drylands. After early seedling stage, sesame is able to tolerant moisture stress and is capable of utilizing stored soil moisture efficiently. The local varieties inspite of their drought tolerance are leafy with excessive vegetative growth. As such they possess low harvest index, long period of growth and poor adaptability. Hence these are caught up in drought at crucial flowering phase leading to very low productivity. Another problem associated with the land races is their higher degree of susceptibility to pests and diseases. Added to all this, there is excessive shattering of nature pods in the local varieties. Another disadvantages associated with this crop are its high theme and photosensitivity. As a result the crop is a failure under delayed onset of monsoon.

Improvement to be done

(1) Earliness in maturity and a high degree of adaptability.

(2) Improvement in harvest index i.e. a higher proportion of siliquae to the total biological production. It can be achieved by evolving a suitable single stemmed plant type with 4-6 capsules per node. This way it would be possible to increase the population density per unit area which ultimately would lead to higher yield levels.

(3) Phyllody: A common type of sterility has been found to be due to phyllody where the floral parts are transformed into various kinds of vegetative (or leaf like) structures. It found that early sowing, if accompanied with a heavy rainfall greatly favours phyllody. There is a urgent need to develop a resistant cultivar to this disease. The suitable varieties identified for the region are T-13, Var 4-2 and TC-25 (Henry and Daulay, 1987, Henry and Kumar, 1991).

Castor (*Ricinus communis* L.)

Traditional varieties cultivated during the *kharif* season are mostly tall and 270-300 days of duration. Such traditional varieties flower in September, about 80 days after planting and some of the varieties even later. The primary spikes mature in about 55 days after flowering. The spikes of secondary, tertiary and latter order appear towards October and thereafter. Since moisture becomes a limiting factor beyond the month of October, the development of the subsequent spikes frequently suffers rendering the yield low and unpredictable.

Improvement Required

The castor variety should be tailored in such a way that the flowering at least up the fourth order spike, could be completed during rainy season i.e. before the end of September. This will help in elevating and stabilizing the yield levels in spite of fluctuating rains. Thus, a plant type, which could produce the first, second, third and fourth spikes at 35, 60, 70 and 80 days respectively, would furnish an ideal edeotype. Some of the new castor varieties satisfy these requirements.

The evaluation of newly developed castor varieties indicated that by growing varieties like GAUCH-1 and Aruna, the higher yield potential can be obtained in arid regions of western Rajasthan (Henry and Daulay, 1985; Daulay, Singh and Henry, 1984).

Sunflower (*Helianthus annus* L.)

The following problems are associated with its production in India:

(i) Poor seed set and filling is one of the important constraints limiting its productivity. This refers to the large percentage of empty and partially filled seeds in the capitulum resulting in low test weight. This affects both seed and quality of the produce.

(ii) The varieties are highly heterogeneous and lack of uniformity appears to be a general feature.

(iii) Susceptibility to diseases like rust, alternaria leaf spot and various root and stem rot.

(iv) Bird damage.

(v) Problems associated with the marketing of the produce.

Improvement Required

Sunflower breeding in Soviet Union has made remarkable progress in improving seed oil content, from about 30 per cent in the early 1920's to almost 50 per cent in the present-day varieties. They have used a modified form of recurrent selection similar to the ear to row method used in U.S. for maize. The discovery of cytoplasmic genetic male sterility has made possible the evolution of hybrids. These hybrids have removed the above-mentioned problems to some extent. Hybrid sunflower has the following special features.

(i) More production stability.

(ii) More self-fertility leading to superior seed set.

(iii) Uniformity in flowering.

(iv) More tolerance to diseases.

(v) Early in maturity and more drought tolerance.

(vi) Suitable of input intensive agriculture. The performance of newly evolved hybrid/ varieties at CAZRI, Jodhpur indicated that hybrid BSH-1 MSFH-9 and MSFH-7 holds the promise for yield potential in the area (Daulay, Singh and Henry, 1984; Henry and Kumar, 1991).

Safflower (*Carthamus tinctorius* L.)

It is a drought hardy crop and can also be grown well under salt affected soil conditions and in light to heavy textured soils. It has a strong tap root which draws water from fairly deep layers of the soil profile and, therefore, is best suited to the arid regions. Compared to other oilseed crops, it is low in oil content (28-32%). Hence in safflower, breeding for high oil content assumes greater importance than any other trait. Another disadvantage associated is that all the native cultivars of safflower are handicapped by very thick pericarp which adversely limits the per hectare yield of oil as well as the quantity and quality of the meal in the crop, thereby, making its cultivation low and less attractive. Apart from their low yield potential, all available germplasm sources of high oil content from outside sources display very poor adaptability to Indian conditions. The incorporation of alleles for low hull and/or high oil content into the genetic background of locally adapted agronomic base should form an important breeding activity in the immediate future. Like sunflower, poor seed filling is also a problem in case of safflower. Karve *et al* (1981) reported that this may be due to the presence of partial self-incompatibility, coupled with differences in insect pollinators.

The evaluation of newly evolved safflower varieties in the region revealed that varieties like Tara, JL28-7 and S-4, APPR-1 are most promising from yield point of view (Henry and Daulay 1990, Henry and Kumar 1991).

Rape and Mustard (*Brassicas* spp.)

Rape and mustard belong to the group of *Brassicas* species. Rape generally refers to the *Brassica campestris* and mustard or Raya to *B. jucoa*. Taramira (*Eruca sativa*) though belongs to a different genus but of the same family, is generally referred under rape and mustard. The following factors are responsible for the low yield of crops viz.

(i) Susceptibility to aphid attack
(ii) Frost damage to the crop
(iii) Lower content of oil in the seed
(iv) Low seed yield potential under conserved moisture.

Improvement Required

Model for increase yield: Drought resistance appeared to be associated with deep roots, waxy leaves and ability to drop most of the leaves under drought stress after the seed pods have been formed. This adaptation to stress appeared to be useful since most of photosynthate used for the production of seeds is produced by the green pods and stems. The mustard species appeared to have greater drought tolerance than rapeseed.

It is indicated that early maturing cultivars should have adequate sink capacity. Sinks have been defined as areas in the plant where the products of photosynthesis are utilized for the formation of grains. The plant should not have more than 12 leaves. The first 3-4 leaves should be fairly large to provide leaf area as quickly as possible and thus provide nutrients for the production of plant structures, such as roots, stems and leaves. The remainder of the leaves should be progressively smaller to provide a pyramidal form which will permit the maximum penetration of sunlight into the plant canopy. The plant probably should not have more than 3-6 branches near the top and should have very little basal branching. Stem strength should be sufficient to prevent early lodging but slight lodging near maturity may be beneficial in reducing wind damage to the crop. The variety should also be resistant to aphid attack and frost damage for realising higher yield potential. This idiotype should result plants which are not highly competitive in a mixture of genotypes.

Large number of genotypes of rape and mustard were evaluated over the years at CAZRI, Jodhpur, under limited moisture supply conditions.

Indian Mustard (*Brassica juncea*)

Indian mustard was indicated that genotype T-59' (Varuna) was a stable genotype and was able to take advantage of the favourable growing conditions and yielded reasonably high under less favourable conditions as compared to rest of the genotypes tried (Henry and Daulay, 1985). In another study, it was revealed that the sowing time together with irrigation level had the major influence on the yield performance of Indian mustard and the genotypes like Pusa Bold, RH-8133, TM-9, RK-8203 and RLC-1012 performed better under high yielding environment. Genotype DIR-147 was found to perform best amongst all under low yielding environments (Henry and Daulay, 1988). Evaluation of newly generated breeding lines indicated that genotypes DIR-153 and RII-827 were promising from the yield point of view and also had unit responses to changes in environmental conditions. The exploitation of above genotypes in breeding programmes is likely to give high yielding genotypes (Henry and Daulay 1990). The aphid resistance in different mustard varieties was worked out by Satya Vir and Henry (1987).

Brown sarson (*Brassica rapa* var. ulti)

Promising genotypes of brown sarson were evaluated for 3 winter seasons with 3, 2 and 1 irrigation, respectively of 5 cm each at critical stages of crop growth. The highest mean seed yield was obtained with 3 irrigation and the lowest with one irrigation. Genotype 'Pusa Kalyani' had indicated god response to changing environmental conditions and was also high yielder. However, genotype 'BID-10' gave the highest mean seed yield and was consistent in its performance, besides being stable (Henry and Daulay, 1988).

Taramira (*Eruca sativa* L.)

Taramira could be grown successfully under conserved moisture or with very meagre irrigation resources of the farmers of Rajasthan including desert districts. It is generally grown under no input management conditions. Five important genotypes were evaluated for 3 winter seasons. The crop was given only one irrigation of 5 cm at the time of flowering during all the seasons. The results revealed that genotype RTM-1 gave consistent high seed yield during all the 3 seasons and was stable. Genotype RTM-2 was found to be responsive especially for adverse growing conditions (Henry and Daulay, 1990).

Indian Rape (*Brassica campestris* L.)

Indian rape or toria is a short duration oilseed crop in rapeseed-mustard group. It is generally grown in the intervening period of rainy season and winter season as a catch crop. The yield potential of the crop is low. The performance of 11 diverse promising genotypes of toria was evaluated for 3 winter seasons. Irrigation levels varied in different seasons. The level of irrigation applied was 5 cm each, and was given at critical stages of crop growth. Under such situations, genotypes 'Sangam' and 'TGC-1' indicated higher yield potential over others. Out of these, TGC-1 was a stable genotype (Henry and Dauley 1990).

Yellow Sarson (*Brassica campestris* sub sp. Sarson prain Var. Yellow Sarson)

Five important genotypes of yellow sarson were evaluated for their performance during the 3 winter seasons under limited moisture supply condition. Genotype YID-1 gave the highest mean seed yield and was suitable for favourable growing conditions. DYS-3 and YID-3 also had high seed yield and had responses near to unity. However, response of YID-3 was only stable (Henry and Daulay, 1990).

The salient features of the important varieties of the oilseed crops are given Table 2.2.

Table 2.2: Salient features of oilseed varieties

Name of crop	Name of the variety	Maturity	Yield (kg/ha)	Remarks
Sesame	RT-46	85-90	400-500	It is a white seeded branched variety. It is found resistant towards oozing disease complex.
	RT-127	80-90	500-600	It is identified for cultivation in Rajasthan state in 1998
	TC-25	85-90	400-500	It is a white seeded branched variety having a wide adaptability
	T-13	90	350-400	It is a white seed branched variety having a wide adaptability
Castor	Variety			
	Aruna (Mutant of HC-6)	115-120	600-700	It is a early maturing variety.
	Bhagya (R 63)	130	500-600	
	Hybrid varieties			
	GCH-3 (TSP-10R)	—	—	It is a high yielding hybrid variety
	GAUCH-1	130	800-900	Green stem, spiny capsule, triple bloom.
	(VP-1XVI-9)		(1500-1600)	
	GCH-2 (VP-IXJI-35)	—	(1700-1800)	Green stem, non-shattering triple bloom
	GCH-4 (All India) (VP-1 X 48-1)	—	(1900-2000)	Wilt resistant and monogamy stem, triple bloom.
	GCH-5 (All India) (Geeta x SH 72)	—	(2600-2700)	Double bloom, wilt resistant and monogamy
	TMVCH-1 (T-N) (LRES 17 X TMV 5)	—	(1100-1200)	Red stem, non-dehiscent, spiny capsule
Sunflower	Variety			
	Morden	75-80	600-800	36-38% oil content. All States.
	EC-68414	100-110	800-1000	40-42% oil content. All States.
	TNAU-SUF-7	90-95	800-1200	38-42% oil content. All States.
	GAU-SUF-15	90-95	800-1200	38-42% oil content. All States.
	Hybrid			
	BSH-1	85-90	1000-1500	40-42% oil content. All States.
	KBSH-1	90-95	1200-1500	42-44% oil content. All States.
	APSH-11	90-95	1000-1500	40-42% oil content. All States.

(Table 2.2 Contd.)

(Table 2.2 Contd.)

Name of crop	Name of the variety	Maturity (Days)	Yield (kg/ha)	Remarks
	PKVSH-27	75-80	800-1100	38-40% oil content. All Maharashtra.
	MSFH-1, MSFH-8, MSFH-10	High yielding hybrids		
Rapeseed/ Mustard Toria				
	T-9	95-100	1000-1200	Oil content 44.3%
	SANGAM	95-105	1200-1500	Oil content 44.2%
	TH 63 (SUMAN)	85-90	1500-1700	Oil content 43.9%
Brown Sarson	Pusa Kalyani	130-135	1300-1500	Oil content 45%
	BSH-1	130-140	1200-1500	Oil content 45%
Yellow Sarson Mustard	Raudys 89-115	100-105	1500-1700	Oil content 41.5%
	T-59 (Varuna)	115-120	1700-1800	Bold seeded
	PUSA BOLD	110-15	1800-1900	Bold seeded
	KRANTI	115-120	1500-1700	Medium bold
	RH-30	120	1500-1700	Medium bold
	Pusa Jaikisan (1993)	110-115	1600-2000	Bold seeded
	Rajat (PCR-7) (1993)	110-115	1800-2000	Bold seeded
	GM-2 (1996) (Guj)	120	200-2400	Oil content 40%
	PBR-97 (1996)	Late	1900-2200	Oil content 42%
Taramira	RTM-314 (1997)	130	600—800	
	T-27	130-140	600-700	Oil content 38%

IMPROVEMENT ASPECTS OF PULSESS

In pulses following are the main constraints to increased productivity

Evolutionary and Ideotype Constraints

- Narrow base of different germplasm sources
- Mostly grown in marginal lands characterised by moisture and fertility stress with the result that they have been selected for such primitive characteristics as a bushy, spreading and in some cases indeterminate growth habit associated with prolonged maturity duration and non-synchronised development.
- Non-uniform partitioning of dry matter both in vegetative and reproductive phase.
- Lack of ideal plant type for inter and multiple cropping systems.
- Flower drop and dehiscence of pods before harvest are known to affect grain yield considerably.

- Biological nitrogen fixing ability of these crops has not been fully exploited. Inoculation of legumes through efficient rhizobium strains under different agroclimatic and soil conditions, though inexpensive, has not made any significant dent. The responses are variable.

Management Technology Constraints

- Technology for increasing productivity on marginal lands
- Insufficient and costly plant protection umbrella
- Weed management is too poor through chemical
- Post-harvest technology is not properly worked out.

According to Asthana and Chandra (1991), the research approach for varietal improvement in pulses broadly consists of:

- Development of high yielding varieties with resistance to important diseases and pests.
- Development of early maturing varieties which can be fitted into suitable multiple cropping systems.
- Development of relatively photo period and thermo-insensitive varieties which can be grown in non-traditional areas and seasons.
- Development of suitable plant types for inter-cropping and pure cropping systems.

The ultimate aim is higher yield per unit time and space and stability in performance.

The important emphasis on evaluation of the new plant types in crops of arid region are as follows:

Mothbean or Dewgram

- To identify/evolve a variety having a semi-erect plant habit, dark green leaves, early maturity and high yield potential and resistance to yellow mosaic virus.
- To identify suitable variety for inter/mixed cropping in pearl millet.
- Breeding programmes for improvement of Mothbean started at CAZRI, Jodhpur in 1975 (Singh *et al*, 1981). In identification of suitable varieties for region indicated variety like T-2 and T-18 (Henry and Daulay, 1983b), IPCMO-880, IPCMO-186 (Henry and Daulay, 1987), IPCMO-371, PLMO-34, PLMO-34, IPCMO-433 (Henry, 1955b) holds promise for the region. In hybridisation programmes breeding lines JMS-1, JMS-2, JMS-3 and JMS-4 indicated superior performance for different growing situation over other varieties (Henry, 1995a). In mutation breeding programme, using chemical mutants EMS on seeds of locally adapted variety 'Jadia' led to the development of promising mutants like JMM-259, JMM-211, JMM-60, which had higher number of pods, early in maturity and higher yield potential than its parent variety 'Jadia' and other checks varieties (Henry and Daulay 1983a, 1983b), out of these mutants, Mutant JMM-259 gave better performance over the check varieties in All India Co-ordinated trails and based on its performance it was released and notified for general cultivation by the Central Seed Committee in Dec. 1988 under the name of Maru Moth-1 (Henry *et al*, 1993). JMM-DBS-1 was another promising mutant evolved (Henry and Daulay 1990). The estimation of genetic diversity in the land races indicated that there exists

enough genetic variability which can be utilised with advantage for breeding high yielding varieties (Henry and Krishna, 1986). The development of high yielding varieties is the basic requirement in Mothbean (Henry and Singh 1985). In evaluation of Mothbean varieties for inter/mixed cropping with pearl millet, it was revealed that variety JMM-259 gave better performance in inter-cropping while var. T-18 gave better performance in mixed cropping.

Clusterbean or Guar

- To identify/evolve a variety, which can be adapted to the agro-climatic conditions of the region at the same time their growth rhythms matching the rainfall pattern of the region.
- To evolve varieties resistant to *Alternaria* blight, bacterial blight and dry root rot diseases.
- To evolve variety having high galactomanan content in the endosperms of the seeds.
- To evolve determinate plant type.

Breeding programmes to bring about genetic improvement in clusterbean was started in erstwhile dry farming centre at CAZRI (Singh and Henry 1981). More than 5 thousand germplasms of this crop are evaluated and characterised for different characteristics. Studies on genetic diversity in the indigenous material indicated that enough genetic variability and diversity are present in the available germplasm for utilisation in the hybridisation programme (Henry *et al*, 1984, Henry *et al*, 1986, Henry and Krishna 1990). High yielding genotypes evolved through crossing programmes are being tested at various locations through All India Co-ordinated Varietal trials. Clusterbean crop has a indeterminate growth habit, where flowering, vegetative growth and pod formation take place simultaneously with the result at maturity upper portion of the plant remains seen and upper portion pods also remains immature and result in dry matter loss at harvesting. To avoid such loss, it was envisaged to evolve a determinate plant type, which is not available in the existing germplasm, through mutation breeding. For the purpose seeds of variety FS-277 were treated with chemical Mutagen, EMS. Variants having determinate growth habit were selected from the segregating generations. These mutants gave better performance over parent-FS-277 (Henry and Daulay, 1987). Its further testing with other promising varieties like HG-75, G-1 indicated at par performance while with vars. like RGC-936, Suvidha, it gave better performance (Henry 1999). In the identification of the suitable varieties, it was indicated that varieties like HG-75, HG 182, D. safed, 2470/12, are suitable ones (Henry and Daulay; 1983, Kackar and Henry, 1984) among medium late maturity group. Among the single stemmed varieties, varieties like HFG-314, PLG-119, PLG-85 proved superior. Among early maturing groups, varieties like HG 7-2 (Naveen) and Suvidha proved superior (Henry and Daulay, 1984; Henry, 1993). Varieties were evaluated against different rhizobial strains (Daulay and Henry, 1997).

Variant selected from the germplasm like 2470/12 gave better performance over check varieties in All India Co-ordinated Varietal trials. Its further testing in adaptive trials and minikit trials indicated its superior performance over the other check varieties tested. Based on its superior performance, it was released and notified for general cultivation by Central Seed Committee in Nov. 1986 under the name Maru-Guar (Henry *et al*, 1992). In evaluation of clusterbean varieties for inter/mixed cropping with pearl millet, it was revealed that var. FS-277 gave better performance in inter-cropping system over others (Daulay *et al*, 1998).

Green Gram (Mungbean)

The major emphasis is on the development of green gram varieties resistant to YMV and other diseases with stable performance and good cooking quality, maturing in 65 to 70 days, so that it can be grown in multiple cropping and in rotation with other crops.

In identification of suitable varieties for the region, it was revealed that variety S-8 gave the most promising performance both under normal and late sown conditions followed by K-851 (Henry and Daulay, 1989). Of late other varieties which are giving promising performance are RMG-11, RMG-131, RMG-146, P-7093, RMG-268, PDM-54 (Henry 1998). The insect-pests scenario on different green gram varieties was worked out in detail by Verma and Henry (1988).

Cowpea (Lobia)

- To identify/evolve varieties which are more compact in their growth and which are relatively erect with a non-trailing habit. This will result in higher seed yield of the crop.
- To evolve varieties resistant to diseases and pests.

Large number of varieties tried in evaluation trials indicated that variety Charodi-1, having brownish seed colour with small medium seed size, gave the most promising performance over the other varieties both under normal and late sown conditions, followed by FS-68 (Henry and Daulay, 1988). Variety K-11 was another variety which gave better performance under late sown conditions only. Efforts were made to improve the most suited variety Charodi-1 through mutation breeding to evolve variants having plant type of Charodi-1 with seeds having white seed coat colour and medium bold size. This led to the development of mutants with above characteristics, viz., CAZC-W, CAZC-9, CAZC-10 and CAZC-11 are being tested in All India Co-ordinated Varietal trials to see its suitability over the various locations (Henry 2001a). CAZC-B was another Mutant having back seed coat colour and was promising over parent variety Charodi-1 (Henry 2001b). Evaluation of different varieties in inter/mixed cropping with pearl millet revealed that variety Charodi-1 also gave better performance in both the inter cropping and mixed cropping system over the other varieties tested. It was found that vars. C-152, V 265 and AC-80 were better for favourable season, while varieties VCM-8, VPC-124, GC-1 were suitable for adverse season (Henry, 1998). Early to medium maturing varieties were found to give better performance in arid areas (Henry, 1999).

Horsegram (*Kulthi*)

This minor pulse crop is being grown in tribal areas of Pali and Jalore districts and the adjoining area of Sirohi. Varietal improvement in this crop is in incipient stage. Efforts were made to collect the land races from its growing areas. High-yielding variant plants were selected subsequently evaluated for various characteristics. Results revealed that genotypes PL-Selection, (brown seed coat colour), SN-selection (dark brown seed coat colour) and SU-

Selection (light dark seed coat colour) were high yielding and stable genotypes for various situations (Henry and Daulay, 1988). Genotype PI-Selection was tested under diverse agro-climatic zones of the country over the years in All India Co-ordinated Varietal trials, and gave the highest mean performance, over the other varieties tested. Based on its superior performance, genotype PL-selection was released and notified for general cultivation by central seed committee in year 1989 under the name Maru-Kulthi-1.

Chickpea (*Gram*)

It is grown to a limited extent in Ganganagar, Nagaur district under limited moisture supply conditions. The objective in varietal improvement of gram in order of priorities are high grain yield with good consumer acceptance parameters (cooking quality, palatability, seed colour, size) and stability in performance by incorporating resistance to diseases like wilt, blight and rust and insect-pests.

The testing of promising varieties of chickpea under limited moisture supply conditions revealed that variety ICCC-42 and Manak (14-16 q ha^{-1}) holds promise or the area.

Pigeonpea (*Arhar*)

With the availability of the extra-early maturity strains in pigeonpea, it was envisaged to evaluate its performance under both rainfed conditions during monsoon season and under limited moisture supply conditions during winter season in the arid regions to see its suitability. The variety should be of short duration and should carry resistance to the three major diseases, viz., wilt, sterility mosaic and *phytophthora* blight as well as to pests particularly pod borer and pod fly.

Table 2.3: The performance of improved var. of grain legumes at CAZRI, Jodhpur

Grain legumes	*Variety*	*Maturity (Days)*	*Mean productivity (kg ha^{-1}) (Long term)*	*State average (kg ha^{-1}) (Long term)*
Clusterbean		Branched medium late varieties		
	Maru Guar HG-75 G-1	97-105	750-850	216
		Branched early varieties		
	Suvidha, Naveen, RGC-936	80-85	800-900	
		Unbranched varieties		
	FS-277, PLG-119, PLG-85, HFG-314	90-95	600-800	
Mothbean		Medium maturing varieties		
	CAZRI Moth-1, Maru Moth-1, Jadia Jwala	80-85	400-600	150
		Medium late varieties		
	IPCMO-880, T-18	90-95	300-500	
Greengram	S-8, K-851	60-65	700-900	194
Cowpea	Charodi-1, FS-68	60-65	700-900	257
Horsegram	Maru-Kulthi-1	100-105	600-700	

The evaluation of extra-early germplasm during monsoon revealed that there exist genetic variability for all the important characters as well as genetic diversity among the genotypes so that the same could be utilised with advantage in crossing programmes (Henry and Krishna 1990, 1992). The performance of different varieties of pigeonpea during winter season revealed that varieties 'Basant' and 'Bahar' are suitable or the post-rainy season cultivation under limited moisture supply conditions (Henry and Daulay, 1988).

The performance of improved var. of grain legumes at CAZRI, Jodhpur are presented in Table 2.3.

REFERENCES

Anonymous (1981). Canola – a new oilseed from Canada. *J. Amer Oil Chem. Soc.* 58 723 A-725A.

Anonymous (1997). The Hindu : Survey of India Agriculture.

Asana R.D. (1969). Plant types for drought conditions. *Symposium on planning for drought areas.* National Institute of Sciences of India. IARI, New Delhi (4-6 May 1971): 31.

Asthana A.N. and Chandra S.S. (1991). Varietal improvement in pulse crops : Current trend and future prospects. In Golden Jubilee Symp. of Genetic Research and Education : Current Trends and the next fifty years. (Eds. Sharma *et al.*) pp. 925-959 *Ind. Soc. Of Gen. Of Pl. Breed.*

Daulay H.S., Henry A. and Bhati T.K. (1991). Studies on suitable genotypes of grain legumes for inter/ mixed cropping with pearlmillet. In *Abst. Vol. II Golden Jubilee Symp. on Genetic Research and Education – current Trends and the next fifty years* Feb. 12-15, 1991. *Indian Soc. of Genel. & Pl. Breed* New Delhi pp. 414-415.

Daulay H.S., Singh H.P. and Henry. A. (1984). A decade of research on oilseeds under arid conditions (1972-73 to 1982-83). *Div. Bull* No. 11, pp. 49.

Daulay H.S. and Henry A. (1997). Response of clusterbean varieties to different rhizobial strains in drylands of western Rajasthan. Forage Research : 23: 105-107.

Daulay H.S., Henry A. and Bhati T.K. (1998). Compatibility of promising varieties of clusterbean for inter/mixed cropping in drylands of western Rajasthan. Forage Research. 24: 13-16.

Daulay H.S., Henry A. and Bhati T.K. (1999). Studies on suitable genotypes of green gram or inter/mixed sowing in pearlmillet on drylands of western Rajasthan. Current Agriculture 23: 93-96.

Daulay H.S., Henry A. and Bhati T.K. (2000). Screening of suitable genotypes of clusterbean, mothbean and green grain in legume based intercropping with pearlmillet. Current Agric. 24: 85-88.

Henry A. and Daulay H.S. (1983a). Performance of induced mutants in mothbean, *Indian Journal of Genetic and Plant Breed* 43: 343-344.

Henry A. and Daulay H.S. (1983b). Genotype x environment interactions for seed yield in mothbean, *Indian Journal of Agricultural Sciences* 53: 469-470.

Henry A. and Daulay H.S. (1983c). Genotype x environment interactions for seed yield in clusterbean, *Indian Journal of Agricultural Sciences* 53: 467-468.

Henry A. and Daulay H.S. (1983d). Genotype x environment interactions for seed yield in mothbean, *Madras Agri. Journal* 70: 705-708.

Henry A., Mathure B.K. and Daulay H.S. (1984). Genetic divergence in clusterbean. *Forage Res.* 10: 89-94.

Henry A., Daulay H.S. and Krishna G.V.S.R. (1986). Correlation, path coefficient analysis and genetic diversity in clusterbean. *Madras Agril. Jour.* 73: 11-16.

Henry A. and Mathure B.K. (1987). Correlation and path coefficient analysis in clusterbean. *Annals of Arid Zone.* 26: 159-162.

Henry A. and Daulay H.S. (1987a). Phenotypic stability for seed yield in mothbean and clusterbean. *Proceeding First Symp. On Crop Imp.* Vol. II pp. 137-145 (Eds.) K.S. Gill *et al. Crop Imp. Soc. of Indian*, PAU, Ludhiana.

Henry A. and Daulay H.S. (1987b). Genotype x environment interactions for seed yield in sesame. *Indian Journal of Agricultural Sciences* 57: 622-624.

Henry A. and Daulay H.S. (1987c). Induced mutants in mothbean and clusterbean. *Abst. First Symp. on Crop Imp.* 23-27 Feb. 1987. *Crop. Imp. Soc. of India* PAU, Ludhiana, pp. 109-110.

Henry A. and Daulay H.S. (1988a). Genotype x environment interactions for seed yield in horsegram. *Indian Journal of Agricultural Sciences* 58: 684-686.

Henry A. and Daulay H.S. (1988b). Relative performance of pigeonpea genotypes in arid regions of western Rajasthan, India. *International pigeonpea Newsletter* No. 7, 8-9.

Henry A. and Daulay H.S. (1988c). Genotype x environment interactions for seed yield in brown season. *Indian Journal of Agricultural Sciences* 58: 634-635.

Henry A. and Daulay H.S. (1998d). Genotype x environment interactions for seed yield in Indian mustard. *Indian Journal of Agricultural Sciences* 58: 794-795.

Henry A. and Daulay H.S. (1988e). Grow cowpea variety Charodi-1 in arid regions. *Indian Farming* 38: 9-10.

Henry A. and Daulay H.S. (1989). Technology for mungbean cultivation in arid regions. *Indian Farming* 39: 12-13.

Henry A. and Krishna G.V.S.R. (1990a). Correlation and path coefficient analysis in pigeonpea. *Madras Agril. J.* 77: 440-442.

Henry A. and Daulay H.S. (1990b). Induced dark brown mutants in mothbean. *Madras Agril. J.* 77: 501-504.

Henry A. and Daulay H.S. (1990c). Phenotypic stability or seed yield in safflower. *Madras Agril. J.* 77: 471-474.

Henry A. and Daulay H.S. (1990d). Phenotypic stability or seed yield in taramira . *Madras Agril. J.* 77: 5100-511.

Henry A. and Daulay H.S. (1990e). Phenotypic stability or seed yield in Indian Rape. *Madras Agril. J.* 77: 522-525.

Henry A. and Daulay H.S. (1990f). Phenotypic stability or seed yield in Indian mustard. *Madras Agril. J.* 77: 533-534.

Henry A. and Daulay H.S. (1990g). Genotype x environment interactions for seed yield in yellow sarson. *Madras Agril. J.* 77: 570-571.

Henry A. and Kumar D. (1991). Performance of some varieties of oilseed crops on drylands of western Rajasthan. *Trans Ind. Soc. Desert Tech.* 16: 113-120.

Henry A. and Krishna G.V.S.R. (1992). Genetic divergence in pigeonpea. *Madras Agril. J.* 79: 41-44.

Henry A., Daulay H.S. and Bhati T.K. (1992). Maru-Guar—a promising clusterbean for arid regions. *Indian Farming* 42: 24-25.

Henry A. (1993). Performance of clusterbean genotypes on dryland of western Rajasthan Forage Research 19: 338-340.

Henry A., Daulay H.S. and Bhati T.K. (1993). Maru-Moth-1 – A promising mothbean variety for arid and semi-arid regions. *Indian Farming* 43: 21-24.

Henry A. (1995a). Phenotypic stability in mothbean for seed yield in drylands of western Rajasthan. Forage Res. 21: 58-61.

Henry A. (1995b). Note on performance of mothbean genotypes on dryland of western Rajasthan Forage Research 21: 149-151.

Henry A. (1997). Role of improved varieties of arid legumes crops for increasing the productivity. Pp. 71-72. Abst. In National Symp. on "Recent advances in management of arid ecosystems" organised by Arid Zone Research Assoc. of India, CAZRI, Jodhpur 3-5 March.

Henry A. (1998). Relative performance Mungbean varieties on drylands of western Rajasthan. Pp. 142-143. Abst. In National Symp. on "Management of Biotic and Abiotic Stresses in Pulse Crops" org. by Indian Soc. Of Pulses Res & Dev. IIPR, June 26-28, 1998 Kanpur (U.P.).

Henry A. (1998). Relative performance cowpea varieties on drylands of western Rajasthan. pp. 151 Abst. In National Symp. on "Management of Biotic and Abiotic Stresses in Pulse Crops" org. by Indian Soc. of Pulses Res & Dev. IIPR, June 26-28, 1998 Kanpur (U.P.).

Henry A. (1999). Performance of different maturity groups of cowpea varieties in dryland of western Rajasthan. Abst. In National Seminar on plant physiology of interface of agri-horticuluture and industry org. by Indian Sco. Of Plant Physiology, New Delhi, RAU, Bikaner and Agril Univ. Udaipur. Dec. 30 January 1, 2000 at RCA, Udaipur pp. 74.

Henry A. (1999). Performance of determinate plant type mutant vis-à-vis indeterminate varieties of clusterbean on dryland of western Rajasthan. Abst. In National Seminar on plant physiology of interface of agri-horticuluture and industry org. by Indian Sco. Of Plant Physiology, New Delhi, RAU, Bikaner and Agril Univ. Udaipur. Dec. 30 January 1, 2000 at RCA, Udaipur pp. 23.

Henry A. (2001). Performance of induced black seed coat mutant in cowpea. *Indian J. Genet.* 61: 381-382.

Henry A. (2001). Performance of induced Mutants in Cowpea. Pp 225. Abst. In Hundred years of post Mendelian Genetics and Plant Breeding—Actrospects and Prospects organised by ISGPB, Nov. 6-9, 2001. New Delhi.

Kackar N.L. and Henry A. (1984). Genotype x environment interaction for seed yield in clusterbean. *Indian J. of Genetic and Pl. Breeding* 44: 465-468.

Kumar P.R. (1996). Rape seed mustard. In fifty years of crop science research (Eds. R.S. Paroda and K.L. Chadha) pp. 330-351.

Satyavir and Henry A. (1987). An assessment of yield loss due to mustard aphid *Lipaphis erisimikalt* in some important varieties of Raya *Pesticides* 21: 30-32.

Satyavir, Singh M.P. and Henry A. (1990). Yield loss in important cultivars of Raya and effect of date of sowing on aphid infestation under arid climate of Rajasthan. *Indian Journal of Entomology*. 52: 541-546.

Verma S.K. and Henry A. (1988). Intensity of insect pests of mungbean cultivars in arid Rajasthan. *Annals of Arid Zone* 27: 71-74.

Virupakshappa K. (1996). Sunflower. In fifty years of crop science research in India (R.S. Paroda and K.L. Chadha) (Eds) pp. 352-365.

CHAPTER 3

Degree of Submergence Stress Tolerance of Lowland Rice at Varying Growth Stages

B. Haloi and P.C. Dey

Abstract

Twelve days of continuous submergence stress imposed upon on lowland rice germplasm at seedling stage and panicle initiation stage indicated varying degrees of submergence stress tolerance within the cultivars and at various phonological stages. The degree of survival varied from 43.8 to 96.0% when submergence stress was imposed upon at seedling stage. However, the degree of survival was low when subjected to same stress at tillering and panicle initiation stages. The cultivars under TTB series exhibited high degree of tolerance capabilities than the resistant check FR13A on being subjected to submergence stress at panicle initiation stage. The desirable traits of maintenance of low profile of the pace of degradation of carbohydrate, chlorophyll content and dry matter accumulation with low tissue water potential during stress period reflected the inherent capabilities of these cultivars to withstand the submergence stress.

Key words: Chlorophyll, carbohydrate content, dry matter, submergence stress, tissue water potential

INTRODUCTION

In south and southeast Asia, rainfed lowland rice is generally affected by submergence during flash floods. The frequency and duration of floods differ from place to place, year to year causing different degrees of rice crop mortality. In Assam, the crop suffers from partial to complete submergence in rainfed lowland areas due to widespread occurrence of floods in wet season in areas where localised excess rainfall and water supplied by nearby rivers or riverlets

Address for correspondence: Regional Agricultural Research Station, Assam Agricultural University, Titabar–785630, Assam, India interact with impaired natural drainage. This situation leads to complete submergence of rice seedling and the situation may continue for 10 to 20 days and even sometimes at advance stages of the crop. Such situation warrants for either need of rice varieties having inherent morpho-physiological make up to tolerate complete submergence stress more so at seedling stage or needs some management practices for avoidance of submergence stress. However, no such systematic studies have been pursued with location specificity in Assam. Therefore, an attempt was made to characterise the physiological behaviour of certain lowland rice varieties in relation to their degree of submergence stress tolerance at varying growth stages.

MATERIALS AND METHODS

The experiment was conducted during *kharif* season, 2001 and 2002 at Regional Agricultural Research Station, Titabar. Thirty days old seedlings of 13 cultivars (Table 3.1) were transplanted in a cemented tank having dimensions 10 m × 7 m × 1 m following recommended package and practices in a randomised block design with three replications. Ten days after seedling establishment, the seedlings were subjected to complete submergence stress for a period of 12 days maintaining a water depth of 80 cm in the tank. The survival count, shoot elongation ability at the relief of submergence stress, regeneration ability at 15 days relief of submergence stress and tissue water potential by water status console (Model 3005, San Dimas, CA USA), change in shoot dry matter and chlorophyll status had been recorded before and after the relief

Table 3.1: Physiological behaviour of *Sali* rice cultivars to submergence stress imposed at different phonological stages (2001).

Cultivars	*Stages at which submergence stress imposed at different phonological stages*											
	Seedling stage				*Tillering stage*				*Panicle initiation stage*			
	1	2	3	4	1	2	3	4	1	2	3	4
IET16481	43.8	6.7	0.22	0.101	55.8	5.8	0.16	0.091	50.0	1.1	0.31	0.081
FR13A (resistant check)	96.0	0.9	0.46	0.230	83.9	1.2	0.51	0.085	20.4	3.1	0.49	0.211
Sabita	83.8	0.9	0.37	0.087	53.8	1.3	0.33	0.015	80.2	0.0	0.32	0.075
Purnendu	81.3	8.0	0.03	0.087	51.2	0.0	0.65	0.141	90.6	0.0	0.67	0.142
TTB202-3	92.8	2.1	0.53	0.174	69.7	7.6	0.14	0.011	90.8	8.2	0.61	0.156
TTB202-4	94.1	5.2	0.66	0.181	69.1	4.1	0.20	0.021	80.3	1.4	0.67	0.174
TTB202-25	89.0	5.2	0.14	0.032	62.8	1.3	0.13	0.016	40.7	1.0	0.28	0.084
TTB202-16	90.3	3.9	0.45	0.044	49.3	4.5	0.14	0.018	80.3	8.3	0.44	0.106
IET16237	77.7	16.6	0.23	0.087	58.2	8.1	0.06	0.000	50.0	6.2	0.06	0.024
IET16473	87.0	11.4	0.23	0.145	19.8	9.6	0.08	0.015	50.8	0.0	0.11	0.011
Bahadur	72.8	13.1	0.15	0.000	49.3	2.5	0.21	0.024	20.2	0.0	0.11	0.011
IET10016	81.9	4.6	0.89	0.430	43.6	1.6	0.53	0.102	30.5	0.0	0.11	0.005
IR-42 (Susceptible check)	0.0	0.0	0.00	0.000	0.0	0.0	0.00	0.000	40.6	1.0	0.00	0.000
$CD_{0.05}$	4.12	2.33	–	–	3.02	1.11	–	–	3.46	0.09	–	–

1. Survival p.c. at 7 days after relief of submergence period
2. Change in shoot elongation (cm) during submergence period
3. Change in shoot dry matter ($g.hill^{-1}$) during submergence period
4. Change in chlorophyll content ($mg.g^{-1}$ .fr.wt) during submergence period

of the submergence stress. The chlorophyll content was determined on the second leaf from the top in submerged plants and also in plants before submergence by measuring tissue extract absorbency in a Spectronic 20 colorimeter following Haloi and Dey (2003). The total carbohydrate content was estimated on plants before and after the submergence on dry weight basis using anthrone reagent and glucose as the standard solution.

Another two sets consisting of 24 pots in each set with two replications were maintained with 30 days transplanted rice seedlings. One set of pots was subjected to complete submergence stress at tillering stage and another set was subjected to submergence stress at panicle initiation stage (60-65 days after transplanting). The submergence status was given for twelve days. The necessary observations were recorded before and after the relief of the submergence stress as at cemented tank.

RESULTS AND DISCUSSION

It has been observed from Table 3.1 and Table 3.2 that when submergence was imposed at seedling stage, all the cultivars except IR42, IET16481 and Bahadur could survive submergence stress to a considerable degree. The p.c. survival of the cultivars under TTB series developed at Regional Agricultural Research Station, Titabar was at par with the resistant check FR13A when stress was imposed at seedling stage. This was attributable to their lesser degree of degradation of chlorophyll content and so much so in their dry matter accumulation and carbohydrate content during submergence period (Table 3.2). Such low rate of depletion of dry matter and for that matter the slow depletion of carbohydrate content during submergence appears to be a desirable trait for tolerance as also envisaged by Chaturvedi *et al* (1994). It was reported by Adak and Dasgupta (2000) that photosynthesis and related characters were adversely affected in rice varieties subjected to submergence stress. The survival ability was low, however, to some extent when submergence was imposed at tillering stage during 2001 due to turbid nature of submerged water. But during 2002 on being submerged at tillering stage, all the cultivars except IET16481 could withstand stress to some extent (Table 3.2). The significant point was that when submergence was imposed at panicle initiation stage, the survival ability decreased to a greater extent in all the cultivars more so during 2002 period. The notable feature of TTB series of cultivars was that they could maintain higher degree of survival and chlorophyll content than the resistant check FR13A on being subjected to submergence at panicle initiation stage.

Although the same trends were observed in dry matter accumulation, change in carbohydrate content and chlorophyll content, TTB series of cultivars exhibited lesser degree of degradation of carbohydrate, chlorophyll content and dry matter accumulation during submergence period in all the phonological stages. Haloi and Dey (2003) reported that these traits to be desired for submergence stress tolerant cultivars. Another important observation was that the change in shoot elongation of the cultivars was rapid during submergence period at seedling stage as compared to other two stages (Table 3.1). Such genotypic traits make physiological make up of the crops to be tolerable to submergence at early stages as also envisaged by Singh *et al*, (1999).

It was interesting to observe that the cultivars with high survival units had the lower water potential value (Table 3.3) at the relief of the submergence stress at all the stages, while the cultivars with higher tissue water potential tended to become susceptible to submerged

Table 3.2: Physiological behaviour of Sali rice germplasm to submergence stress at different growth stages (2002)

Germplasm	*Stages at which submergence stress was imposed upon*														
	Seedling stage					*Tillering stage*					*Panicle initiation stage*				
	1	2	3	4	5	1	2	3	4	5	1	2	3	4	5
IET16481	31.2	0.105	2.1/3.8	0.14	70.5/29.5	38.1	0.085	2.8/4.2	0.23	63.4/36.6	17.0	0.044	–	0.08	35.2/64.8
FR13A (resistant check)	83.1	0.242	2.6/2.9	0.55	82.6/17.4	88.4	0.175	3.6/4.8	0.61	76.3/23.7	22.2	0.122	–	0.34	34.3/65.7
Sabita	59.7	0.083	2.6/3.8	0.41	79.7/20.3	60.1	0.065	3.0/5.1	0.34	68.2/31.8	37.0	0.024	–	0.15	31.6/68.4
Purnendu	56.2	0.021	2.2/3.5	0.08	68.4/31.6	55.3	0.012	3.6/4.1	0.14	60.8/39.2	52.6	0.014	–	0.06	32.4/67.6
TTB202-3	74.1	0.181	3.2/3.6	0.58	85.9/14.1	74.5	0.174	4.2/5.8	0.52	74.5/25.5	43.9	0.126	–	0.31	40.0/60.0
TTB202-4	86.8	0.207	3.1/3.5	0.72	70.5/29.5	80.3	0.188	4.4/5.4	0.58	71.2/28.8	39.2	0.132	–	0.28	33.7/66.3
TTB202-25	66.8	0.122	2.7/3.4	0.22	77.2/22.8	70.5	0.102	3.8/4.9	0.16	64.6/35.4	30.4	0.114	–	0.09	24.0/75.8
TTB202-16	75.5	0.145	2.7/3.8	0.48	80.1/19.9	77.2	0.108	4.5/5.1	0.43	70.4/29.6	35.8	0.102	–	0.28	21.3/78.7
IET16237	64.5	0.016	1.8/2.8	0.24	74.0/26.0	60.3	0.007	2.1/3.7	0.21	58.3/41.7	68.9	0.004	–	0.02	25.6/74.4
IET16473	60.3	0.115	1.8/3.1	0.30	68.3/31.7	62.8	0.101	2.2/4.0	0.26	52.1/47.9	66.0	0.092	–	0.11	0.0/100.0
Bahadur	39.5	0.000	2.6/3.7	0.18	83.6/17.3	50.3	0.045	3.9/5.4	0.24	60.8/39.2	48.0	0.035	–	0.08	4.3/95.7
IET10016	26.4	0.065	2.1/3.4	0.44	82.7/17.3	41.4	0.021	2.6/4.7	0.19	71.4/28.6	0.0	0.000	–	0.00	0.0/0.0
$CD_{0.05}$	11.2	0.014	–	0.16	6.01/5.63	17.31	0.012	–	0.18	8.44/8.82	9.03	0.021	–	0.14	10.27/14.0

1. Survival p.c. at 7 days after relief of submergence stress
2. Change in chlorophyll content ($mg.g^{-1}$ .fr.wt) during submergence period
3. Numerator = Carbohydrate content (p.c.) on dry wt. basis after submergence
   Denominator = Carbohydrate content (p.c.) on dry wt. basis before submergence
4. Change in shoot dry matter ($g\ hill^{-1}$) during submergence
5. Numerator = p.c. good grain
   Denominator = p.c. chaffy grain

situation leading to low regeneration ability than the tolerable cultivars. Such effects are due to the fact that the osmotic potential of fully submerged leaves when reached above a critical point leads to lower stomatal conductance and intercellular CO_2 concentration, retards carboxylation process as also supported by Hirasawh *et al*, (1996).

It was further observed from Table 3.2 that when submergence imposed at PI stage, chaffy

Table 3.3: Tissue water potential and regeneration ability of lowland rice under submergence pressure (2001)

Genotype	*Water potential (kpa) before submergence stress*			*Water potential (kpa) after submergence stress*			*Regeneration ability*
	S.S	T.S	P.I	S.S	T.S	P.I	S.S
IET16481	2000	2000	2020	2260	2200	2200	32.3
FR13A (resistant check)	1920	1900	1920	1400	1420	1500	80.4
Sabita	2540	2500	2520	2700	2720	2540	72.1
Purnendu	2460	2500	2400	2500	2560	2500	60.5
TTB202-3	1940	1900	1900	1560	1520	1560	64.3
TTB202-4	2000	1960	2020	1620	1600	1620	68.7
TTB202-25	3000	2860	2860	2620	2800	2700	58.3
TTB202-16	3000	3000	3020	2800	2500	2660	62.7
IET16237	2300	2240	2300	2500	2260	2300	55.2
IET16473	2200	2240	2200	2420	2320	2320	47.3
Bahadur	2320	2300	2320	2240	2300	2360	23.8
IET10016	2420	2400	2460	2300	2400	2500	42.6
IR-42 (Susceptible check)	2420	2520	2520	–	–	–	–

S.S.: Seedling Stage, T.S.: Tillering Stage, P.I.: Panicle initiation Stage.

Table 3.4: Grain yield and good grain p.c. in relation to submergence stress

Entries	*Grain yield (t ha^{-1})*		*Good grain p.c.*	
	Normal water regime	*Submergence stress*	*Normal water regime*	*Submergence stress*
IET16481	2.2	3.1	57.3	70.5
FR13A(resistant check)	3.1	4.3	97.7	82.6
Sabita	3.2	3.3	83.3	79.7
Purnendu	3.5	3.1	85.1	68.4
TTB202-3	2.9	3.2	60.1	85.9
TTB202-4	2.8	3.1	75.6	70.5
TTB202-25	2.8	5.3	76.2	77.2
TTB202-16	2.1	4.2	67.0	80.1
IET16237	2.0	3.0	63.5	74.0
IET16473	3.0	3.3	54.9	68.3
Bahadur	3.3	3.0	75.8	83.6
IET10016	2.7	2.4	86.0	82.7
CD(0.05)	0.67	0.81	7.22	6.01

grain formation was accelerated in all the cultivars. However, good grain p.c. and grain yield were satisfactory in TTB series of cultivars when submerged at early two stages (Table 3.4).

Thus, the study revealed that the degree of submergence tolerance at various phonological stages of lowland rice crop depend on several physiological traits such as low degradation of carbohydrate, low tissue water potential, minimum degradation of chlorophyll and having potential regeneration ability etc.

REFERENCES

Adak M.K. and Dasgupta D.K. (2000). Prolonged waterlogging on photosynthesis and related characters in rice. *Indian J. Plant Physiol.* 5(4): 380-382.

Chaturvedi G.S, Ram P.C., Singh A.K., Ram P., Lugram K.T., Singh B.B., Singh RK. and Singh V.P. (1994). Carbohydrate status of rainfed lowland rice in relation to submergence, drought and shade tolerance. *Proceedings of the international conference on stress physiology of rice*, 28 Feb to 5 March, 1994, Lucknow, UP, India: 103-119.

Haloi B. and Dey P.C. (2003). Seedling size and varietal characteristics as the tools for avoidance to submergence stress at seedling stage in rice. *Indian J. Plant Physiol.* 8(1): 38-43.

Hirasawh T., Iida Y. and Ishihara K. (1996). The relationship between environmental factors and behaviour of stomata in rice plant. *Japanese J. Crop Sci.* 58: 383-389.

Singh D.N., Singh S. Singh R.K. and Sarkarung S. (1999). Screening for submergence tolerance using double haploid rice lines for rainfed lowlands Oryza. 36(1): 38-41.

CHAPTER 4

Effect of SO_2 on Growth, Photosynthesis and Antioxidant Enzyme Activities in Blackgram (*Vigna mungo* L. Hepper)

Sarvajeet Singh and Nafees A. Khan*

Abstract

Blackgram (*Vigna mungo* L. Hepper) cultivars T-9 and PU-19 exhibited reduction in growth and photosynthesis, increased intercellular CO_2 and malondialdehyde level when exposed to 0.1 and 0.2 ppm SO_2 for 3h on alternate days for 6 weeks. Maximal reduction was noted with PU-19. Activities of antioxidant enzymes responded differentially in the two cultivars. Superoxide dismutase (SOD) increased with increasing SO_2 in T-9 but remained unaltered in PU-19, whereas peroxidase (APX) increased in both the cultivars. Results indicated that lesser decrease in growth and photosynthesis in T-9 was maintained by higher activities of SOD and APX.

Key words: Chlorophyll; Dry mass; Lipid peroxidation; Peroxidase; Superoxide dismutase

INTRODUCTION

Reduction in growth and photosynthesis has been reported as a result of various physiological and biochemical imbalances induced by sulphur dioxide (SO_2) (Barrett and Benedict 1970). After entry in the plant system, SO_2 produces superoxide anions (O_2^-), hydrogen peroxide (H_2O_2) and hydroxyl radicals (Asada 1980, Norman *et al*, 2001) by the oxidation of sulphite and bisulphite ions which decrease enzymatic protein functions (Alsher 1984), membrane organisations (Shimazaki *et al*, 1980, Barrett and Benedict 1970), photosynthetic rate and dry mass production (Khan and Malhotra 1977, Sugahara 1984, Anderson *et al*, 1987, Singh and

*Address for correspondence: Department of Botany, Aligarh Muslim University, Aligarh 202 002, India
Tel.: +91-571-2702016, fax: +91-571-2702016, E-mail address: naf9@lycos.com (N.A. Khan)

Singh 2003). Increased activities of peroxidase (APX) and superoxide dismutase (SOD) have been reported in SO_2 exposed plants (Qian and Yu 1991, Rao 1992, Ranieri *et al*, 1997, Bernardi *et al*, 2001). However, the response of plants to SO_2 may depend on their sensitivity and their potential of resistance depends on the behaviour of cultivars. Therefore, the present study was undertaken to compare the growth and photosynthetic response of T-9 and PU-19 cultivars of blackgram (*Vigna mungo* L. Hepper) exposed to SO_2 and associated changes in the activities of APX and SOD.

MATERIALS AND METHODS

Plant Material and Culture Conditions

An experiment was conducted in the net house of the Botany Department, Aligarh Muslim University, Aligarh, (27°52′N latitude, 78°51′E longitude and 187.45 m altitude), India under natural day/night conditions. A mixture of compost and soil (3:1) with sandy loam in texture and neutral in reaction (pH = 7.7 – 7.8) was used for the study. Blackgram (*Vigna mungo* L. Hepper) cultivar (cv. T-9 and PU-19) plants were raised from seeds in 23 cm diameter clay pots. The average temperature and relative humidity during the experiment varied from 32–35°C and 62–70% respectively.

Gas Exposure System

Exposure chamber of 90 × 90 × 120 cm dimensions made of transparent fibre glass, with an exhaust duct on the top (20 × 20 cm) and double walled perforated bottom were used. SO_2 was produced in a generator by the reaction of sodium sulphite (Na_2SO_3) and dilute sulphuric acid (10% H_2SO_4). The outlet (0.4 cm diameter) of SO_2 generator (Standard Appliances, Varanasi, India) was connected to the blower inlet of the chambers. The blower assembly, which was run at a constant speed, mixed the gas with ambient air and dispensed into the chamber. Desired concentrations of SO_2 i.e. 0.0, 0.1 and 0.2 ppm were obtained by preparing Na_2SO_3 solutions of different concentrations. During each exposure, an air sampler (Kimoto Electricals, Japan) was kept inside the chamber to estimate actual concentration of the gas. Sodium tetra chloromercurate was used as an absorbing media for SO_2. After sampling, the solution was analysed colorimetrically (Anonymous 1986). The air flow rate was measured at different points in the exhaust duct of the chamber using an electronic anemometer during exposure. An air flow rate of 1.9 ms^{-1}, which was obtained with 210 voltage supply to the blower of the chamber, was maintained. The air (with or without gas) inside the chamber was replaced approximately seven times in 1 min.

Exposure Schedule

Plants were exposed to various concentrations of SO_2 ranging from 0.1 to 1.0 ppm. A 50% reduction in growth was observed at 0.25 ppm SO_2. Therefore, 0.1 and 0.2 ppm concentrations were taken as the experimental treatments. Two-week old plants of each cultivar were exposed to 0.1 and 0.2 ppm SO_2 for 3 hours on alternate days for 6 weeks. Untreated plants served as control which were also kept for the same duration in exposure chambers without SO_2. Each treatment was replicated 10 times.

Growth Determinations

Five plants of each cultivar and each treatment were uprooted at 56 days after the start of first exposure and plant length, shoot and root dry masses, shoot/root dry mass ratio were determined. Leaf area was measured with a leaf area metre (*LA 211*, Systronics, India). Dry masses were taken after drying the samples in an oven at 80°C till constant weight.

Leaf Gas-exchange Measurements

Photosynthetic rate (P_N), stomatal conductance (g_S) and intercellular CO_2 (Ci) on fully expanded uppermost leaves of plants were measured using portable photosynthesis system (*LICOR 6200*, Nebraska, USA) at about 900 µmol $m^{-2}s^{-1}$ saturating photosynthetic active radiation between 1100-1200 hours.

Chlorophyll Determination

The total chlorophyll content of leaves was determined according to the method of Arnon, 1949 by extracting the pigment in 80% acetone.

Lipid Peroxidation

The level of lipid peroxidation products in leaf samples was expressed as malondialdehyde (MDA) content (Heath and Packer 1968). 200 mg leaf tissues was ground in 0.25% 2-thiobarbituric acid (TBA) in 10% trichloroacetic acid (TCA) using mortar and pestle. After heating at 95°C for 30 min, the mixture was quickly cooled in an ice bath and centrifuged at 10,000 × g for 10 min. The absorbance of the supernatant was read at 532 nm and corrected for unspecific turbidity by subtracting the absorbance of the same at 600 nm. The blank was 0.25% TBA in 10% TCA. The concentrations of lipid peroxides together with oxidatively modified proteins of plants were thus quantified in terms of MDA level using an extinction coefficient of 155 mM^{-1} cm^{-1}.

Enzyme Determinations

The activity of SOD (EC 1.15.1.1) was assayed by monitoring the inhibition of photochemical reduction of nitroblue tetrazolium (NBT), according to the method of Giannopolitis and Ries, 1977 and Beyer and Fridovich, 1987. Leaf samples were homogenised in a pre-chilled mortar and pestle for 2 min with 1.5 g of quartz sand and 10 ml of homogenising solution containing 50 mM HEPES buffer and 0.1 mM Na_2EDTA (pH 7.6). The homogenate was centrifuged at 15,000×g for 15 min, and then filtered through Whatman 42 filter paper to produce the crude extract, which was used for SOD assay. A 5 ml reaction mixture containing 50 mM HEPES (pH 7.6), 0.1 mM EDTA, 50 mM Na_2CO_3 (pH 10.0), 10 mM methionine, 0.025 % (m/v) triton X-100, 63 µmol NBT, 1.3 µmol riboflavin and an enzyme extract was illuminated for 15 min (360 µmol m^{-2} s^{-1}) and a control set of reaction mixture was not illuminated to correct for background absorbance. A unit of SOD activity was defined as the amount of enzyme required to cause 50% inhibition of the reduction of NBT as monitored at 560 nm.

Peroxidase (EC 1.11.1.7) was extracted and determined according to Quesada *et al*, 1992. Fresh leaf material was homogenised in 0.05 M acetate buffer (pH 4.0). The homogenate was filtered through three layers of cheese cloth and the filtrate was incubated overnight at 4°C with polyvinylpyrrolidone and then centrifuged at 15,000 × g for 40 min. The supernatant was used for soluble peroxidase. Following incubation of the extracts with 0.26 mM *0*-dianisidine, 8.8 mM H_2O_2 in 20 mM phosphate buffer (pH 6.0 at 25°C), absorbance at 460 nm was recorded. A unit of enzyme activity was defined as the one-increment increase in absorbance at 460 nm per min.

The protein content in leaf extracts was determined using bovine serum albumin (BSA) as standard (Lowry *et al*, 1951).

Statistical Analysis

Data were subjected to ANOVA test (SPSS Ver. 10 Inc., Chicago, USA). Means were compared by least significant difference (LSD). Comparison at $P < 0.05$ was considered significantly different.

RESULTS

Effect of SO_2 on Growth Characteristics

Growth characteristics were maximally reduced by 0.2 ppm SO_2 in T-9 (Table 4.1). Plant length decreased significantly by 27% and 34% in T-9 and PU-19, respectively in 0.2 ppm SO_2 compared to control. SO_2 decreased shoot and root dry masses significantly. However, cultivar T-9 exhibited a decrease of 25% in shoot dry mass and 31% in root dry mass, whereas PU-19 showed the decrease of 38% and 42%, at 0.2 ppm SO_2, over respective controls. The shoot/root dry mass increased in T-9 and PU-19 with increasing SO_2 concentration indicating a larger decrease in root. Leaf area of cultivar T-9 was decreased by 29% whereas of PU-19 by 36% at 0.2 ppm SO_2.

Effect of SO_2 on Leaf Gas Exchange and Chlorophyll Content

The P_N and g_S of both the cultivars were reduced by SO_2 (Table 4.2) but maximum decrease of 35% was observed in PU-19 at 0.2 ppm SO_2. SO_2 significantly increased the Ci by 20% in PU-19 and by 17.4% in T-9 at 0.2 ppm SO_2, over respective controls. In T-9, 0.2 ppm SO_2 led to 28% decrease in total chlorophyll content, whereas PU-19 registered maximum decrease of 42%.

Effect of SO_2 on Lipid Peroxidation

SO_2 exposure led to a significant increase in the level of MDA content in both the cultivars at 0.1 and 0.2 ppm (Table 4.3). The MDA accumulation in PU-19 was higher than T-9, indicating a higher degree of lipid peroxidation in PU-19. In PU-19, MDA content increased by 46.8 and 69.7% and in T-9 by 33.6 and 43.3% at 0.1 and 0.2 ppm SO_2 over respective controls.

Table 4.1: Effect of sulphur dioxide on growth characteristics of blackgram (*Vigna mungo* L. Hepper) cv. T-9 and PU-19.

Parameters	*T-9*			*PU-19*		
	SO_2 concentration (ppm)					
	0.0	*0.1*	*0.2*	*0.0*	*0.1*	*0.2*
Plant length (cm $plant^{-1}$)	46.9 ± 1.38^a	37.3 ± 1.21^{ab}	33.9 ± 0.48^b	45.2 ± 1.50^a	32.3 ± 1.03^b	29.8 ± 0.28^b
Shoot dry mass (g $plant^{-1}$)	4.7 ± 0.45^a	4.1 ± 0.22^{ab}	3.5 ± 0.13^b	4.8 ± 0.28^a	3.9 ± 0.30^b	3.3 ± 0.04^b
Root dry mass (g $plant^{-1}$)	0.78 ± 0.07^a	0.61 ± 0.04^{ab}	0.48 ± 0.05^b	0.75 ± 0.12^a	0.59 ± 0.04^b	0.43 ± 0.02^b
Shoot/Root dry mass	6.0 ± 0.47^a	6.7 ± 0.28^a	7.1 ± 0.34^a	6.4 ± 0.28^b	6.6 ± 0.34^b	7.6 ± 0.7^a
Leaf area (cm^2 $plant^{-1}$)	421.0 ± 8.08^a	317.9 ± 2.54^b	296.0 ± 8.66^b	425.0 ± 9.2^a	305.0 ± 4.6^b	272.0 ± 13.2^c

Values represent mean ± standard error. Means within a row followed by the same letter are not significantly different by the Duncan's multiple range test at 5% probability level.

Table 4.2: Effect of sulphur dioxide on the rate of photosynthesis, stomatal conductance, intercellular CO_2 and total chlorophyll content of blackgram (*Vigna mungo* L. Hepper) cv. T-9 and PU-19.

Parameters	*T-9*			*PU-19*		
	SO_2 concentration (ppm)					
	0.0	*0.1*	*0.2*	*0.0*	*0.1*	*0.2*
Net photosynthetic rate (μ mol CO_2 $m^{-1}s^{-1}$)	12.5 ± 0.11^a	11.0 ± 0.17^b	10.2 ± 0.055^b	12.7 ± 0.06^a	10.0 ± 18.0^b	8.2 ± 0.5^c
Stomatal conductance (mol $m^{-2}s^{-1}$)	0.26 ± 2.8^a	0.24 ± 3.4^a	0.24 ± 4.04^a	0.23 ± 0.00^a	0.19 ± 0.1^b	0.16 ± 0.01^b
Intercellular CO_2 (ppm)	286.0 ± 10.3^a	309 ± 9.23^{ab}	336 ± 2.3^a	290.0 ± 8.08^a	327.8 ± 4.61^b	348.0 ± 3.46^c
Chlorophyll (mg g^{-1} fresh mass)	1.80 ± 0.02^a	1.64 ± 0.05^{ab}	1.28 ± 0.04^b	1.87 ± 0.02^a	1.69 ± 0.3^b	1.14 ± 0.02^c

Values represent mean ± standard error. Means within a row followed by the same letter are not significantly different by the Duncan's multiple range test at 5% probability level.

Table 4.3: Effect of sulphur dioxide on lipid peroxidation and activities of superoxide dismutase and peroxidase of blackgram (*Vigna mungo* L. Hepper) cv. T-9 and PU-19.

Parameters	*T-9*			*PU-19*		
	SO_2 concentration (ppm)					
	0.0	*0.1*	*0.2*	*0.0*	*0.1*	*0.2*
MDA (n mol mg^{-1} proteins)	21.48 ± 0.72^b	28.7 ± 1.84^a	30.1 ± 1.11^a	22.55 ± 0.94^c	32.4 ± 0.79^b	38.27 ± 0.52^a
SOD (Units g^{-1} proteins s^{-1})	4.8 ± 0.47^c	5.6 ± 0.11^b	6.2 ± 0.11^c	4.9 ± 0.57^a	4.9 ± 0.75^a	5.1 ± 0.77^a
APX (Units g^{-1} proteins s^{-1})	22.3 ± 0.46^a	28.5 ± 0.09^b	31.5 ± 0.46^c	18.8 ± 0.53^c	23.1 ± 0.46^b	26.3 ± 0.63^a

Values represent mean ± standard error. Means within a row followed by the same letter are not significantly different by the Duncan's multiple range test at 5% probability level.

Effect of SO_2 on Antioxidative Enzymes Activities

Both the cultivars exhibited significant increase in APX activity with increasing SO_2 concentration, whereas SO_2 had no impact on SOD activity of PU-19 (Table 4.3). In T-9, 0.1 and 0.2 ppm SO_2 increased the SOD activity by 16 and 29%, respectively, in comparison to the control.

APX activity increased with increasing SO_2 concentration being maximum in T-9. In T-9, APX activity increased by 24% and 41% and in PU-19 by 22.8% and 39.8% at 0.1 and 0.2 ppm SO_2 over respective controls.

DISCUSSION

SO_2 decreased the growth characteristics to a larger extent in PU-19 than T-9. Root growth was restricted due to reduction in the availability and translocation of photosynthates from shoot to root. The available lesser photosynthates were used by the plants for maintenance in SO_2 exposed conditions. The adverse effect of SO_2 on the growth characteristics have also been reported by other workers (Lee *et al*, 1997, Ampily 1999, Deepak and Agrawal 1999, Raza and Gouri 2000, Singh and Singh, 2004).

Total chlorophyll content was reduced more in PU-19 due to SO_2 exposure. It has been suggested that chlorophyll destruction by SO_2 is caused either by conversion to pheophytin or by the production of superoxide radicals by the reaction of sulphite with chlorophyll under illumination (Shimazaki *et al*, 1980). Photosynthesis is particularly sensitive to SO_2 (Wellburn 1985, Winner *et al*, 1985). Both the cultivars PU-19 and T-9 showed decreased rate of photosynthesis due to reduction in stomatal conductance. This led to increase in intercellular CO_2 concentration in SO_2 exposed plants which may be due to competition between SO_3^{2-} and CO_2 for binding sites on the rubisco enzymes (Ziegler 1972).

In the present study, increased activities of SOD and APX were observed which could be indicative of increased production of ROS and a build up of a protective mechanism to reduced oxidative damage triggered by stressed plant. SO_2 exposure led to a significant increase in SOD and APX activities in T-9, however, SOD activity remained unchanged in PU-19 (Table 4.3). Similar results on the activities of these enzymes have been reported in other plants subjected to SO_2 exposure (Nandi *et al*, 1990, Qian and Yu, 1991, Lee 2002). In tolerant plants species, SOD and APX activities were found to be higher, enabling plants to protect themselves against oxidative stress (Tanaka and Sugahara 1980, Varshney and Varshney 1985, Scalet *et al*, 1995), whereas such activities were not observed in sensitive plants (Peters *et al*, 1989). The SO_2^- induced SOD and APX activities in T-9 indicated that it had a higher capacity for the decomposition of O_2^- and H_2O_2. However, in PU-19, unaltered SOD activity may not be enough to scavenge O_2^-, therefore, highly reactive hydroxyl radicals were formed which were not then eliminated and caused more oxidative stress in PU-19. The results reported here show higher degree of accumulation of MDA in PU-19 indicating a high rate of lipid peroxidation due to SO_2. The lesser degree of membrane damage (as indicated by low MDA content) and the higher activity of SOD and APX observed in SO_2^- exposed T-9 plants indicated that this cultivar had a high capacity for scavenging of reactive oxygen species (ROS) generated by SO_2 stress. The antioxidative enzymes, SOD and APX, function as effective quenchers of ROS and their level may also determine the sensitivity of plants to lipid peroxidation (Kanazawa *et al*, 2000).

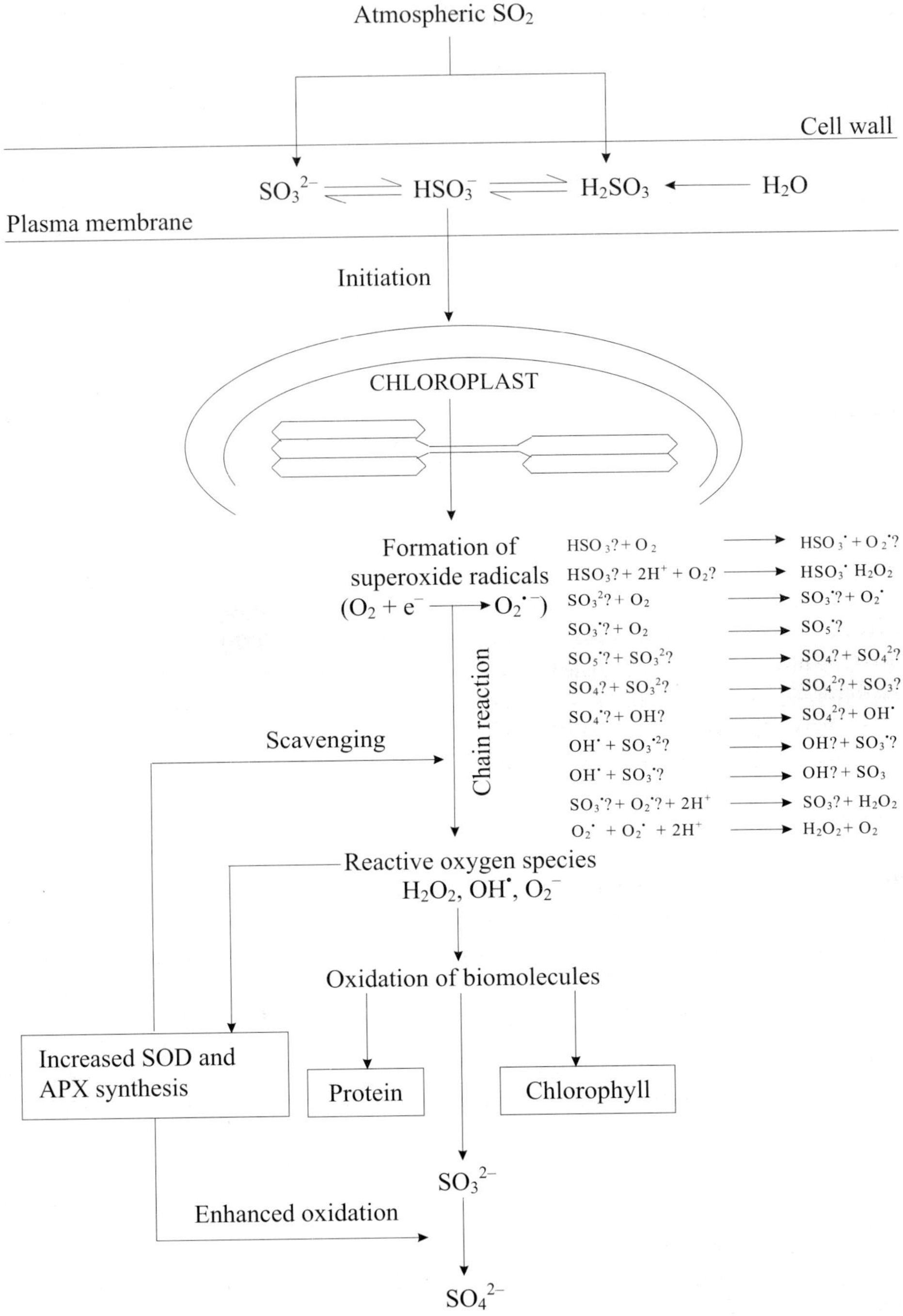

Fig. 4.1: SO_2 interaction in cell and its detoxification

Deficiency of these activities may result in the enhancement of the free radical mediated lipid peroxidation (Manohar and Balasubramaniam, 1986).

Fig. 4.1. represents a general model for SO_2 interaction in cell and its detoxification. In the cell walls the SO_2 is first dissolved in water [$SO_2.H_2O$] and its reaction products, HSO_3^- and SO_3^{2-}, are then distributed inside the cell. The free radicals of sulphur i.e. SO_3^{2-} and HSO_3^- ions having lone pair of electrons (Malhotra and Hocking 1976, Hippeli and Elstner 1996) promotes the formation of O_2^-, which through a series of free radical chain reaction initiates oxidation of SO_3^{2-} in illuminated chloroplast (Asada and Kiso, 1973). During this chain reaction, oxidising species such as H_2O_2 and OH^- are formed from O_2^-. These are highly toxic and if plants initially fail to detoxify them, these free radicals then disturb the metabolic pathways and cause changes in protein and lipids in the cell walls and membrane organisations (Alsher 1984, Norman *et al*, 2001). Therefore, many groups of compounds need scavenging the free radicals. SOD catalyses dismutation of superoxides and active oxygen species. Similarly, APX provide measure to reduce levels of peroxides. Additionally, increased level of these enzymes may induce oxidation of sulphite to sulphate (Varshney and Varshney 1985).

CONCLUSION

This study showed that plants having high activities of SOD and APX seem to be more tolerant to SO_2. Blackgram cv. T-9 which has a high level of antioxidative enzymes activities is relatively less sensitive to SO_2 than PU-19. Varietal difference in growth and photosynthesis was related to the differences in the activities of antioxidative enzymes. Unaltered activity of SOD in PU-19 unabled the plants to protect them against superoxide anions which caused more damage. Whereas, increased SOD and APX activity in T-9 protects the chloroplast from O_2^- and H_2O_2, resulted in less decrease in photosynthetic rate and dry matter production than PU-19.

REFERENCES

Alsher R. (1984). Effects of SO_2 on light-modulated enzyme reactions, in: M.J. Kozoil, F.R. Whatley (Eds.), Gaseous air pollutants and plant metabolism. Butterworths, London pp. 181-200.

Ampily R. (1999). Interaction of SO_2 and *Fusarium solani* Socc. on growth characteristics of *Cicer arietinum* cv. U-24, *J. Ecotoxic. Environ. Monit.* 9: 167-175.

Anderson L.E., Mushinez G. & Marques I. (1987). Effects of SO_2 and sulphite on stromal metabolisms, in: S. Schulte-Hostede, N.M. Darrell, L.W. Blank, A.R. Welburn (Eds.), Air Pollution and Plant Metabolism, Elsevier Applied Science, London pp. 134-147.

Anonymous (1986). Air quality standards, in: Technical Mannual, Envirotech Industries, Okhla-New Delhi, India.

Arnon D.I. (1949). Copper enzymes in isolated chloroplast polyphenol oxidase in *Beta vulgaris*, Plant Physiol. 24: 1-15.

Asada K. (1980). Formation and scavenging of superoxides in chloroplasts, with reaction to injury by sulphur dioxide, Natl. Inst. Environ. Stud. 11: 165-179.

Asada K. & Kiso K. (1973). Initiation of aerobic oxidation of sulphite by illuminated spinach chloroplasts, Eur. J. Biochem. 33: 253-257.

Barrett T.W. & Benedict H.M. (1970). Sulphur dioxide, in: J.S. Jacobsen and A. Dydetull (Eds.), Recognition of air pollution injury to vegetation: a pictorial atlas, Air Pollution Control Association, Pittsburg, Pa.

Bernardi R., Nali C., Gargiulo R., Pugliesi C., Lorenzini G. & Durante M. (2001). Protein pattern and Fe superoxide dismutase activity of bean plants under sulphur dioxide stress, J. Phytopathol. 149: 477-480.

Beyer W.F. & Fridovich I. (1987). Assaying for superoxide dismutase activity: some large consequences of minor changes in conditions, Anal. Biochem. 161: 559-566.

Deepak S.S. & Agrawal M. (1999). Growth and yield response of wheat plants to elevated levels of CO_2 and SO_2 singly and in combination, Environ. Pollut. 104: 411-419.

Giannopolitis C.N. & Ries S.K. (1977). Superoxide dismutase occurrence in higher plants. Plant Physiol. 59: 309-314.

Heath R.L. & Packer L. (1968). Photoperoxidation in isolated chloroplasts. I. Kinetics and Stoichiometry of fatty acid peroxidation, Arch. Biochem. Biophys. 125: 180-198.

Hippeli S. & Elstner E.F. (1996). Mechanism of oxygen activation during plant stress: Biochemical effects of air pollutants, J. Plant Physiol. 148: 249-257.

Kanazawa S., Sano S., Koshiba T. & Ushimaru T. (2000). Changes in antioxidative enzymes in cucumber cotyledons during natural senescence: comparison with those during dark-induced senescence, Physiol. Plant 109: 211-216.

Khan A.A. & Malhotra S.S. (1977). Effects of aqueous sulphur dioxide on pine needle glycolipids, Phytochem. 16: 539-543.

Lee E.H., Pausch R.C., Rowland R.A., Mulchi C.L. & Rudorff B.F.T. (1997). Responses of field grown soybean (cv. Essex) to elevated SO_2 under two atmospheric CO_2 concentrations, Environ. Exp. Bot. 37: 85-93.

Lee M. (2002). Effects of Na_2SO_3 on the activities of antioxidant enzymes in geranium seedlings, Phytochem. 59, 493-499.

Lowry O.H., Rosebrough N.J., Farr A.L. & Randall R.J. (1951). Protein measurement with folin-phenol reagent, J. Biol. Chem. 193: 265-295.

Malhotra S.S. & Hocking D. (1976). Biochemical and cytological effects of sulphur dioxide on plant metabolism, New Phytol. 76: 229-237.

Manohar M. & Balasubramaniam K.A. (1986). Antioxidant enzymes in gastrointestinal tract, Ind. J. Biochem. Biophys. 193: 265-275.

Nandi P.K., Agrawal M., Agrawal S.B. & Rao D.N. (1990). Physiological responses of *Viccia faba* plants to sulphur dioxide, Ecotoxic. Environ. Safety 19: 64-71.

Norman H.A., Krizek D.T. & Mirecki R.M. (2001). Changes in membrane lipid and free fatty acid composition during low temperature preconditioning against SO_2 injury in *Coleus*, Phytochem 58: 263-668.

Peters J.L., Castillo F.J. & Heath R.H. (1989). Alteration of extra cellular enzymes in Pinto bean leaves upon exposure to air pollutants, ozone and sulphur dioxide. Plant Physiol. 89: 159-164.

Qian Y.C. & Yu S.W. (1991). Adaptive responses of soybean plants to SO_2, Acta Phytophysiologica Sinica 17: 232-238.

Quesada M.A., Sanchez-Roldan C., Heredia A, Valpuesta V. & Bukovac M.J. (1992). Peroxidase and IAA oxidase activities and peroxidase isoenzymes in the percarp of seeded and seedless Red haven peach fruit, J. Plant Growth Regul. 11: 1-6.

Ranieri A., Castagna A., Lorenzini G. & Soldatmi G.F. (1997). Changes in thylakoid protein patterns and antioxidant levels in two wheat cultivars with different sensitivity to sulphur dioxide, Environ. Exp. Bot. 37: 125-135.

Rao M.V. (1992). Cellular detoxifying mechanism determine the age dependent injury in tropical trees exposed to SO_2. J. Plant Physiol. 140: 733-740.

Raza S.H. & Gouri K. (2000). Studies on the impact of certain oxidising pollutants and growth regulating substances on *Oryza sativa* var. Tulsi, in: M. Yunus, N. Singh, L.J. de Kok (Eds.), Environmental Stress: Indications, Mitigation and Eco-Conservation pp. 165-171.

Scalet M., Federice R., Guido M.C. & Manes F. (1995). Peroxidase activity and polyamine changes in response to ozone and simulated acid rain in Aleppo pine needles, Enviorn. Exp. Bot. 35: 417-425.

Shimazaki K., Sakaki T., Kondo N. & Sughara K. (1980). Active participation in chlorophyll destruction and lipid peroxidation in sulphur dioxide fumigated leaves of Spinach, Plant Cell Physiol. 27: 1193-1204.

Singh L.P. & Singh S. (2003). Single and joint effect of SO_2 and O_3 in plant growth, photosynthesis and macroelements of mustard, Proc. XII Natl. Symp. Environ. (NSE-12), Tehri, U P, India pp. 320-324.

Singh S. & Singh L.P. (2004). Response of two cultivars of wheat to sulphur dioxide, Asian J. Microbiol. Biotech. Environ. Pollut. 6: 73-76.

Sugahara K. (1984). Effects of air pollutants on light reaction in chloroplasts, in: M.J. Koziol, F.R. Whatley (Eds.), Gaseous Air Pollutants and Plant Metabolism, Butterworths, London pp. 169-180.

Tanaka K. & Sugahara K. (1980). Role of superoxide dismutase in defense against SO_2 toxicity and an increase in superoxide dismutase activity with SO_2 fumigation, Plant Cell Physiol. 21: 601-611.

Varshney S.R.K. & Varhsney C.K. (1985). Response of peroxidase to low levels of SO_2, Environ. Exp. Bot. 25: 107-114.

Wellburn A.R. (1985). SO_2 effects on stromal and thylakoid function, in: W.E. Winner, H.A. Mooney, R.A. Goldstein (Eds.), Suplhur dioxide and Vegetation, Stanford University Press, Stanford, CA pp. 133-147.

Winner W.E., Mooney H.A., Williams K. & Caemmerer S.V. (1985). Measuring and assessing SO_2 effects on photosynthesis and plant growth, in: W.E. Winner, M.A. Mooney, R.A. Goldstein (Eds.), Sulphur dioxide and vegetation, Stanford University Press, Stanford, CA pp. 118-132.

Ziegler (1972). The effect of SO_3^{2-} in the activity of ribulose 1,5-diphosphate carboxylase in isolated spinach chloroplasts, Plant 103: 155-163.

CHAPTER 5

Stimulatory Effect of Herbicide (Atrataf 50W) on Pollen Germination and Tube Growth of Successive Flowers of Apocynaceae: A Critical Review

S.A. Salgare

Abstract

All the different concentrations ($10^{-17} - 10^{-2} - 10^{-3}$, 1, 5, 10, 20-20-100 mg/ml) of atrataf 50W tried were found to be toxic for the germination of pollen of F-24 series of red-flowered cultivar of *Nerium odorum* and F-72 series of pink-flowered cultivar of *Catharanthus roseus*. The widest range of concentrations of atrataf 50W was proved to be 10-17 – 100 mg/ml which stimulated the pollen germination and tube growth of successive flowers of Apocynaceae. The said concentration of the herbicide stimulated the pollen germination and tube growth of F-24 series of white-flowered cultivar of *C. roseus*. Atrataf 50W produced as high as 412.50% stimulation in pollen germination of successive flowers of Apocynaceae which was noted in optimum concentration of atrataf 50W. However, 375.00% stimulation proved to be highest produced by the herbicide in the pollen tube growth of successive flowers of Apocynaceae.

Key words: Environmental Sciences, Growth substances, Physiology of Pollen.

INTRODUCTION

Pollen physiology has attracted the attention of plant breeders and horticulturists ever since the discovery of pollen tube by Amici (1824). Horticulturists and plant breeders often fail to get

Address for correspondence: Salgare Research Foundation Pvt. Ltd., Prathamesh Society, Shivaji Chowk, Karjat-410 201 (MS), India; E-mail: drsalgare@rediffmail.com fertile seeds in spite of all the care taken during artificial pollination. Unless sterility is the main cause, failure of seed setting may be due to slow growth of the pollen tube or its early degeneration in the style.

MATERIALS AND METHODS

Pollen of the following 5 cultivars of Apocynaceae were used: *Nerium odorum* Soland. - red-, pink- and white-flowered cultivars and *Catharanthus roseus* (L.) G. Don. – pink- and white-flowered cultivars. To find out the germination potential of pollen in the bud stage of floral development and their response to the various concentrations ($10^{-17} - 10^{-2} - 10^{-3}$, 1, 5, 10, 20-20-100 mg/ml) of the herbicide (atrataf 50W), the flower buds of various sizes marking the various stages of development and the open flowers were plucked at the same time soon after the dehiscence of the anthers (in open flowers). The range of dehiscence of the anthers was observed from 8:30 a.m. to 9:30 am in *N. odorum* from 6:20 a.m. to 6:55 a.m. in *C. roseus*. Series of successive flowers used - F, F-24, F-48, F-72 series i.e. open flowers and the flower buds which required 24, 48, 72 hours to open respectively. Germination of pollen grains of successive flowers was studied by standing-drop technique in an optimum concentration of sucrose as well as in an optimum concentration of sucrose supplemented by the wide range of concentrations ($10^{-17} - 10^{-2} - 10^{-3}$, 1, 5, 10, 20-20-100 mg/ml) of atrataf 50W (see Tables 5.1 and 5.2). Pollen grains were incubated soon after the dehiscence of anthers. The cultures then transferred to a moist fiher chamber, stored at room temperature (23.6-32.4°C) with RH 84% in the diffused laboratory light at the Institute of Science, Mumbai. The experiments were run in triplicate and average results were recorded. For each experiment a random count of 100 grains was made (from different fields on the slide) to determine the percentage of pollen germination. For measurement of length of pollen tubes, 50 pollen tubes were selected randomly and measured at a magnification of 100 ×. The observations on pollen germination and tube growth were made after 24 hours of sowing.

RESULTS AND DISCUSSION

As a rule the percentage of pollen germination is always less than the pollen viability. However, Banerji and Gangulee (1937) and Dharurkar (1971) reported higher percentage of pollen germination than the pollen viability in *Eichhornia crassipes*. The claim of Banerji and Gangulee (1937) and Dharurkar (1971) is challenged by Salgare (1986c, 1995, 2000d) who stated that the observations of Banerji and Gangulee (1937) and Dharurkar (1971) are exaggerating.

In the present investigation the potentiality of pollen germinability was noted in F series of all the 5 cultivars of the Apocynaceae studied. It was the pollen of F-24 series of red-flowered cultivar of *Nerium odorum* and both the cultivars of *Catharanthus roseus* found germinated in the optimum concentrations of sucrose. It should be pointed out that the pollen of F-48 and F-72 series of pink-flowered cultivar of *C. roseus* showed their germination in the optimum concentrations of sucrose. Thus the potentiality of pollen germinability in Apocynaceae was noted 10 out of 20 series investigated (Tables 5.1 and 5.2). Potentiality of pollen germinability was noted in all the 5 forms of the Solanaceae in F and F-24 series, while the pollen of F-48 series of *Brunfelsia americana* and both the forms of *Datura fastuosa* showed germination in sucrose

Table 5.1: Pollen germination and tube growth of successive flowers of Apocynaceae in sucrose medium

Species	Series	%PV	S%	%G	mm	V/O
N. odorum pink-flowered	F	80	50	35	1485	12.38
N. odorum red flowered	F	74	20	20	1250	11.36
N. odorum white-flowered	F	62	50	20	0675	05.63
C. roseus pink-flowered	F	90	20	60	1575	0.750
C. roseus white-flowered	F	88	20	40	1256	06.28
N. odorum red-flowered	F-24	74	20	06	0486	04.41
C. roseus pink-flowered	F-24	90	50	28	0240	01.14
C. roseus white-flowered	F-24	88	50	16	0248	01.24
C. roseus pink-flowered	F-48	90	50	14	0095	00.45
C. roseus pink-flowered	F-72	90	80	10	0065	00.31

%G, % pollen germination; %PV, % pollen viability; S% concentrations of sucrose in %; V/O, *in vitro* pollen tube length in compare to *in vivo* in %; μm, pollen tube length in μm.

culture medium. It should be noted that the pollen of F-72 series of white-flowered form of *D. fastuosa* showed their germination in the optimum concentration of sucrose. Thus, the potentiality of pollen germinability in the 5 forms of the Solanaceae was reported in 14 out of 20 series investigated (Ram Indar, 1981). Sharma (1984) stated that the potentiality of pollen germinability in 5 cultivars of *Petunia grandiflora* studied was reported in 13 out of 20 series investigated. Pollen of F and F-24 series of all the 5 cultivars showed their germination in the optimum concentrations of sucrose. However, pollen of F-48 series of all the 3 cascades found germinated in the optimum concentrations of sucrose. Singh (1985) reported that the potentiality of pollen germinability in F and F-24 series of all the 5 cultivars of *Solanum melongena* studied. Thus, the potentiality of pollen germinability in 5 cultivars of *S. melongena* was reported in 10 out of 20 series investigated. Salgare (1986a) observed the potentiality of pollen germinability in F and F-24 series of all the 5 cultivars of *Petunia axillaris*. However, it was the pollen of F-48 series of white-flowered cultivar of *P. axillaris* showed their germination in the optimum concentration of sucrose. Thus, the potentiality of pollen germinability in 5 cultivars of *P. axillaris* was found to be in 11 out of 20 series investigated. Johri and Chhaya Roy Chowdhury (1957) stated that in *Citrullus colocynthis*, where pollen grains 'mostly' remained attached in tetrads, satisfactory germination is observed. Kwack (1965) stated that the pollen of *Mimosa pudica* does not germinate in the absence of calcium in the culture medium. However, the Kwack's (1965) argument is challenged by Salgare (1986e, 97b, 2000c). Salgare (1986e, 1997b, 2000c) observed that the germination of pollen of *M. pudica* even *in situ*, tap water, distilled water. Salgare also observed that the germination of pollen of *M. pudica* in the culture medium of glucose and sucrose in the absence of calcium. This proves the superficial and misleading observations of Kwack (1965).

It should be pointed out that even the lowest concentration (10^{-17} mg/ml) of atrataf 50W tried found to be toxic for the germination of pollen of F-24 series of red-flowered cultivar of *Nerium odorum* and F-72 series of pink-flowered cultivar of *Catharanthus roseus*. As a result of which cent per cent suppression of the germination of pollen was caused in the said series. This proves that the pollens of the said series are highly sensitive and act as an ideal indicator of pollution. Thus, it is confirmed that pollen development and activity are more sensitive indicators of adverse factors in the botanical environment and the use of an entire vascular plant (Berg, 1973; Brandt, 1974; Vick and Bevan, 1976; Rasmussan, 1977; Navara, Horvath and Kaleta, 1978; Mhatre, 1980; Mhatre, Chaphekar, Ramani Rao, Patil, Haldar, 1980; Shetye, 1982 and Giridhar, 1984) as an indicator of pollution is a very crude method and rather a wrong choice. There is no evidence of any entire vascular plant exhibiting this much degree of sensitivity. It should be pointed out that Salgare (1983, 84b, 85a, c-d, 86g-i, 2000g, 2001a; Salgare and Theresa Sebastian 86b) and Salgare's Research Group (Ram Indar, 1981; Sharma, 1984 and Singh, 1985) have already proved that the pollen is an ideal indicator of pollution.

The widest range of concentrations of atrataf 50W was proved to be 10^{-17} – 100 mg/ml which stimulated the pollen germination and tube growth of successive flowers of Apocynaceae. The said concentration of the herbicide stimulated the pollen germination and tube growth of F-24 series of white-flowered cultivar of *C. roseus*. Atrataf 50W produced as high as 412.50% stimulation in pollen germination of successive flowers of Apocynaceae which was noted in optimum concentration of atrataf 50W. However, 375.00% stimulation proved to be highest produced by the herbicide in the pollen tube growth of successive flowers of Apocynaceae (Table 5.2).

Thus, the present investigation proves that the herbicides can be most successfully used as the growth substance which is very economical (Table 5.2). This is already proved earlier by Salgare (1983, 86a, 87, 88, 89, 91, 95, 2000a-b) and his Research Group (Ram Indar, 1981; Sharma, 1984; Singh, 1985).

The pollen tube length *in vitro* is 12.31% in F series of red-flowered cultivar of *Nerium odorum* of the tube length found *in vitro* is the longest of all the cultivars investigated of the Apocynaceae (Table 5.2). This proves that though atrataf 50W stimulate tube length, however, it cannot be produced as long as to that found in nature.

Table 5.2: Stimulatory effect of atrataf 50W on pollen germination and tube growth of successive flowers of Apocynaceae

Pollen germination	In optimum concentration of atrataf 50W									
	Pollen tube growth							ROCOHFS		
Species	Series	H	%G	%S	H	μm	%S	V/O	RG	RT
N. odorum pink-flowered	F	10^{-17}	28	Nil	10^{-17}	1053	Nil	08.78	Nil	Nil
N. odorum red-flowered	F	10^{-17}	36	080.00	10^{-17}	1354	008.00	12.31	10^{-17}–10^{-7}	10^{-17}–10^{-13}
N. odorum white-flowered	F	10^{-13}	30	050.00	10^{-11}	0527	Nil	04.39	10^{-17}–10^{-7}	Nil
C. roseus pink-flowered	F	10^{-15}	82	036.67	10^{-9}	1338	Nil	06.37	10^{-17}–10^{-7}	Nil
C. roseus white-flowered	F	10^{-15}	72	080.00	10^{-5}	2170	072.78	10.86	10^{-17}–40	$10–17^{-5}$
N. odorum red-flowered	F-24	Ng	Ng	Ng	Ng	Ng	Ng	Ng	Ng	Ng
C. roseus pink-flowered	F-24	10^{-17}	84	200.00	10^{-7}	0682	184.17	03.25	10^{-17}–80	10^{-17}–80
C. roseus white-flowered	F-24	10^{-15}	82	412.50	10^{-9}	1178	0375.00	05.89	10^{-17}–100	10^{-17}–100
C. roseus pink-flowered	F-48	10^{-9}	28	100.00	10^{-11}	0382	302.11	01.82	10^{-17}–10	10^{-17}–1
C. roseus pink-flowered	F-72	Ng	Ng	Ng	Ng	Ng	Ng	Ng	Ng	Ng

%G, % pollen germination; H, concentration of atrataf 50W in mg/ml; RG, range of concentrations of atrataf 50W for stimulation of pollen germination; RT, range of concentrations of atrataf 50W for stimulation of pollen tube growth; ROCOHFS, range of concentrations of atrataf 50W for stimulation of pollen germination and tube growth; %S, % stimulation; V/O, *in vitro* pollen tube length in compare to *in vivo* in %; μm, pollen tube length in μm.

Although several attempts have been made to improve pollen tube growth on artificial media, no method is yet known which gives rise pollen tube growth as good as that obtained in nature. As a rule, the length of the pollen tube obtained *in vitro* is significantly shorter than that seen *in vivo* (Tables 5.1 and 5.2).

Atrataf 50W stimulated pollen germination in 7 series and tube growth in 5 series of the Apocynaceae (Table 5.2). This shows that atrataf 50W acts as a better stimulator for pollen germination than tube growth.

Sudhakaran (1967) stated that in *Vinca rosea* besides pollen grains which produced single pollen tube, it has also been noticed that tetraploid grains frequently produce more than one pollen tube. Pollen tubes are branched quite frequently. Aberrations of this type in the pollen tube development are not observed in diploid pollen tubes, but quite frequently met with the pollen grains of irradiated plants. Extensive work of Salgare (1979a, 83, 86b) made it very clear that Sudhakaran (1967) had failed to trace out the branched pollen tubes and polysiphonous condition which is fairly common even in diploid pollen grains. Apart from this Sudhakaran (1967) was not able to report the various types of pollen tube deformities either with diploid or tetraploid grains.

The delay in pollen germination was interpreted by Saoji and Chitaley (1972) as being due to the grains not being mature enough to effect pollination, immediately after being shed from the anther. Further they stated that 4-5 hours are required for the complete maturation of pollen grains. It was Salgare (1983) for the first time proved that the pollen grains require resting period before germination and it was the failure of Saoji and Chitaley (1972) who misinterpreted the resting period for pollen maturity (Salgare, 1983, 84a, 85b, 86a, j and Salgare and Theresa Sebastian, 1986c). This is further confirmed even with the extensive work of Salgare (1983, 84a, 85b, 86a, j) and Salgare and Theresa Sebastian (1986c) and Salgare's Research Group (Ram Indar, 1981; Sharma, 1984; Singh, 1985). This resting period of pollen is altered by the effect of herbicides (Salgare, 1983, 84a, 85d, 86a, j; Salgare and Theresa Sebastian, 1986c and Salgare's Research Group - Ram Indar, 1981; Sharma, 1984; Singh, 1985).

Pollen germination and tube elongation are two distinct processes differing in their sensitivity to different concentrations of the herbicide was also confirmed with this extensive work. However, Nair, Nambudiri and Thomas (1973) stated that it has been significant that the optimum percentage of germination and tube length were attained in the same growth medium. With the present work it could be concluded that the observations of Nair, Nambudiri and Thomas (1973) are superficial and misleading. This is also confirmed earlier by Salgare (1979b, 83, 86a, f) and his Research Group (Ram Indar, 1981; Sharma, 1984; Singh, 1985).

The weed control by chemical means has been replacing or supplementing traditional cultural methods in most of the agriculturally advanced countries. Due attention is being paid to the problem in these countries since the economic return from the crop is dependent upon it to a considerable extent. A large number of chemicals, inorganic, organic and auxin herbicides have been used for this purpose. Khosla (1967) stated that no lethal doses could be determined for *Achyranthes aspera* and *Cassia tora* since maleic hydrazide recrystalised at room temperature beyond 5400 ppm. In fact it is apparent that the dosage required to kill the weeds through its conductive tissue is often sufficient to kill the crop as well. Thus, Crafts (1946) reported that spray of 4% solution of sodium arsenite destroyed all barley plants and weeds growing in the same plot. Mitra (1952) and Shivapuri and Sinha (1953) noted some adverse effect of higher concentrations of 2,4-D on crop plants such as abnormalities and sterility of some of the florets and consequent reduction in the yield of wheat. Salgare (1996b, 98a, c, 2000e, 2001 b) doubted whether the chemical weed control is justified?

Alternative method of weed control is suggested by Singhal and Sen (1981) where they recommended that instead of killing the weed, suppress the seed germinability. It should be pointed out that Singhal and Sen's (1981) theory is challenged by Salgare (1986d, 96a, 98b, 2001c) and Salgare and Ram Indar (1995) stating that Singhal and Sen (1981) have forgotten that it is the reproductive stage which consumes maximum nutrients from the soil. Salgare (1986d, 96a, 98b, 2001c) and Salgare and Ram Indar (1995) recommended that instead of killing the weed flowering should be suppressed with very low doses of herbicides. It should be pointed out that Salgare (1996b, 98a, c, 2000e, 2001b) doubted the practice of chemical weed control itself.

Levi and Crafts (1952), Molero and Blackhurst (1956) and Carroll (1957) have reported that MH does not show any residual effect on the soil. It should be pointed out that the findings of these workers are challenged by Salgare (1997a, 2000f), Salgare and Theresa Sebastian (1986a, 90a-c, 91, 95) and Theresa Sebastian and Salgare (1990, 91) also observed the residual effect of dalapon, sodium arsenite, sodiumpentachlorophanate, MH, 2,4-D and 2,4,5- T on the soil and on the succeeding crops. This shows that we must minimise the use of the higher concentrations of the herbicides.

REFERENCES

Amici G.B. (1824). Observations microscopiques sur diverses especes de plantes. Ann. Sci. Nat Bot. 2: 41-70, 211-248.

Banerji I. and Gangulee H.C. (1937). Spermatogenesis in *Eichhornia crassipes* Solam. J. Indian Bot. Soc. 16: 289-296.

Berg H. (1973). Plants as indicators of air pollution. Toxicol. 1: 79-89.

Brandt C.C. (1974). Plants as indicators of air quality. in: Indicators of Environmental quality Ed: Thomas W A Plenum Press, New York p. 101-107.

Carroll R.N. (1957). Literature summary on maleic hydrazide. 1957-1963. Compiled by J. W. Zukel.

Crafts, A.S. 1946. Selectivity of herbicides. Plant Physiol. 21: 345-361.

Dharurkar R.D. (1971). Effect of herbicides on the cytomorphology of *Eichhornia crassipes* Solam (Mart). Ph.D. Thesis, Univ. Bombay.

Giridhar B.A. (1984). Study of interactions between industrial air pollutants and plants. Ph.D. Thesis, Univ. Bombay.

Johri B.M. and Chhaya Roy Chowdhury (1957). A contribution to the embryology of *Citraullus colocynthis* Schare. and *Melothria maderaspatana* Cong. New Phytol. 56: 51-60.

Khosla S.N. (1967). Effect of herbicides on cytomorphology of weeds. Ph.D. Thesis, Univ. Bombay.

Kwack B.H. (1965). The effect of calcium on pollen germination. Proc. Amer. Soc. Hort. Sci. 86: 818-823.

Levi E. and Crafts A. S. (1952). Hilgardia 21: 431.

Mhatre G.N. (1980). Studies in responses to heavy metals in industrial Environment. Ph.D. Thesis, Univ. Bombay.

Mhatre G.N., Chaphekar S.B., Ramani Rao I.V., Patil M.R. and Haldar B.C. (1980). Effect of industrial pollution on the Kalu river ecosystem. Environ. Pollut. Series A 23: 67-78.

Mitra A.K. (1952). A note on measures to check weeds in field of food crops by chemical means. Proc. 8th Meet. Crop. and Soil Wing Bd. Agri. and Ann. Husb. p. 152-153.

Molero F.J. and Blackhurst H.T. (1956). Proc. Amer. Soc. Hort. Sci. 67: 416.

Nair P.K. K., Nambudiri E.M.V. and Thomas M.K. (1973). A note on pollen germination at various stages of development of flower buds of Balsam (*Impatians balsam*). J. Palyno. 4: 29-33.

Navara J., Horvath I. and Kaleta M. (1978). Contribution to the determination of limiting of values of SO_2 for vegetation in the region of Bratislava. Environ. Pollut. Series A 16: 249-262.

Ram Indar (1981). Effect of herbicides on plants of the Solanaceae – I. M.Sc. Thesis, Univ. Bombay.

Rasmussan L. (1977). Epiphytic bryophytes as indicators of changes in the background levels of air-borne metals from 1951 to 1975. Environ. Pollut. Series A 14: 34-45.

Salgare S.A. (1979a). Reinvestigation of Palynology of *Vinca rosea* Linn. Proc. 66th Session of Indian Sci. Congr. Botany Section, Abstract No. 130.

Salgare S.A. (1979b). A Criticism of Nair, Nambudiri and Thomas' Paper entitled, 'A Note on pollen germination at various stages of development of flower buds of Balsam (*Impatiens balsam*).' Proc. 66th Session of Indian Sci. Congr. Botany Section, Abstract No. 131.

Salgare S.A. (1983). Pollen physiology of angiosperms – I. Ph.D. Thesis, Univ. Bombay.

Salgare S.A. (1984a). A Criticism on the Paper of Saoji and Chitaley (1972) entitled, Palynological studies in *Bauhinia variegata* Linn.' Proc. 4th Indian Palynological Conf., held on March 19-21, 1984 at Andhra Univ., Visakhapatnam. Abstract No. S III-05.

Salgare S.A. (1984b). Further Evidence of a Criticism on the Hypothesis of Mhatre (1980), Mhatre, Chaphekar, Ramani Rao, Patil, Haldar (1980) and Shetye (1982). Proc. 4th Indian Palynological Com., held on March 19-21,1984 at Andhra Univ., Visakhapatnam. Abstract No. S III-06.

Salgare S.A. (1985a). A Criticism on the Findings of Mhatre (1980), Mhatre, Chaphekar, Ramani Rao, Patil, Haldar (1980) and Shetye (1982). Proc. 2nd Ann. Com. of Nat. Environ. Sci. Acad., held on May 25-27, 1985 at Awadh Univ., Faizabad. Abstract No. 2.

Salgare S.A. (1985b). A Criticism on the Hypothesis of Saoji and Chitaley (1972). Proc. 2nd Ann. Com. of Nat. Environ. Sci. Acad., held on May 25-27,1985 at Awadh Univ., Faizabad. Abstract No. 72.

Salgare S.A. (1985c). Further Evidence of a Criticism of the Findings of Mhatre (1980), Mhatre, Chaphekar, Ramani Rao, Patil, Haldar (1980), Shetye (1982) and Giridhar (1984). Proc. Nat. Symp. on Environ. Sci. and Warm Water Aquculture-Finfish, held on November 6-9, 1985 at M.J.K.(P.G.) College, Bettiah. Abstract No. 2.

Salgare S.A. (1985d). A Criticism on the Findings of Mhatre (1980), Mhatre, Chaphekar, Ramani Rao, Patil, Haldar (1980) and Shetye (1982). Proc. Nat. Symp. on Environ. Sci. and Warm Water Aquculture-Finfish, held on November 6-9,1985 at M.J.K.(P.G.) College, Bettiah. Abstract No. 73.

Salgare S.A. (1986a). Effect of herbicides on pollen physiology of *Petunia axillaries* BSP. Ph.D. Thesis, World Univ.

Salgare S.A. (1986b). A Criticism on Ph.D. Thesis of Sudhakaran (1967) entitled, 'Cytogenetic studies in Vinca rosea Linn.' Proc. 1st Nat. Symp. on Environ. Biol., held on December 30-31, 1986 at Sri Krishnadevaraya Univ., Anantapur. Abstract No. 5.

Salgare S.A. (1986c). A Criticism on Ph.D. Thesis of Dharurkar (1971) entitled, 'Effect of Herbicides on the Cytomorphology of *Eichhornia crassipes* Solam (Mart.).' Proc. 1st Nat. Symp. on Environ. Biol., held on December 30-31, 1986 at Sri Krishnadevaraya Univ., Anantapur. Abstract No. 6.

Salgare S.A. (1986d). A New Approach in Weed Control and a Criticism on the Findings of Singhal and Sen (1981). Proc. 1st Nat. Symp. on Environ. Biol., held on December 30-31, 1986 at Sri Krishnadevaraya Univ., Anantapur. Abstract No. 10.

Salgare S.A. (1986e). A Criticism on the Findings of Kwack (1965). Proc. 1st Nat. Symp. on Environ. Biol., held on December 30-31, 1986 at Sri Krishnadevaraya Univ., Anantapur. Abstract No. 12.

Salgare S.A. (1986f). A Criticism on the Findings of Nair, Nambudiri and Thomas (1973). Proc. 1st Nat. Symp. on Environ. Biol., held on December 30-31, 1986 at Sri Krishnadevaraya Univ., Anantapur. Abstract No. 13.

Salgare S.A. (1986g). A Criticism on Ph.D. Thesis of Giridhar (1984) entitled, 'Study of Interactions between Industrial Air Pollutants and Plants.' Proc. 1st Nat. Symp. on Environ. Biol., held on December 30-31, 1986 at Sri Krishnadevaraya Univ., Anantapur. Abstract No. 14.

Salgare S.A. (1986h). A Criticism on the Hypothesis of Berg (1973), Brandt (1974), Vick and Bevan (1976), Rasmussan (1977), Navara, Horvath and Kaleta (1978), Mhatre (1980), Mhatre, Chaphekar, Ramani Rao, Patil, Haldar (1980), Shetye (1982) and Giridhar (1984). Proc. 1st Nat. Symp. on Environ. Biol., held on December 30-31, 1986 at Sri Krishnadevaraya Univ., Anantapur. Abstract No. 15.

Salgare S.A. (1986i). A Criticism on Ph.D. Thesis of Shetye (1982) entitled, 'Effect of Heavy Metals on Plants.' Proc. 1st Nat. Symp. on Environ. Biol., held on December 30-31, 1986 at Sri Krishnadevaraya Univ., Anantapur. Abstract No. 16.

Salgare S.A. (1986j). Further Evidence of a Criticism on the Findings of Saoji and Chitaley (1972). Proc. 1st Nat. Symp. on Environ. Biol., held on December 30-31, 1986 at Sri Krishnadevaraya Univ., Anantapur. Abstract No. 17.

Salgare S.A. (1987). Use of herbicides as growth substances. Bull. Environ. Sci. 4:22-25.

Salgare S.A. (1988). Stimulatory effect of some herbicides on pollen tube growth of successive flowers of Apocynaceae. Ad. Plant Sci. 1(2 suppl.): 255-261.

Salgare S.A. (1989). Stimulatory effect of gramoxone and dalapon on pollen tube growth of successive flowers of Apocynaceae. Pesticides 23: 46-48.

Salgare S.A. (1991). Stimulatory effect of herbicides (atrataf 50W, basalin EC, femoxone) on pollen tube growth of successive flowers of Apocynaceae. Bioved 1: 183-188.

Salgare S.A. (1995). Interesting observations on the physiology of pollen of *Eichhornia crassipes* Solam. (Mart). Proc. 13th Nat. Symp. on Life Sci., Proceeded with Ann. Session of Indian Soc. of Life Sci., held on December 30-31, 1995 & January 1, 1996 at CH. Charan Singh Univ., Meerut. Abstract No. 83.

Salgare S.A. (1996a). A new method of weed control. Proc. 5th APSI Scientists Meet-Cum-Nat. Seminar on Scope of Microbiology in Agri. & Industry & Information on Medicinal plants, held on October 4-5, 1996 at Gurukul Kangri Univ., Hardwar. Abstract No. 64.

Salgare S.A. (1996b). Is chemical weed control justified? Proc. 5th APSI Scientists Meet-Cum-Nat. Seminar on Scope of Microbiology in Agri. & Industry & Information on Medicinal plants, held on October 4-5, 1996 at Gurukul Kangri Univ., Hardwar. Abstract No. 68.

Salgare S.A. (1997a). Effect of persisting herbicides on seed germination of succeeding crops. Proc. 14th Nat. Symp. of Indian Soc. of Life Sci. on Recent Trends in Life Sci., held on October 23-25, 1997 at Panjabi Univ. Patiala. Abstract No. 74.

Salgare S.A. (1997b). A fresh light on the observations of Kwack (1965). Proc. Nat. Symp. on Recent Advances in Biosciences, Proceeded with Ann. Session of Biosciences, held on November 3-5, 1997 at Maharshi Dayanand Univ. Rohtak. Abstract No. 239.

Salgare S.A. (1998a). Is chemical weed control justified? Proc. Nat. Symp. on Combating Pollutant Accumulation in Eco-system for Sustainable Agri., held on October 27-28, 1998 at Allahabad Agricultural Institute, Allahabad.

Salgare S.A. (1998b). A new method of weed control. Proc. Inter. Conf. on Pest and Pesticide Management for Sustainable Agri., held on December 11-13, 1998 at C.S. Azad Univ. of Agri. and Technology, Kanpur.

Salgare S.A. (1998c). Is chemical weed control justified? Proc. 23rd Ann. Conf. of Orissa Bot. Soc., held on December 29-30, 1998 at P. N. College, Khurda. Abstract No. 24.

Salgare S.A. (1999). Use of herbicide (acrolein) as growth substance–A Critical Review-I. Biojournal 11: 15-27.

Salgare S.A. (2000a). Use of herbicide (dalapon) as growth substance. Him. J. Environ. Zoo. 14: 69-82.

Salgare S.A. (2000b). Use of herbicide (Sodium arsenite) as growth substance – A Critical Review – I. J. Ecotoxicol. Environ. Monit. 10: 221-232.

Salgare S.A. (2000c). Germination of pollen of the sensitive plant, *Mimosa pudica* Linn. In glucose. Bionotes 2: 35.

Salgare S.A. (2000d). A Criticism on the findings of Banerji and Gangulee (1937) and Dharurkar (1971): Him. J. Environ. Zoo. 14: 159-160.

Salgare S.A. (2000e). Is chemical weed control justified? (Use of MH) — A Critical Review. Proc. 1st Nat. Com. on Recent Trends in Life Management, held on October 22-23, 2000 at Bipin Bihari (P.G.) College, Jhansi. Abstract No. 81.

Salgare S.A. (2000f). Further Evidence of a Criticism on the Hypothesis of Levi and Crafts (1952), Molero and Blackhurst (1956) and Carroll (1957) – A Critical Review. Proc. 1st Nat. Conf. on Recent Trends in Life Management, held on October 22-23, 2000 at Bipin Bihari (P.G.) College, Jhansi. Abstract No. 88.

Salgare S.A. (2000g). Additional Evidence of a Criticism on the Hypothesis of Berg (1973), Brandt (1974), Vick and Bevan (1976), Rasmussan (1977), Navara, Horvath and Kaleta (1978), Mhatre (1980-Ph.D. Thesis), Mhatre, Chaphekar, Ramani Rao, Patil, Haldar (1980), Shetye (1982-Ph.D. Thesis), and Giridhar (1984-Ph.D. Thesis) – A Critical Review. Proc. 1st Nat. Com. on Recent Trends in Life Management, held on October 22-23, 2000 at Bipin Bihari (P.G.) College, Jhansi. Abstract No. 92.

Salgare S.A. (2001a). Monitoring of herbicide (Nitrogen) toxicity by using pollen as Indicators – A Critical Review – I. Website www.microbiologyou.com U.S.A.'s Publication.

Salgare S.A. (2001b). Is chemical weed control justified? (Use of MH) – A Critical Review – I. Proc. 88th Session of Indian Sci. Congr., New Delhi, January 3-7, 2001, Botany Section, Abstract No. 72.

Salgare S.A. (2001c). A New method of weed control – A Critical Review. Proc. 88th Session of Indian Sci. Congr., New Delhi, January 3-7, 2001, Botany Section, Abstract No. 110.

Salgare S.A. and Ram Indar (1995). A New Method of weed control. Proc. Nat. Symp. on Recent Advances in Biosciences, Proceeded with Ann. Session of Biosciences, held on November 3-5, 1997 at Maharshi Dayanand Univ., Rohtak. Abstract No. 115.

Salgare S.A. and Theresa Sebastian (1986a). A Criticism on the hypothesis of Levi and Crafts (1952), Molero and Blackhurst (1956) and Carroll (1957). Proc. 1st Nat. Symp. on Environ. Biol., Proceeded with 3rd Ann. Conf. of Nat. Environ. Sci. Academy, held on December 30-31, 1986 at Sri Krishnadevarays Univ., Anantapur. Abstract No. 8.

Salgare S.A. and Theresa Sebastian (1986b). A Criticism on the hypothesis of Berg (1973), Brandt (1974), Vick and Bevan (1976), Rasmussan (1977), Navara, Horvth and Kaleta (1978), Mhatre (1980), Mhatre, Chaphekar, Ramani Rao, Patil, Haldar (1980), Shetye (1982) and Giridhar (1984). Proc. 1st Nat. Symp. on Environ. Biol., Proceeded with 3rd Ann. Conf. of Nat. and Environ. Sci. Academy, held on December 30-31, 1986 at Sri Krishnadevaraya Univ., Anantapur. Abstract No. 15.

Salgare S.A. and Theresa Sebastian (1986c). Further evidence of a criticism on the findings of Saoji and Chitaley (1972). Proc. 1st Nat. Symp. on Environ. Biol., Proceeded with 3rd Ann. Conf. of Nat.

Environ. Sci. Academy, held on December 30-31, 1986 at Sri Krishnadevaraya Univ., Anantapur. Abstract No. 17.

Salgare S.A. and Theresa Sebastian (1990a). Effect of persistent 2,4,5-T in soil on seed physiology of *Vigna mungo* (L.) Hepper. Glimpes of Research Activities of the Govt. Institute of Sci., Bombay, held on January 25, 1990 at Govt. Insti. of Sci., Bombay. Abstract No. Bot. 11.

Salgare S.A. and Theresa Sebastian (1990b). Residual toxicity of sodium arsenite on seed. physiology of *Vigna mungo* (L.) Hepper. Proc. Inter. Symp. on Environ. Influences on seed and germination mechanism – Recent advances in Research and Technology, held on January 27-29, 1990 at Univ. Jodhpur, Jodhpur. Abstract No. 25.

Salgare S.A. and Theresa Sebastian (1990c). Role of persisting dalapon in soil on seed germination. Proc. Inter. Symp. on Environ. Influences on seed and germination mechanism – Recent advances in Research and Technology, held on January 27-29, 1990 at Univ. Jodhpur, Jodhpur. Abstract No. 26.

Salgare S.A. and Theresa Sebastian (1991). Residual toxicity of MH on productivity of *Vigna mungo*. Nat. Symp. on Environ. and Life, held on April 27-28, 1991 at D.A.V. College, Kanpur. Abstract No. 33.

Salgare S.A. and Theresa Sebastian (1995). Effect of persisting herbicides in soil on germination of seeds of succeeding crops. Proc. Nat. Symp. on Recent Advances in Biosciences, Proceeded with Ann. Session of Biosciences, held on November 3-5, 1997 at Maharshi Dayanand Univ., Rohtak. Abstract No. 114.

Saoji A.A. and Chitaley S.D. (1972). Palynological studies in *Bauhinia variegata* Linn. Botanique 3: 7-10.

Sharma R. I. (1984). Effect of herbicides on plants of the Solanaceae – II. Ph.D. Thesis, Univ. Bombay.

Shetye R.P. (1982). Effect of heavy metals on plants. Ph.D. Thesis, Univ. Bombay.

Shivapuri T.N. and Sinha R.P. (1953). Studies on weed eradication in U. P. Proc. 40th Session of Indian Sci. Congr. Botany Section 3: 146.

Singh S.R. (1985). Effect of herbicides on pollen physiology of Brinjal. M.Sc. Thesis, Univ. Bombay.

Singhal B.K. and Sen D.N. (1981). A new method of weed control. Curr. Sci. 50:414-415.

Sudhakaran, I.V. 1967. Cytogenetic studies in *Vinca rosea* Linn. Ph.D. Thesis, Univ. Bombay.

Theresa Sebastian and Salgare S.A. (1990). Residual toxicity of dalapon on productivity. Proc. Nat. Seminar on Agro- Technology and Bioenergy for Sustained Crop and Agro-Industrial Productivity, held on May 10-11, 1990 at Vasantdada Sugar Institute, Manjari (Bk.). Abstract No. A. I (10).

Theresa Sebastian and Salgare S.A. (1991). Residual toxicity of MH on productivity of *Vigna mungo*. Proc. Nat. Symp. on Environ. and Life, held on April 27-28, 1991 at Kanpur. Abstract No. 33.

CHAPTER 6

Abiotic Stresses and Physiology of Drought Resistance in Chickpea

S.C. Gupta

INTRODUCTION

Plants grow and develop through a well defined process of growth and differentiation. Normal functioning of these processes requires an optimum supply of water, minerals, oxygen, carbon dioxide, suitable temperature, light etc. Scarcity or excess supply of any of these abiotic factors limit the normal functioning of plant processes. In normal plant life cycle the major plant processes are germination, seedling growth, vegetative growth, leaf area development, dry matter accumulation, flowering, fruiting and senescence—all of which are sensitive to environmental conditions. Any unusual, natural or anthropogenic aberration in the environment, thus affect the plant growth, its development and productivity.

The estimates show that world over the arable land is about 1474 million hectares out of which the unirrigated area is tremendous (about 1247 million hectares, see: Kumar 2001). In this unirrigated area drought is an important abiotic factor limiting the crop production. The drought can be primary stress produced by water deficiency in the environment surrounding the plant and/or a secondary stress induced by either chilling, freezing, heat or salt stress (Levitt, 1980). In India about 36% of the land area is arid or semi-arid while the rest 64% is exposed to temporary drought during the crop season.

ABIOTIC STRESS

Abiotic stress is usually defined as an external factor that exerts a disadvantageous influence on plant processes resulting in reduced growth and productivity. The external abiotic factors causing stress are inadequacy or excess supply of:

Address for correspondence: Rajasthan Agricultural University, Agricultural Research Station, Durgapura-Jaipur, India

I. Water
II. Temperature
III. Salts
IV. Solar radiation
V. Atmospheric pollutants

Out of these the first three are the common factors causing stress to crop plants. A brief account of the effect of these stress factors is described here.

I. WATER

Water is the most important and essential ingredient required for various plant processes during the life cycle of a plant. The important functions of water in plants are:

(i) It acts as a solvent and provides a medium for chemical reactions.
(ii) It provides medium for the transport of inorganic and organic solutes.
(iii) It gives turgor to plant cells, essential to promote cell enlargement.
(iv) It causes hydration and neutralisation of charges on colloidal molecules. For enzymes, water of hydration helps to maintain structure and facilitates catalytic functions.
(v) It is a raw material for photosynthesis.
(vi) It helps in monitoring plant temperature through transpirational cooling of plant surfaces.

Looking to the ubiquitous involvement of water in plant life any aberration in its availability due to drought will affect all the plant processes and in turn the growth and yield of crops.

Drought Stress

Scarcity of available water in plants results into a situation well known as drought. The availability of water during the whole plant life is determined not only by rainfall alone, but by the soil characteristics and evaporative demand also. Hence, the intensity and duration of drought become highly location specific. The eratic pattern of rainfall distribution further adds to the uncertainty of stage specificity to the drought. Thus a working definition of drought would be "the inadequacy of water availability, including precipitation and soil moisture storage capacity, in quantity and distribution, during the life cycle of a crop to restrict the expression of its full genetic yield potential". Generally drought manifests itself when various combinations of physical and environmental factors cause internal water deficit in crops sufficient to reduce their productivity. Arnon (1975) differentiated between "atmospheric drought" and "soil drought", because the response of plants to these two types of droughts may be different.

Atmospheric drought: Low air humidity, frequently accompanied by hot, dry winds results into atmospheric drought. It may occur even under conditions of relatively high soil moisture.

Soil drought: When the availability of soil moisture to the plant is less than evapotranspiration, it results in soil drought. It is usually gradual and progressive.

Pulses often experience two types of drought :

(i) Intermittent drought may occur during any of the growth stages due to soil moisture deficiency.

(ii) Terminal drought may occur during crop maturity stage due to receding soil moisture, more crop demand and increasing air temperatures.

Effects of Drought

Water deficiency adversely affects the plant growth (Ali *et al*, 1998) and has effects at different plant levels. These can be summarised as follows:

Morphological effects: At whole plant level:

(i) The whole plant stature is reduced.

(ii) There is reduction in dry matter accumulation (Fig. 6.1)

(iii) Seedling vigour and crop establishment are adversely affected (Singh 1990).

(iv) Crop productivity is reduced (Fig. 6.2).

(v) The crop survival becomes difficult under severe water deficiency.

At organ level:

(i) There is reduction in stem elongation.

(ii) Leaf enlargement is reduced (Fig. 6.3A and 6.3B).

(iii) The fruits do not develop to their full size.

(iv) There is poor nodulation.

At cellular level:

(i) Cell divisions are adversely affected.

(ii) There is reduction in cell enlargement.

(iii) The cell differentiation is poor.

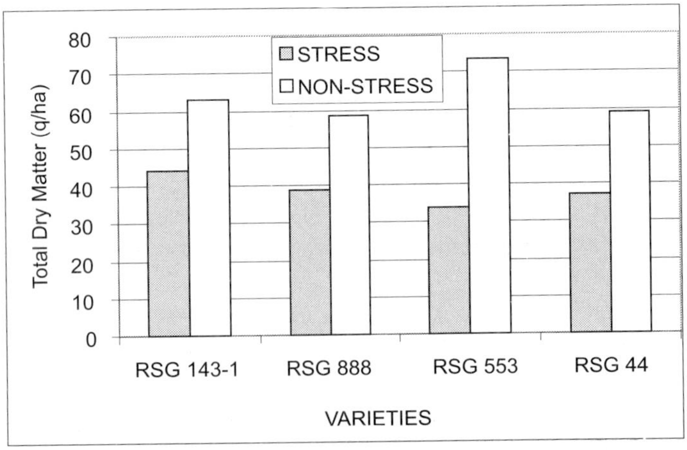

Fig. 6.1: Effect of moisture stress on above ground total dry matter production in some chickpea genotypes as compared to the non-stress conditions in fields.

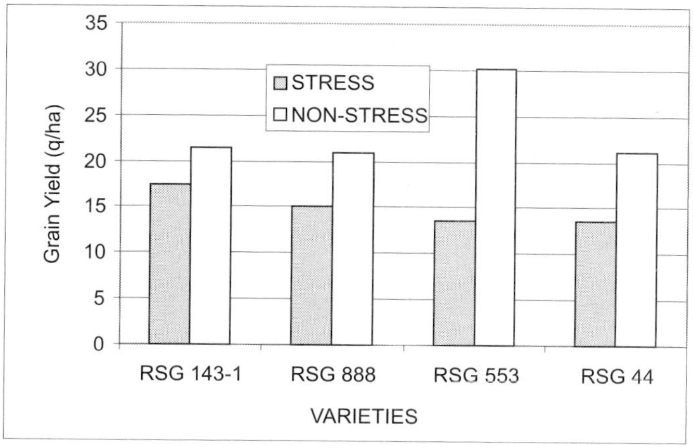

Fig. 6.2: Effect of moisture stress on grain productivity in some chickpea genotypes as compared to the non-stress conditions in fields.

Physiological effects: There is a reduction in:

(i) mineral uptake due to air filled pores between the soil particles and there being no direct pathway from the soil to the root surface (Nye and Tinker, 1977; see : Pugnaire *et al*, 1999),

(ii) translocation of solutes,

(iii) plant water potential (Garg *et al*, 2001),

(iv) stomatal conductance,

(v) carbon dioxide diffusion, thus adversely affecting photosynthesis (Fig. 6.4),

(vi) nitrate reductase activity (Garg *et al*, 2001),

(vii) chlorophyll contents (Garg *et al*, 2004),

(viii) specific leaf area (Fig. 6.5).

There is an increase in:

(i) respiratory breakdown of reserve solutes and peroxidase activities (Singh and Kumar 1992),

(ii) hydrolysis of starch to sugars which in turn may adversely affect the photosynthetic carbon fixation (Garg *et al*, 2001),

(iii) production of reactive oxygen forms (O_2^-, superoxide) which cause damage to the enzymes, the chloroplast pigments, the membrane lipids and proteins, and alterations of their structural integrity (Anderson *et al*, 1990; Navari – Izzo *et al*, 1994),

(iv) ABA level in roots, which is translocated to leaves and results in increased stomatal resistance (Zhang *et al*, 1987),

(v) ethylene production, which results in early abscission of leaves, flowers and fruits (see: Taiz and Zeiger 1998),

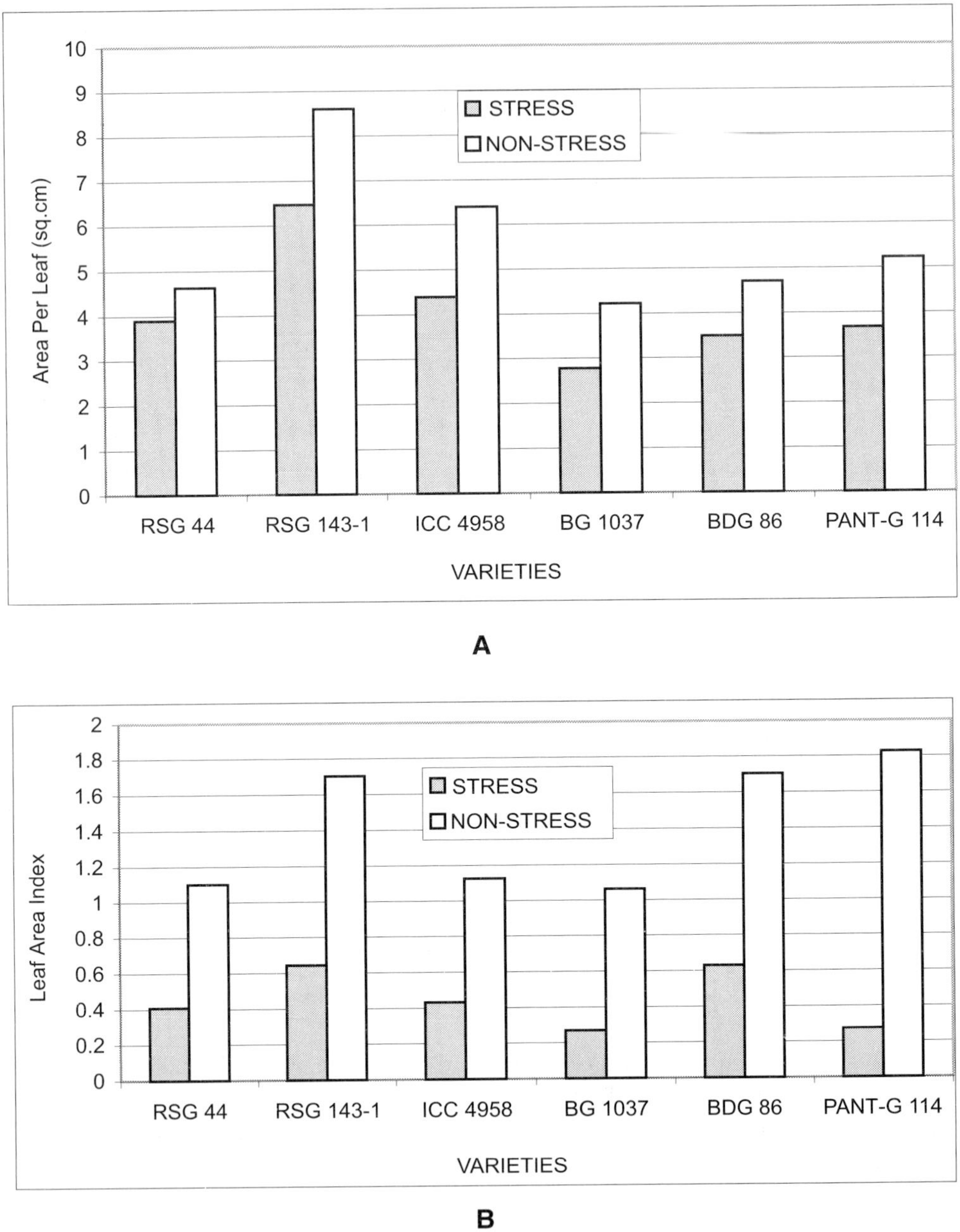

Fig. 6.3: Effect of moisture stress on leaf area development in some chickpea genotypes as compared to the non-stress conditions in fields. (A) Effect on individual leaf (B) Effect on canopy expressed as leaf area index i.e. the ratio of leaf area to land area in a plot.

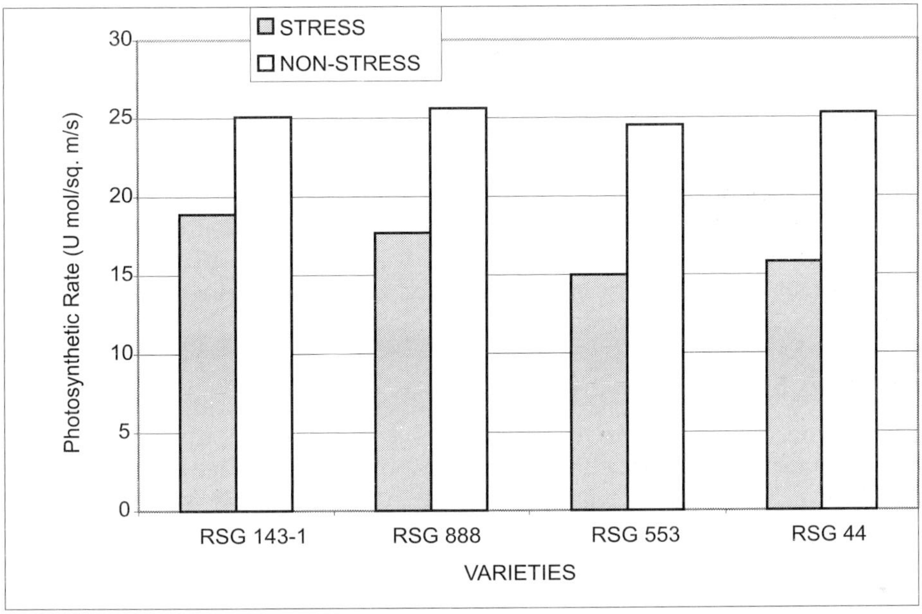

Fig. 6.4: Effect of moisture stress on photosynthetic rate in some chickpea genotypes as compared to the non-stress conditions in fields. See that the tolerant genotype RSG 143-1 maintains comparatively higher photosynthetic rate under stress.

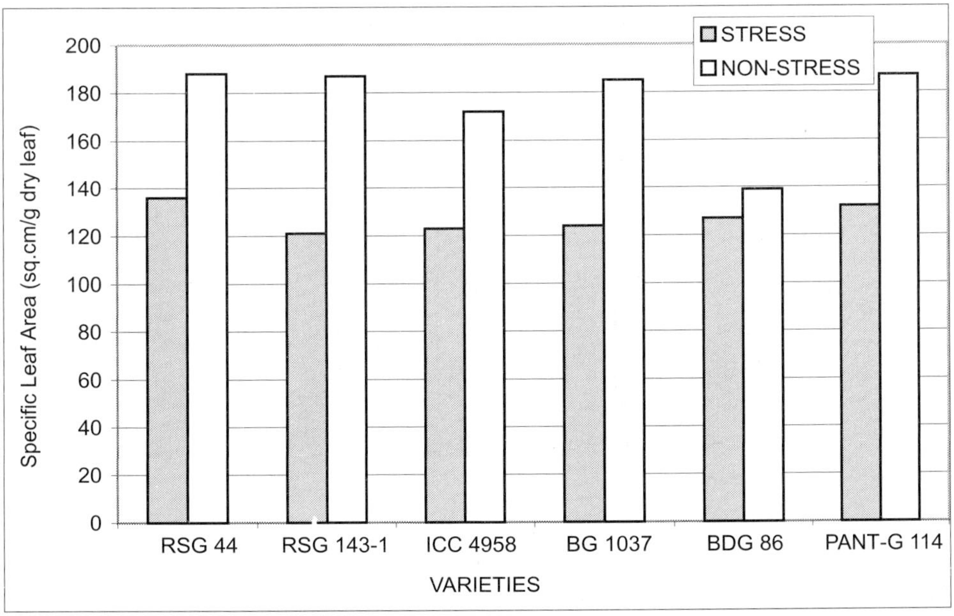

Fig. 6.5: Effect of moisture stress on specific leaf area i.e. the ratio of leaf area to its dry weight in some chickpea genotypes as compared to the non-stress conditions in fields.

(vi) proteolytic activities, with increased protein hydrolysis it results into accumulation of toxic products like ammonia, which may cause tissue injuries (see: Taiz and Zieger 1998),

(vii) soluble osmolytes (Gupta *et al.* 2000, Ghorbanli *et al*, 2003; Fig. 6.6),

(viii) canopy temperature (Fig. 6.7).

In general the metabolic activities including nucleic acid and protein synthesis are impaired under stress. The increased respiration combined with reduced photosynthesis may cause starvation. Plants are usually more sensitive to the stress during late vegetative and reproductive stages because of the competition for assimilates between roots and developing flowers and fruits (Silim and Saxena, 1993).

Waterlogging Stress

Excess of water in the soil results in waterlogging stress which adversely affects the plant processes and in turn crop production. Rainy season crops may face such type of stress when the precipitation is more and the water is stagnant in the fields or is poorly drained. In flooded soils, water fills the pores and blocks the diffusion of oxygen in the gaseous phase. Dissolved oxygen diffuses very slowly in the stagnant water. It is only a few centimetres of soil near the surface that gets oxygenated. Thus, waterlogging in the root zone causes poor soil aeration or the anaerobic conditions thereby inducing unhealthy physiological changes in plants and subsequently influences growth and development. When temperatures are low and plants are

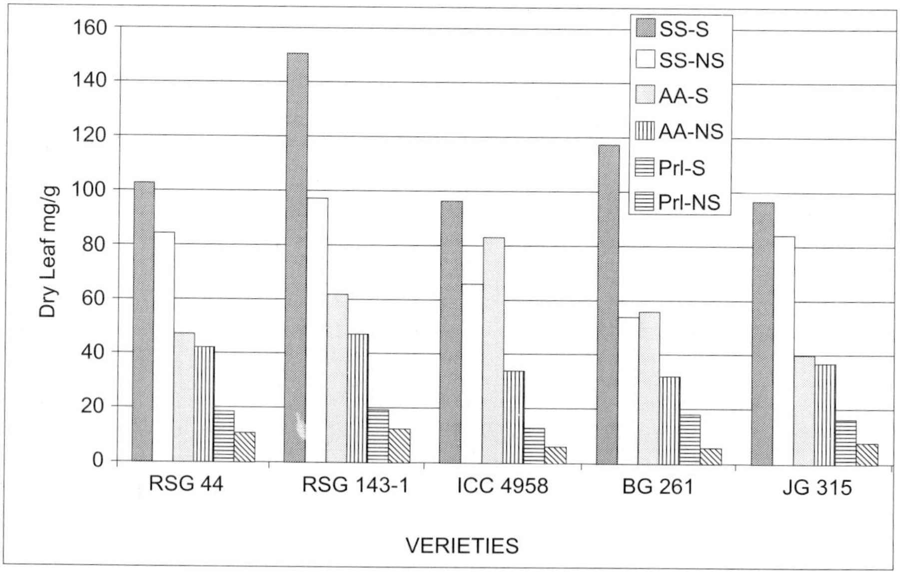

Fig. 6.6: Effects of moisture stress on accumulation of osmolytes in leaf tissue in some chickpea genotypes as compared to the non-stress conditions in fields. See that the accumulation of osmolytes under stress is comparatively higher in the tolerant genotype RSG 143-1 (SS = Soluble Sugar, AA = Soluble Amino Acids, Prl = Proline, -S = Stress, -NS = Non-Stress; the values of Amino acids and proline in the figure are ten times the original values)

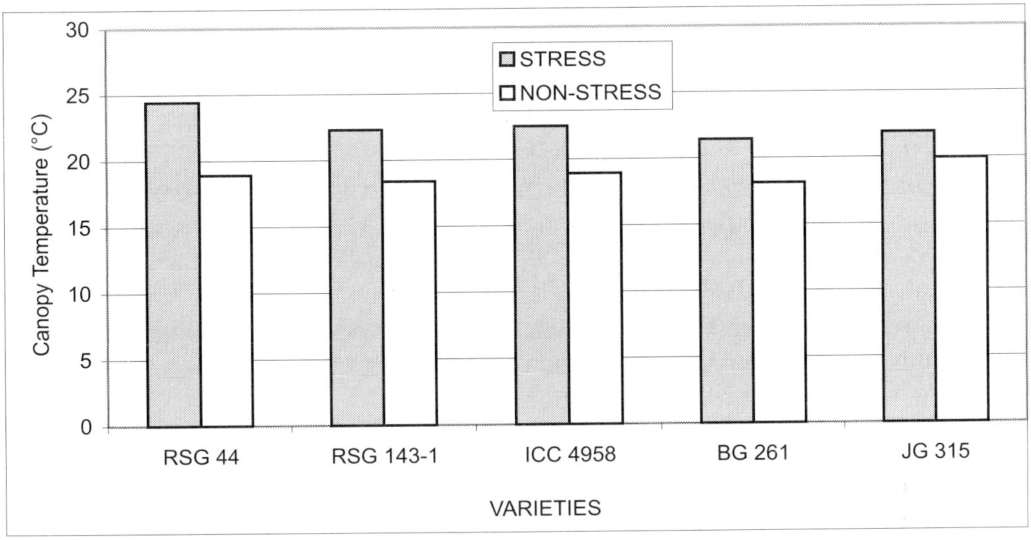

Fig. 6.7: Effect of moisture stress on canopy temperature in some chickpea genotypes as compared to the non-stress conditions in fields.

relatively dormant, oxygen depletion is very slow and the consequences of stress are less harmful. However, when the temperatures are higher (>20°C), oxygen consumption by plant roots, soil fauna and soil microorganisms can totally deplete oxygen from the bulk of the soil - water in as little as 24 hours. The growth and survival of many plant species are greatly depressed under such conditions and crop yields are severely reduced. The problem of salinity is often accompanied with waterlogging which further deteriorates the growth conditions. In India more than six million hectare area suffers with the problem of waterlogging stress.

Effects of Waterlogging Stress

Waterlogging resulting into oxygen deficiency which may cause the following aberrations in plant processes:

(i) Failure of roots to absorb nutrient ions and transport them to the xylem and shoot leading to a shortage of ions within the developing tissues.

(ii) Production and translocation of ABA by hypoxic roots and may result in stomatal closure (Zhang and Zhang, 1994).

(iii) Production of anaerobic stress proteins.

(iv) Anaerobes in the soil reduce sulphates to hydrogen sulphide, which is a respiratory poison.

(v) Anaerobes reduce ferric iron to ferrous stage which is more soluble and can rise to the toxic level if soils are anaerobic for several weeks.

(vi) Bacteria release metabolites such as acetic acid and butyric acid in soil water, which along with reduced sulphur compounds account for the unpleasant odour and are toxic to plants at high concentrations (Drew and Lynch, 1980).

(vii) Anaerobes derive their energy from the reduction of nitrate to nitrite to nitrous oxide and finally to molecular nitrogen (denitrification), which is then lost into the atmosphere.

(viii) Cytoplasmic acidosis due to leaking of protons across the tonoplast into the cytosol, which irreversively disrupts metabolism in the cytoplasm.

(ix) Tricarboxylic acid cycle may become non-operative due to lack of oxygen.

(x) Ceasing of electron transport and oxidative phosphorylation in mitochondria.

(xi) ATP production only by fermentation of pyruvic acid to lactate through the action of lactate dehydrogenase.

(xii) Lower ATP production (2 moles per mole of hexose sugar entering in glycolysis) is insufficient to drive the essential metabolic processes (Drew 1997).

(xiii) Reduction in dry matter accumulation.

(xiv) Under severe conditions the total plant loss may occur.

II. TEMPERATURE

High Temperature Stress

Arid environments are generally hot and dry. Short periods of high temperature (>35°C) are common, particularly during the grain filling period of rabi crops limiting the genetic yield potential. The duration of such episode of high temperature may vary from short, less destructive periods to prolonged highly destructive periods (Blumenthal *et al,* 1991). Further global warming due to the impact of greenhouse gases is accentuating the degree of high temperature stress. The projected temperature changes of global level (Climate Change 1992) are 1.5–4.0°C rise annually, which is much greater than any climatic change. Thus, high temperature stress is becoming a commonly experienced problem hampering the plant productivity.

Effects of High Temperature Stress

High temperatures, depending on duration and the stage of crop growth, reduce crop growth and productivity through the following effects:

(i) Reduction in growth period and grain filling period.

(ii) Ceasing plant growth due to starvation. The optimum temperature for respiration is higher than that for photosynthesis. Thus reduced photosynthesis with increased respiration cause starvation (Björkman *et al,* 1980).

(iii) Inhibition of specific enzyme reactions and breakdown of proteins.

(iv) Accumulation of toxic substances like ammonia, causing injury to plant tissues.

(v) Adverse effect on seed germinability due to heat stress during seed maturation.

(vi) Excessive fluidity of membranes resulting into the loss of metabolic integrity.

(vii) Untimely closure of stomata and adverse effect on gaseous exchange (Lu *et al,* 1998).

(viii) More water loss does not keep pace with absorption and hence results into protoplasmic plasmolysis and reduction in growth.

(ix) Synthesis of specific proteins called heat shock proteins which may sometimes be helpful in protection from the stress damage (Sachs and Ho, 1986; Vierling, 1991; Bonk *et al*, 1996).

Low Temperature Stress

Tropical and subtropical plants generally show chilling injury when temperatures are too low for normal growth but not low enough for ice to form. However, when the temperature falls below the freezing point of water, it causes freezing injury due to ice formation. Following may be the effects of chilling or freezing stress on plants:

(i) Poor germination and seedling vigour (Özdemir 1996; O'-Toole *et al*, 2001).

(ii) Slow plant growth due to low rate of metabolic activities and poor yields (Singh *et al*, 1993; Erskine *et al*, 1994; Toker and Cagiran, 1996; Croser *et al*, 2003).

(iii) Discolouration or lesions appear on leaves.

(iv) Foliage appears water soaked.

(v) Chilling of roots adversely affects absorption and translocation processes thus resulting into desiccation and wilting of plants.

(vi) Changes occur in the protoplasmic colloidal system.

(vii) Adverse changes occur qualitatively and quantitatively in chlorophyll, organic and inorganic contents (Rao and Sahu, 1991).

(viii) Formation of intracellular ice, first in the cytoplasm and then in the vacuoles, causes membranes to become permeable and may result in bursting of cells.

(ix) Extracellular ice formation lowers the water potential and thus drives outwards movement of water causing dehydration.

(x) When the temperatures is low but well above 0°C, the lipids in the bilayer of membranes having higher percentage of saturated fatty acid chains tend to solidify into a semicrystalline stage, thus making the membrane less fluid and disrupting the functioning of the protein component, which in turn inhibit several enzyme dependent metabolic activities (Williams *et al*, 1988; Palta *et al*, 1993; Sharma and Lavanya, 2002).

(xi) Poor pod set due to floral abortion, aberration in gamete development during early reproductive growth and pollen malfunctioning occurs (Srinivasan *et al*, 1999; Dua, 2001; Sandhu and Arasakesary, 2003).

III. SALTS

Excess of salts in the rooting medium cause osmotic or ionic stress while their deficiency results in mineral deficiency stress. However, the excess salt stress is of common occurrence in arid and semi-arid tropics. The increased concentration of salts in the soil to the extent that limits the expression of plant's full genetic yield potential is called salinity stress. Physiologically the salt stress may be due to osmotic stress or ionic stress.

(i) Osmotic stress: It is caused by the decrease in the external osmotic potential of soil solution due to the higher concentration of dissolved salts.

(ii) Ionic stress: It is caused by the specific toxic effects of cations and anions like Na^+, Cl^-, SO_4^{-2}, etc. on physiological and biochemical processes in plant cells.

The problematic soils may be either saline or alkaline. The saline soils predominantly contain soluble salts particularly sulphate and chloride salts of sodium, calcium, magnesium etc. The electrical conductivity of soil solution of saline soils is more than 4.0 ds/m and pH is less than 8.3. The alkaline soils are predominated by carbonate and bicarbonate salts with their exchangeable sodium percentage (ESP) and pH more then 15 and 8.5, respectively. Chickpea is very sensitive to salinity stress and in chloride dominated salinity of 2.7 to 5.3 ds/m shows about 50% reduction in yield.

Effects of Salt Stress

Plants vary in their sensitivity to salts (Maas and Hoffman, 1977; Greenway and Munns, 1980). Usually there is a threshold concentration of salts specific to a plant species above which the crop plants begin to show the salt induced injuries. Such effects may be:

(i) The general water balance of plants is disturbed due to higher concentrations of salts in the soil because, leaves need to develop a lower water potential to maintain a "downhill" gradient of water potential in soil-plant continuum. Thus, the effect of dissolved solutes, in this sense, is similar to that of a soil water deficit reducing conductance and related physiological processes (Muhammad Ashraf and Alweena Bashir, 2003).

(ii) The ionic balance in the protoplasm is disrupted with growth retardation (Mamo *et al,* 1996).

(iii) An abnormally high ratio of sodium and potassium ions and higher concentration of total salts in the protoplasm inactivate enzymes, inhibit protein synthesis and influence the activities of photorespiratory enzymes (Murumkar *et al,* 1985; Uberoi *et al,* 2001; Singh *et al,* 2001a, 2001b).

(iv) At a higher concentration, sodium ion can displace calcium ion from the plasma membranes of root hairs causing a change in membrane permeability and leakage of potassium ions from the cells (Cramer *et al,* 1985).

(v) Free sterols and triglycerides in root cells are increased (Zenoff *et al,* 1994)

(vi) Higher concentration of sodium and/or chloride ions in chloroplasts disrupts the metabolic status and inhibits photosynthesis (Sharma, *et al,* 1990a; Singh and Singh, 1999a).

(vii) The overall effects on plants are leaf discolouration, growth inhibition, loss of dry weight and decreased productivity (Neera Garg *et al,* 2002).

(viii) Germination is hampered and the seedling growth is adversely affected (Özdemir and Engin, 1994; Singh and Singh, 1999b; Al Mutawa, 2003; Katerji *et al,* 2003).

(ix) Nodulation and nitrogen fixation are also reduced (Sharma *et al,* 1990b; Soussi *et al.* 2001; Neera Garg and Garg, 2002).

(x) Sodium-potassium ratio in the plant tissue increases along with an increase in the proline, betaine and free amino acid contents (Singh and Singh, 1999c; Muhammad Ashraf and Alweena Bashir, 2003).

(xi) A reduction in number of pods per plant and seed weight occurs (Francois *et al*, 1990).

(xii) Plants show lower leaf weight ratio (Han *et al*, 1989).

(xiii) Phosphorylation efficiency in mitochondria is reduced (Melo *et al*, 1994).

(xiv) Chlorophyll content is lowered and the chloroplast integrity is adversely affected (Hernandez *et al*, 1995).

PHYSIOLOGY OF DROUGHT RESISTANCE

Drought resistance may be defined as the mechanism(s) minimising the loss of yield in a water deficient environment relative to the maximum yield in a water constraint free management of the crop. Arnon (1975) grouped the plant adaptations to dry habitats as follows:

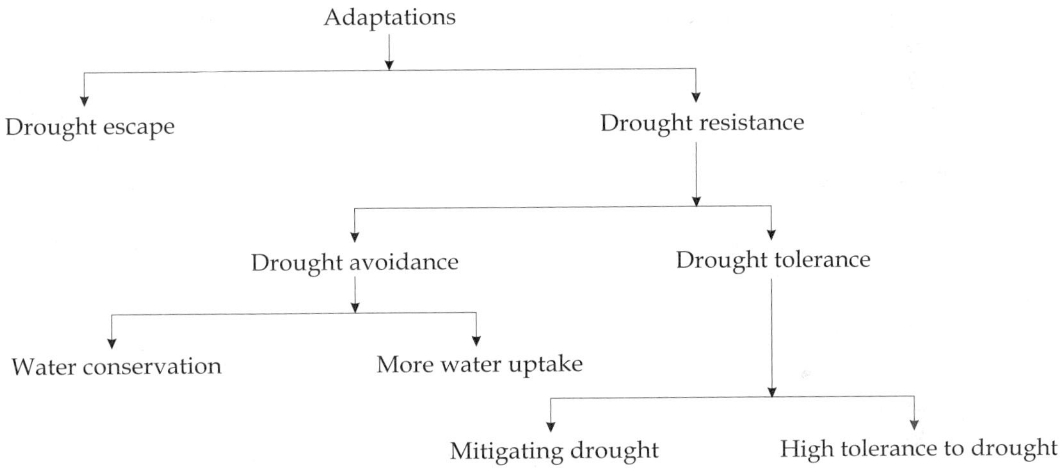

Kramer (1980) also proposed the mechanisms that operate in different plant systems which were further elaborated by Turner (1986) and Blum (1988). According to them, the three drought resistance categories are:

(i) Drought escape

(ii) Dehydration postponement

(iii) Dehydration tolerance.

Drought Escape

It is an adaptation of matching the phenology to the water supply and allows to complete the plant's life cycle during favourable environment before the commencement of drought situation. In chickpea, such adaptation has been of limited use only in the sense that in response to water deficit stress the plants show early flowering and forced maturity and it is at the cost of plant productivity. There has been some success reports through selections for early flowering under environments prone to terminal drought (Hall *et al*, 1978; Hall and Patel, 1985; Subbarao *et al*, 1995). At International Crops Research Institute for Semi Arid Tropics, Hyderabad, India, some short duration chickpea germplasm lines like ICCV-2, ICCC-37, ICCV-10 and KAK-2 have been developed which effectively escape end-of-season drought (Dar *et al*, 2001; Serraj *et al*, 2003). Also, phenological plasticity has been considered an important characteristic in drought prone environments (Turner, personal communications). However, the problem of unpredictable and intermittent drought cannot find a solution in early flowering types.

Dehydration Postponement

It is the mechanism of maintaining turgor during drought period through efficient water uptake, reduction in water loss or osmotic adjustment. The plant water indices like water retention and water content in fresh samples have been suggested as screening tools for tolerance to drought in chickpea (Pannu *et al*, 1993).

Efficient water uptake is largely a function of rooting system of the plant for increasing soil water extraction. There have been reports of genotypic variations in rooting depth and its ability to extract water at different soil profiles having impact on seed yield (Gregory, 1988; Mathews *et al*, 1988; White and Castillo, 1990; Wright *et al*, 1991). Supporting the earlier findings (Gupta *et al*, 1987), Saxena *et al*, (1990) have reported that denser and longer root system in chickpea is advantageous under drought stressed environments. The chickpea genotypes ICC-4958 and FLIP 87-59C have been identified as drought tolerant lines with higher root biomass (Saxena *et al*,1994; Singh *et al*, 1996) and are presently being used as donor parents for this trait. Another genotype showing drought tolerance is RSG 143-1 (Gupta *et al*, 2003), which has the ability of having more root biomass in the deeper soil depths (>60 cm), thus increasing the ability to extract more moisture from the deeper soil profiles.

Several techniques have been used to screen the genotypes for their rooting characteristics. The technique of Robertson *et al*, (1985) of introducing selective herbicides at depth has been used by Khalfaoui and Havard (1993) for identifying the deep rooting lines. Sand culture technique for the similar purpose in chickpea has been used by Saxena *et al*, (1994) and Krishnamurthy *et al* (1996). Other techniques like aeroponics, root pulling resistance, tensiometer etc. have also been used for screening the varieties (Subbarao *et al*, 1995, Mia *et al*, 1996a, 1996b). In rice the molecular markers have been identified to characterise the root morphology (Champoux *et al*, 1995; Ray *et al*, 1996). Similar studies in grain legumes will be of immense help in screening a large number of genotypes for their rooting behaviour.

There have been debates (Passioura 1994) whether the higher root biomass will be a beneficial characteristic (Jordan *et al*, 1983), or the reduced underground growth (Passioura, 1983) will increase seed yield under water stress conditions. The work at ICRISAT (Saxena *et al*, 1994; Krishnamurthy *et al*, 2003) suggests that the genotypes with higher root biomass perform better under limited water conditions.

Reduction in water loss is mainly accomplished through regulation of stomatal closure under water stress conditions. The moisture status of leaves and the ambient humidity have the major influence on stomatal conductance (Turner 1991). Variations in stomatal conductance in response to varying leaf water potential have been reported in grain legumes (Lawn, 1982; Muchow, 1985, Flower and Ludlow, 1986). There are, however, reports of decreasing stomatal conductance in response to drying soils, even when leaf hydration is maintained (Blackman and Davies 1985; Gollan *et al*, 1986). Thus, a stimulus may be transmitted to the leaves which control the closure of stomates. Abscisic acid (ABA) has been reported to provide such kind of stimulus in plants (Henson *et al*, 1989; Turner and Henson, 1989; Davies *et al*, 1994).

Stomatal conductance controls not only the water loss by the plant, but the rate of photosynthesis as well through regulating the gaseous exchange. Therefore, any reduction in conductance will reduce both the water loss as well as the carbon fixation by the plant. In chickpea the tolerant genotypes make a compromise whereby the conductance is reduced to decrease the water loss to the extent that stomates remain partially open to allow sufficient carbon dioxide entry into the leaves and a lesser reduction in photosynthetic rate is observed (Gupta *et al*, 2000).

Variations in the ABA production ability among the genotypes occur in response to water stress. However, it has been reported that the genes for ABA production under drought are located on chromosomes other than the chromosomes bearing the genes contributing to yield, which very clearly suggests that the drought tolerance in terms of higher ABA production in response to water stress may not have high correlation with grain productivity (Quarrie *et al*, 1995). Hence, to develop drought tolerance it is necessary to explore the genetic variability for ABA production and introduce this trait into a higher yielding cultivar.

Leaflet closure has been shown to reduce water loss in groundnut (Ramesh Babu and Rao, 1983). Such adaptation protects the plants from exposure to excessive solar radiations and thus provides a mechanism to tolerate the stress. Similar reduction in leaflet angle has been observed in chickpea also in response to moisture stress (Choudhary, 2002).

Osmotic adjustment is the active accumulation of solutes in plant cells in response to water deficiency to maintain turgor or to reduce the rate of turgor loss, with a decrease in water potential (Turner and Jones, 1980; Morgan, 1984). It involves the active accumulation of solutes, such as sugars, amino acids, organic acids, NO_3^-, Cl^-, and K^+ in response to drought which lowers osmotic potential and maintains cell turgor at lower soil water potential (Khanna Chopra *et al*, 1995). Osmotic adjustment has been shown to be beneficial in imparting drought tolerance (Wery *et al*, 1993; Choudhary, 2002) at low water potential, in maintaining stomatal conductance and photosynthetic rate (Jones and Rawson, 1979; Ludlow, 1980, 1987), in delaying leaf senescence (Hsiao *et al*, 1984; Flower and Ludlow, 1986, 1987), in reducing flower abortion (Morgan and King 1984) and in enhancing root growth and water uptake (Morgan and Condon, 1986). In chickpea, more lowering of leaf osmotic potential is shown by the tolerant genotypes (Gupta *et al*, 2000). Higher leaf moisture retention ability has also been observed in such genotypes in our laboratory and elsewhere as well (Ali *et al*, 1998). An osmotic adjustment of 0.1 to 1.6 MPa has been reported in different grain legumes (see: Subbarao *et al*, 1995). In chickpea, Morgan *et al*, (1991) reported that an osmotic adjustment of 0.6 MPa in leaves, and showed to be correlated with yield under drought conditions Flower and Ludlow (1986, 1987) explained that in grain legumes the osmotic adjustment delays leaf senescence and increases the remobilisation of food reserves, thus enhances yield under water limited conditions.

Genotypic variability has also been found to occur in maintaining canopy temperature under stress conditions. Chauhan and Singh (2003) have proposed that lower canopy temperature under water stress might be attributed to higher water use efficiency and osmoregulation in the tolerant chickpea genotypes.

There have been genetic studies identifying the trait loci for osmotic adjustment, mostly in cereals (Lilley *et al*, 1996; Morgan *et al.* 1986). Identification of the molecular markers like the proteins, dehydrins, synthesised in response to water deficit (Bartels *et al*, 1993) will simplify the selection of genotypes for complex morphological, physiological and biochemical traits from a large population. Also new tolerant genotypes can be developed through DNA transfer technology application in grain legumes (Christou, 1994).

Dehydration Tolerance

Dehydration tolerance is the ability of plants to maintain cellular metabolism at low water potential. Generally, when the water deficit or plant dehydration reaches a critical level (for many legumes leaf water potential -1.8 to -8.2 MPa, Flower and Ludlow, 1987), the plants die due to disordered membrane functions. The membrane fluidity for ion leakage has been used to select out the plants having more stable membranes and thus tolerance to high temperature and water deficiency (Srinivasan *et al*, 1996; Deshmukh and Kushwaha, 2002). In chickpea, the genotypes showing tolerance to drought have lower membrane injury in response to heat shock treatment (Gupta *et al*, 2000, Fig. 6.8). Wery *et al*, (1993) have also emphasised on high membrane stability under stress for tolerance to drought.

However, there are reports that no relationship exists between electrolyte leakage from dehydrated leaves and crop performance under drought (Blum, 1988) and hence the technique is not in much use in the breeding programmes. But, seeing the active role of membranes in plant metabolism, one can expect the importance of membrane integrity under stress conditions and thus more work is required to quantify this characteristic as a selection criterion for drought tolerance. Sampling of leaves for measurement of ion leakage is important for uniformity of results and their valid comparisons. For example in chickpea, the 5th, 6th and 7th fully expanded healthy leaves of the main shoot from the top can serve as good samples for physiological analyses.

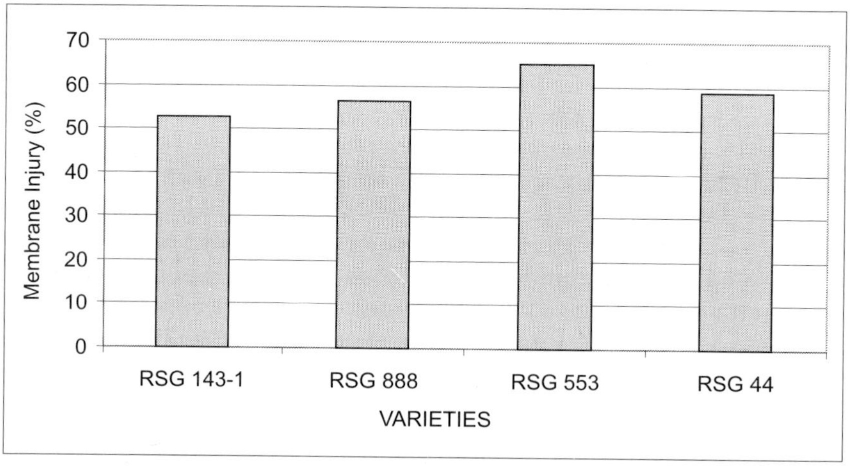

Fig. 6.8: Variations in leaf membrane injury due to stress in some chickpea genotypes. See that the tolerant genotype RSG 143-1 shows comparatively the lowest membrane injury.

The tolerance of grain legumes to the lowest water potential experienced by the viable leaves has been reported to vary from -8.2 to -1.8 MPa (Shackel and Hall, 1983; Flower and Ludlow, 1986, 1987; Sinclair and Ludlow, 1986). However, osmotic adjustment has been reported to play an important role in dehydration tolerance (Hsiao *et al*, 1984) but, the factors other than osmotic adjustment cannot be ruled out as there existed a poor correlation between osmotic adjustment and the lethal osmotic potential for the leaves (Lilley *et al*, 1996). More lowering of osmotic potential in response to water stress has been reported in the drought tolerant chickpea genotypes (Gupta *et al*, 2000, Fig. 6.9) and has been associated with mitigation of the deleterious effects of moisture stress (Turner and Begg, 1981; Morgan, 1984). Accumulation of cellular osmolytes like soluble sugars, free amino acids, proline, glycine betaine etc. has been shown to occur as a general feature under moisture stress conditions (Gupta *et al*, 1994, 2000; Jha and Singh 1997). Out of these osmolytes, proline has been amply implicated to provide an energy source for utilisation during plant recovery from stress (Aspinall and Paleg, 1981; Paleg *et al*, 1981). The physiological significance of proline accumulation has been debatable as in some cases it has been reported to be an stress injury effect (Hanson *et al*, 1977) while in others, a protective agent of enzymes and cellular structures (Schobert and Tschesche, 1978) and still in other cases as a storage compound which is used up during rapid growth after the stress is released (Barnett and Naylor, 1966). In chickpea it has been reported that proline accumulation increases under water stress and the accumulation is more in drought tolerant genotypes (Gupta *et al*, 2000). PEG-6000 induced tolerance to drought in callus cell lines has also been reported as due to higher levels of K^+ and proline (Gulati and Jaiwal, 1994).

Silim and Saxena (1993) in their studies on chickpea reported that deep root system, high leaf water potential at dawn and large number of seeds per plant were associated with drought tolerance. Early seedling vigour has also been shown to play some role for drought tolerance

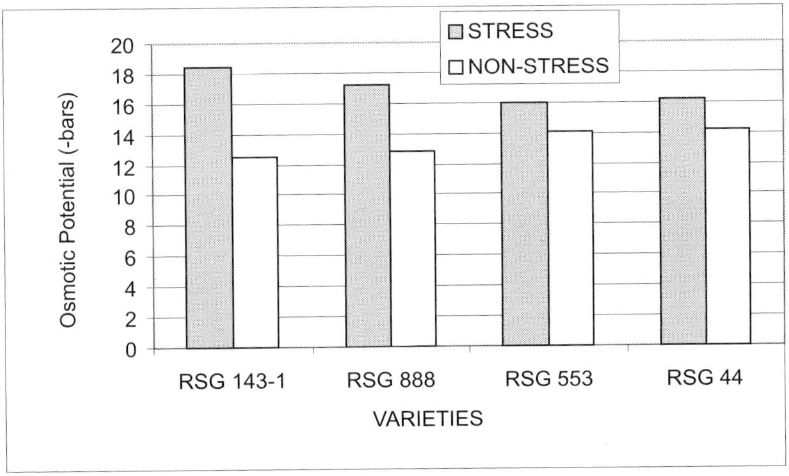

Fig. 6.9: Effect of moisture stress on leaf osmotic potential of some chickpea genotypes as compared to the non-stress conditions in fields. See that the tolerant genotype RSG 143-1 shows comparatively the lowest osmotic potential under stress.

under moisture stress. Harris *et al*, (1999, 2001) found that 10-hour seed soaking in water prior to sowing in fields resulted in faster emergence, better stands, more vigorous plants, better drought tolerance, earlier flowering, earlier harvest and higher grain yield in chickpea. In our experiments under moisture stress conditions, we also found early emergence and more vigorous plants with higher grain yields due to seed priming with water favouring plant performance under moisture stress. About 9% higher seedling vigour index was observed in primed seeds (Fig. 6.10).

Remobilisation of carbon and nitrogen from vegetative parts to developing seed is important in chickpea because the pod filling occurs when net photosynthesis and nitrogen fixation fall due to the onset of terminal drought (Davies *et al*, 2000). Desi genotypes show better drought tolerance, partly due to their ability of better remobilisation and higher pod number.

The various physiological adjustments in response to moisture stress allows a tolerant chickpea genotype to maintain lower drought susceptibility index (DSI, Gupta *et al*, 2000), which is a good measure for drought tolerance in chickpea. Lower DSI value signifies that the genotype has yielded sustainability under the defined drought stress conditions.

POTENTIAL AND LIMITATIONS OF USING PHYSIOLOGICAL BASE FOR DEVELOPMENT OF DROUGHT TOLERANCE

Physiological traits critically explain how a plant system economically manages the water use and the metabolic processes under water stress conditions. Upto a certain limit such economic water use may not affect the grain production appreciably but under severe water stress, survival of life naturally becomes more important for physiological processes, which on priority manage to protect the life of the plant irrespective to the loss in grain yield. Struggle for

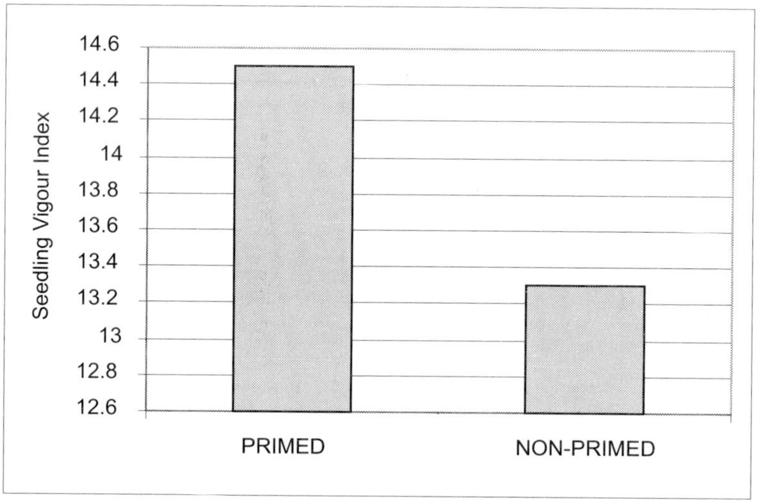

Fig. 6.10: Effect of seed priming prior to sowing on seedling vigour index in chickpea (var. RSG 888) as compared to non-primed seeds.

the survival is a natural phenomenon while production oriented crop manipulations are all anthropogenic fighting against the rule of nature. Hence, significant correlations are expected to exist among the physiological traits involved in water economy management, and not with grain yield under drought conditions, which in true sense signifies the drought tolerance of a system. However, incorporation of the physiological traits responsible for drought tolerance into a high yielding genotype with good agronomic background may be a feasible solution to overcome the problem of water stress and sustainable crop production. Mutation breeding has also shown some success in developing the tolerant systems (Ahloowalia *et al,* 2004). The specific genes encoding for enzymes required for the biosynthesis of various osmoprotectants or modifying membrane lipids etc. are being identified and efforts are in progress to transfer the same in high yielding cultivars (Sharma and Lavanya, 2002). Alfin 1 has been shown to be an essential gene for root growth and in transgenic plants with its over expression confers a manifold increase in root growth, an advantageous trait for stress conditions (Winicov, 2000).

REFERENCES

Ahloowalia B.S., Maluszynski M. & Nichterlein K. (2004). Global impact of mutation derived varieties. Euphytica 135(2): 187-204.

Al Mutawa M.M. (2003). Effect of salinity on germination and seedling growth of chickpea (*Cicer arietinum* L.) genotypes. International Journal of Agriculture and Biology 5(3): 226-229.

Ali Y., Haq M.A., Alam S.S. & Ahmad N. (1998). Variation in drought tolerance in ten chickpea varieties. II : Studies on some physiological characteristics and their relation to growth. Pakistani Journal of Biological Sciences 1(2): 109-110.

Anderson J.V., Hess J.L. & Cheione B.J. (1990). Purification, characterisation and immunological properties for two isoforms of glutathione reductase from eastern white pine needles. Plant Physiol. 94: 1402-1409.

Arnon I. (1975). Physiological principles of dryland crop production. In: Physiological aspects of dry land farming (Ed. U.S. Gupta). Oxford and IBH publishing Co. pp. 3-145.

Aspinal D. & Paleg L.G. (1981). Proline accumulation: Physiological aspects. In: The physiology and Biochemistry of Drought Resistance in Plants (Eds. L.G. Paleg and D. Aspinal), Sydney, Acadmic Press, pp. 205-241.

Barnett N.N. & Naylor A.N. (1966). Amino acid and protein metabolism in Bermuda grass during water stress. Plant Physiol. 41: 1222-1230.

Bartels D., Alexander R., Schneider K., Elster R., Velasco R., Alamillo J., Bianchi G., Nelson D. & Salamini F. (1993). Desiccation related gene products analysed in a resurrection plant and in barley embryos. In: Plant Responses to Cellular Dehydration during Environmental Stress (Eds. T.J., Close & E.A. Bray, Rockville, M.D., American Society of Plant Physiologists, pp. 119-127.

Björkman O., Badger M.R. & Armond P.A. (1980). Response and adaptation of photosynthesis to high temperatures. In: Adaptation of plants to water and high temperature stress (Eds. N.C. Turner & P.J. Kramer). Wiley, New York. Pp. 233-249.

Blackman P.G. & Davies W.J. (1985). Root to shoot communication in maize plants – the effects of soil drying. J. Exp. Bot. 36: 39-48.

Blum A. (1988). Plant breeding for stress environments. Boca Raton, F.L.: CRC Press.

Blumenthal C.S., Batey I.L., Bekes F., Wrigley C.W. & Barlow E.W.R. (1991). Seasonal changes in wheat-grain quality associated with high temperatures during grain filling. Australian Journal of Agricultural Research 42: 21-30.

Bonk M., Tadros M., Vandekerckhove J., Al-Babili S. & Beyer P. (1996). Purification and characterisation of chaperonin 60 and heat shock proteins 70 from chromoplasts of *Narcissus pseudonarcissus*. Involvement of heat-shock protein 70 in a soluble protein complex containing phytoene desaturase. Plant Physiol. 111: 931-939.

Champoux M.C., Wang G. Sarkarung S., MacKill D.J., O'Toole J.C., Huang N. & McCouch S.R. (1995). Locating genes associated with root morphology and drought avoidance in rice via linkage to molecular markers. Theoret. Appl. Genet. 90: 969-981.

Chauhan S.S. & Singh D.N. (2003). Genotypic variation among chickpea genotypes in canopy temperature and relationship to grain yield under soil moisture deficit conditions. Abstracts. International chickpea conference, Raipur, Chhattisgarh, India, January 20-22, 2003, pp. 76-77.

Choudhary P. (2002). Physiological and biochemical studies on some drought tolerant and susceptible genotypes of chickpea. Ph.D. Thesis, Deptt. of Botany, University of Rajathan, Jaipur, India.

Christou P. (1994). The biotechnology of crop legumes. Euphytica 74: 165-185.

Climate Change (1992). The IPCC Scientific Assessment Supplementary Report (Eds. J.T. Roughton, B.A,. Callender, S.K.Varnery). Cambridge University Press, pp. 365.

Cramer G.R., Lauchli A. & Polito V.S. (1985). Displacement of Ca^{2+} by Na^{+} from the plasmalemma of root cells. A primary response to salt stress. Plant Physiol. 79: 207-211.

Croser J.S., Clarke H.J., Siddique K.H.M & Khan T.N. (2003). Low temperature stress: Implications for chickpea (*Cicer arietinum* L.) improvement. Critical Reviews in Plant Sciences 22 (2): 185-219.

Dar W.D., Kumar J., Saxena K.B., Upadhyaya H.D. & Crouch J.H. (2001). International efforts on Improvement of Chickpea and Pigeonpea. In: Souvenir, National Symp. On Pulses for sustainable Agriculture and Nutritional Security, ISPRD, Kanpur, pp. 1-7.

Davies S.L., Turner N.C., Palta J.A., Siddique K.H.M. & Plummer J.A. (2000). Remobilisation of carbon and nitrogen supports seed filling in chickpea subjected to water deficit. Australian Journal of Agricultural Research 51(7): 855-866.

Davies W.J., Tardieu F. & Trejo, C.L. (1994). How do chemical signal work in plants that grow in drying soil? Plant Physiol. 104: 309-314.

Deshmukh P.S. & Kushwaha S.R. (2002). Variability in membrane injury index in chickpea genotypes. Indian Journal of Plant Physiol. 7(3): 285-287.

Drew M.C. (1997). Oxygen deficiency and root metabolism: Injury and acclimation under hypoxia and anoxia. Annu. Rev. Plant Physiol. Plant Mol. Biol. 48: 223-250.

Drew M.C. & Lynch J.M. (1980). Soil anaerobiosis, microorganisms and root function. Annu. Rev. Phytopathol. 18: 27-66.

Dua R.P. (2001). Genotypic variations for low and high temperature tolerance in gram (*Cicer arietinum* L.). Indian Journal of Agricultural Sciences 71 (9): 561-566.

Erskine W., Tufail M., Russell A., Tyagi M.C., Rahman M.M., Saxena M.C., Muehlbauer F.J. & Kaiser W.J. (1994). Current and future strategies in breeding lentil for resistance to biotic and abiotic stresses. In: Expanding the production and use of cool season food legumes. Proceedings of the Second International Food Legume Research Conference on Pea, Lentil, faba bean, chickpea and grasspea. Cairo, Egypt, 12-16 April 1992. (Ed. F.J. Muechlbauer). Kluwer Academic Publ., Dordrecht, Netherland, pp. 559-571.

Flower D.J. & Ludlow M.M. (1986). Contribution of osmotic adjustment to the dehydration tolerance of water stressed pigeonpea [*Cajanus cajan* (L.) Millsp.] leaves. Plant Cell Environment 9: 33-40.

Flower D.J. & Ludlow M.M. (1987). Variation among accessions of pigeonpea (*Cajanus cajan*) in osmotic adjustment and dehydration tolerance of leaves. Field Crops Research 17: 229-243.

Francois L.E., Donovan. T.J. & Maas E.V. (1990). Salinity effects on emergence, vegetative growth and seed yield of guar. Agronomy Journal 82(3): 587-592.

Garg B.K., Burman U. & Kathju S. (2004). Effect of water stress on mothbean (*Vigna aconitifolia* (Jacq.) Marechal) genotypes. Indian J. Plant Physiol. 9(1) N.S.: 30-36.

Garg B.K., Kathju S. & Burman U. (2001). Influence of water stress on water relations, photosynthetic parameters and nitrogen metabolism of mothbean genotypes. Biologia Plantarum 44(2): 289-292.

Ghorbanli M., Farzami Sepehr M.& Farbodnia T. (2003). The changes in total soluble sugars, starch and protein contents at drought stress in two cultivars of Iranian Cicer (*Cicer arietinum* L.). Abstract. International Chickpea Conference, Raipur, Chhattisgarh, India, January 20-22, 2003, pp. 67.

Gollan T., Passioura J.B. & Munns R. (1986). Soil water status affects the stomatal conductance of fully turgid wheat and sunflower leaves. Australian Journal of Plant Physiol. 13: 459-464.

Greenway H. and Munns R. (1980). Mechanisms of salt tolerance in nonhalophytes. Annu. Rev. Plant Physiol. 31: 149-190.

Gregory P.J. (1988). Root growth of chickpea, faba bean, lentil and pea and effects of water and salt stress. In: World Crops: Cool Season Food Legumes (Ed. R.J. Summerfield), Dordrecht, The Netherlands, Kluwer Academic Publishers, pp. 857-867.

Gulati A. & Jaiwal P.K. (1994). *In vitro* selection and characterisation of *Vigna radiata* cell lines resistant to PEG induced drought stress. Acta Physiologiae Plantarum 16(1): 53-60.

Gupta S.C., Bhatia L.K. & Singh A.K. (1994). Role of proline and cellular osmotic agents in drought tolerance in cowpea. In: Plant Productivity under Environmental Stress (Eds. K. Singh and S.S. Purohit), Agro Botanical Publ., Bikaner, India, pp. 237-240.

Gupta S.C., Rathore A.K., Sharma S.N. and Saini R.S. (2000). Response of chickpea cultivars to water stress. Indian J. Plant Physiol 5(3) N.S.: 274-276.

Gupta S.C., Sharma S.N. & Jhabar Singh (2003). Root growth characteristics of some chickpea genotypes under moisture stress. Abstract National Symposium on Improving Crop Production in an Eco-Friendly Environment.: Physiological and Molecular Approaches, October 15-17, 2003, Pantnagar, India.

Gupta S.N., Dahiya B.S. & Singh V.P. (1987). Genetic variability for drought tolerance in chickpea. Abstracts. First Symposium on Crop Improvement, 23-27 February, 1987, Ludhiana, India. Crop Improvement Society of India, pp. 128-129.

Hall A.E., Foster K.W. & Waines J.G. (1978). Crop adaptation to semi-arid environments. In: Agriculture in semi-arid environments (Eds. G.H. Cannell, A.E. Hall & H. Lawton), New York, N.Y., USA, Springer-Verlag, pp. 148-179.

Hall A.E. & Patel P.N. (1985). Breeding for resistance to drought and heat. In: Cowpea Research, Production and Utilisation (Eds. S.R. Singh & K.O. Rachie), Chrichester, USA, John Wiley, pp. 137-151.

Han M.S., Kim J.G., Lee S.B. & Han H.J. (1989). Studies on salt tolerance, dry matter yields and nutritive components of temperate grasses grown on newly reclaimed tidal lands. The Research Reports, The Rural Development Administration, Livestock (Korea R.), 31(4): 36-40.

Hanson A.D., Nelson C.E. & Everson E.H. (1977). Evaluation of free proline accumulation as an index of drought resistance using two contrasting barley cultivars. Crop Sci. 17: 720-726.

Harris D., Joshi A., Khan P.A., Gothkar P. & Sodhi P.S. (1999). On-farm seed priming in semi-arid agriculture: Development and evaluation in maize, rice and chickpea in India using participatory methods. Experimental Agriculture 35(1): 15-29.

Harris D., Pathan A.K., Gothkar P., Joshi A., Chivasa W. & Nyamudeza P. (2001). On-farm seed priming using participatory methods to revive and refine a key technology. Agricultural Systems 69(1-2): 151-164.

Henson I.E., Jensen C.R. & Turner N.C. (1989). Leaf gas exchange and water relations of lupins and wheat. III. Abscisic acid and drought induced stomatal closure. Australian Journal of Plant Physiol. 16: 429-442.

Hernandez J.A., Olmos E., Corpas F.J., Sevilla F., Rio, L.A. del & Del Rio L.A. (1995). Salt induced oxidative stress in chloroplasts of pea plants. Plant Science Limerick 105(2): 151-167.

Hsiao T.C., O'Toole J.C., Yambao E.B. & Turner N.C. (1984). Influence of osmotic adjustment on leaf rolling and tissue death in rice (*Oryza sativa* L.). Plant Physiol. 95: 338-341.

Jha B.N. & Singh R.N. (1997). Physiological responses of rice varieties to different levels of moisture stress. Indian J. Plant Physiol. 2: 81-84.

Jones M.M & Rawson H.M. (1979). Influence of the rate of development of leaf water deficits upon photosynthesis, leaf conductance, water use efficiency and osmotic potential in sorghum. Physiol. Plant. 45: 103-111.

Jordon W.R., Dugas W.A. & Shouse P.J. (1983). Strategies for crop improvement for drought prone regions. Agric. Water Manage. 7: 281 – 299.

Katerji N., Hoorn J.W. Van, Hamdy A., Mastrorilli M. & VanHoorn J.W. (2003). Salinity effect of crop development and yield, analysis of salt tolerance according to several classification methods. Agricultural Water Management 62(1): 37-66.

Khalfaoui J.L.B. & Havard M. (1993). Screening peanut cultivars in the field for root growth: A test by herbicide injection in the soil. Field Crops Research 32: 173-179.

Khanna Chopra R., Moinuddin, Sujata Vasudev, Maheswari M., Amita Srivastava, Bahukhandi D., Vasudev S. & Srivastava A. (1995). K^+, osmoregulation and drought tolerance—an overview. Proceedings of the Indian National Science Academy – Part B, Biological Sciences 61(1): 51-56.

Kramer P.J. (1980). Drought stress and the origin of adaptations. In: Adaptation of plants to water and high temperature stress (Eds. N.C. Turner & P.J. Kramer). New York, N.Y. Academic Press pp. 7-20.

Krishnamurthy L., Johansen C. & Ito, O. (1996). Genotypic variation in root system development and its implications for drought resistance in chickpea. In: Dyanamics of Roots and Nitrogen in Cropping Systems of the Semi-Arid Tropics. (Eds. O. Ito, C. Johansen, J.J. Adu-Gyamfi, K. Katayama, J.V.D.K., Kumar Rao & T.J. Rego), Tsukuba, Japan, Japan International Research Center for Agricultural Sciences, pp. 235-250.

Krishnamurthy L., Kashiwagi J. & Serraj R. (2003). Genetic variation of chickpea (*Cicer arietinum* L.) root traits and their influence on yield stability under drought. Abstracts. International Chickpea Conference, Raipur, Chhattisgarh, India, January 20-22, 2003, pp. 87.

Kumar D. (2001). Breeding approaches in arid legumes for abiotic stresses. In: Pulses for sustainable agriculture and nutritional security (Eds. Masood Ali, S.K. Chaturvedi & S.N. Gurha)—a proceeding of National Symposium, ISPRD, IIPR, Kanpur, pp. 73-81.

Lawn R.J. (1982). Responses of four grain legumes to water stress in south eastern Queensland. I. Physiological response mechanisms. Australian Journal of Agricultural Research 33: 481-496.

Levitt J. (1980). Response of plants to environmental stresses. Vol. 1, 2nd Ed. Academic Press, New York.

Lilley J.M., Ludlow M.M., McCouch S.R. & O'Toole J.C. (1996). Locating QTL for osmotic adjustment and dehydration tolerance in rice. J.Exp.Bot. 47: 1427-1436.

Lu Z.M., Percy R.G., Qualset C.O. & Zeiger E. (1998). Stomatal conductance predicts yields in irrigated Pima Cotton and bread wheat grown at high temperatures. J.Exp.Bot. 49: 453-460.

Ludlow M.M. (1980). Adaptive significance of stomatal responses to water stress. In: Adaptation of Plants to water and High Temperature Stress (Eds. N.C. Turner & P.J. Kramer), New York, N.Y., John Wiley Publ., pp. 123-128.

Ludlow M.M. (1987). Contribution of osmotic adjustment to the maintenance of photosynthesis during water stress. In: Progress in Photosynthesis Research (Ed. J. Biggins) Vol. 4, Dordrecht, The Netherlands, Martinus Nijhoff Publ., pp. 161-168.

Maas E.V. & Hoffman G.J. (1977). Crop salt tolerance-current assessment. J.Irrig. Drainage Div. Am.Soc. Civ. Eng. 103: 115-134.

Mamo T., Richter C. & Heiligtag B. (1996). Salinity effects on the growth and ion contents of some chickpea (*Cicer arietinum* L.) and lentil (*Lens culinaris* Medic.) varieties. Journal of Agronomy and crop science 176(4): 235-247.

Mathews R.B., Harris D., Rao R.C.N., Williams J.H. & Wadia K.D.R. (1988). The physiological basis for yield differences between four genotypes of groundnut (*Arachis hypogaea*) in response to drought. I. Dry matter production and water use. Experimental Agriculture 24: 191-202.

Melo D.F. de, Jolivet Y., Facanha A.R., Gomes Filho E., Lima M.S., Dizengremel P., Fernandes de Melo D. & Silva Lima M. (1994). Effect of salt stress on mitochondrial energy metabolism of *Vigna ungiculata* cultivars differing in NaCl tolerance. Plant Physiology and Biochemistry 32(3): 405-412.

Mia M.W., Yamanchi A. & Kono Y. (1996a). Root system structure of six food legume species: Inter- and intraspecific variations. Jpn. J.Crop.Sci. 65: 131-140.

Mia M.W., Yamanchi A. & Kono Y. (1996b). Plasticity of tap root elongation growth of several food legume species. Jpn. J. Crop. Sci. 65: 368-378.

Morgan J.M. (1984). Osmoregulation and water stress in higher plants. Annu.Rev. Plant Physiol. 35: 299-319.

Morgan J.M. & Condon A.G. (1986). Water use, grain yield and osmoregulation in wheat. Australian Journal of Plant Physiol, 13: 523-532.

Morgan J.M. & King R.W. (1984). Association between loss of leaf turgor, abscisic acid levels and seed set in two wheat cultivars. Australian Journal of Plant Physiol. 11: 143-150.

Morgan J.M., Hare R.A., & Fletcher R.J. (1986). Genetic variation in osmoregulation in bread and durum wheats and its relationship to grain yield of field environments. Australian Journal of Agricultural Research 37: 449-457.

Morgan J.M., Rodriguez-Maribona B. & Knights E.J. (1991). Adaptation to water deficit in chickpea breeding lines by osmoregulation: Relationship to grain –yields in the field. Field Crops Research 27: 61-70.

Muchow R.C. (1985). Stomatal behaviour in grain legumes grown under different soil water regimes in a semi-arid tropical environment. Field Crops Research 11: 291-307.

Muhammad Ashraf & Alweena Bashir (2003). Salt Stress induced changes in some organic metabolites and ionic relations in nodules and other plant parts of two crop legumes differing in salt tolerance. Flora Jena. 198(6): 486-498.

Murumkar C.V., Rajmane N.A., Karadge B.A. & Chavan P.O. (1985). Influence of salt stress on photorespiratory enzymes in legumes differing in salt tolerance. Photosynthetica. 19(2): 244-246.

Navari-Izzo F., Pinzino C., Quartacci M.F., Sgherri C.L.M. & Izzo, R. (1994). Intracellular membranes: Kinetics of superoxide production and changes in thylakoids of resurrection plants upon dehydration and rehydration. Proc. R.Soc. Edinburgh, 102 B: 187-191.

Neera Garg & Garg N. (2002). Salinity stress induced changes in key metabolites in the nodules of *Glycine max* L.(soybean) and *Cicer arietinum* L. (chickpea) and the manoeuvrability of their response through plant growth regulators. Journal of Plant Biology 29(2): 137-142.

Neera Garg, Jasleen Chawla, Garg N., Chawla J. & Hemantaranjan, J. (2002). Variable response of legume species and their cultivars to salinity stress during various stages of growth and metabolism. Advances in Plant Physiology 4: 133-164.

Nye P.H. & Tinker P.B.H. (1977). Solute movement in the soil-root system. Oxford, U.K., Blackwell.

O' Toole N., Stoddard F.L. & O'Brien L. (2001). Screening of chickpea for adaptation to autumn sowing. Journal of Agronomy and Crop Science 186(3): 193-207.

Özdemir S. (1996). Effect of temperature on germination of chickpea. International Chickpea and Pigenopea Newsletter No. 3: 27-29.

Özdemir S. & Engin M. (1994). Effect of NaCl concentration on chickpea (*Cicer arietinum* L.) germination and seedling growth. Turkish Journal of Agriculture and Forestry 18(4): 323-328.

Paleg L.G., Douglas T.J., Van Daala & Kecch B.D. (1981). Proline, betaine and other organic solutes protect enzymes against heat inactivation. Australian Journal of Plant Physiol. 8: 107-114.

Palta J.P., Whitaker B.D. & Weiss L.S. (1993). Plasma membrane lipids associated with genetic variability in freezing tolerance and cold acclimation of *Solanum* species. Plant Physiol. 103: 793-803.

Pannu R.K., Singh D.P., Singh P., Choudhary B.D. & Singh V.P. (1993). Evaluation of various plant water indices for screening the genotypes of chickpea under limited water environment. Haryana Journal of Agronomy, 9(1): 16-22.

Passioura J.B. (1994). The yield of crops in relation to drought. In: Physiology and determination of crop yield (Eds. K.J. Boote, J.M. Bennett, T.R. Sinclair & G.M. Paulsen), Madison, WI: Crop Science Society of America, pp. 343-359.

Passioura J.P. (1983). Roots and drought resistance. Agric. Water Manage, 7:265-280.

Pugnaire F.I., Serrano L. & Pardos J (1999). Constraints by water stress on plant growth. In: Hand book of plant and crop stress (Ed. M. Pessarkli), Marcel Dekker, Inc. New York. Basel. pp. 271-283.

Quarrie S.A., Peki S., Lazi-Jani V., Ivanovi M. & Vasi G. (1995). ABA and drought tolerance in maize: future prospects. In: Maize Breeding, Production and Utilisation: 50 years of Maize Research Institute, Zemun Polje (Ed. Z. Videnovic), Belgrade, Maize Research Institute, pp. 121-137.

Ramesh babu V. & Rao D.V.M. (1983). Water stress adaptations in the groundnut (*Arachis hypogaea* L.). Foliar characteristics and Adaptation to moisture stress. Plant Physiology and Biochemistry 10(1): 64-80.

Rao S.S. & Sahu M.P. (1991). Effect of sulphur and foliar applied chemicals on cold tolerance in chickpea (*Cicer arietinum* L.) I. Physiological changes. Journal of Agronomy and Crop Science 167(5): 320-325.

Ray J.D., Yu L.X., McCouch S.R., Champoux M.C., Wang G. & Nguyen H.T. (1996). Mapping quantitative trait loci associated with root penetration ability in rice (*Oryza sativa* L.). Theoret. Appl. Genet. 42: 627-636.

Robertson B.M., Hall. A.E. & Foster K.W. (1985). A field technique for screening genotypic differences in root growth. Crop Science 25: 1084-1090.

Sachs M.M. & Ho D.T.H. (1986). Alternation gene expression during environmental stress in plants. Annu. Rev. Plant Physiol. 37: 363-376.

Sandhu J.S. & Arasakesary S.J. (2003). Evaluation of Chickpea genotypes for cold tolerance. International Chickpea and Pigeonpea Newsletter No.10, pp. 9-12.

Saxena N.P., Johansen C., Rheenen H.A. Van, Saxena M.C., Walby B.J. & Hall S.D. (1990). Chickpea ideotypes for genetic enhancement of yield stability in South Asia. In: Chickpea in the nineties : Proceedings of the second international workshop on chickpea improvement, 4-8 December 1989. (Eds. H.A.Van Rheenen, M.C. Saxena, B.J. Walby & S.D. Hall), ICRISAT, India pp. 81-85.

Saxena N.P., Krishnamurthy L. & Johansen C. (1994). Registration of a drought resistant chickpea germplasm. Crop Science 33: 1424.

Schobert B. & Tschesche H. (1978). Unusual solution properties of proline and its interaction with proteins. Biochem. Biophysactata. 541: 270-277.

Serraj R., Bidinger F.R., Chauhan Y.S., Seetharma N., Nigam S.N., Saxena N.P., Kijne J.W., Barker R. & Molden D. (2003). Management of drought in ICRISAT cereal and legume mandate crops. In: Water Productivity in Agriculture: Limits and Opportunities for Improvement. (Eds. J.W. Kijne & R. Barker), CABI Publi., Wallingford, U.K., pp. 127-144.

Shackel K.A. and Hall A.E. (1983). Comparison of water relations and osmotic adjustment in sorghum and cowpea under field conditions. Australian Journal of Plant Physiol. 10: 423-435.

Sharma K.K. & Lavanya M. (2002). Recent developments in transgenics for abiotic stress in legumes of the semi-arid tropics. In: Genetic engineering of crop plants for abiotic stress. Proceedings of an APEC-JIRCAS Joint Symposium and Workshop 3-7 September, 2001, JIRCAS-Working Report 2002, No. 23: 61-73.

Sharma K.D., Datta K.S. & Varma S.K. (1990a). Effect of chloride and sulphate types of salinity on some metabolic drifts in chickpea (*Cicer arietinum* L.). Indian Journal of Experimental Biology 28(9): 890-892.

Sharma K.D., Datta K.S. & Varma S.K. (1990b). Effect of chloride and sulphate salinity on nodulation and nitrogen fixation in chickpea. Annals of Biology 6(1): 13-20.

Sharma K.K. & Lavanya M. (2002). Recent developments in transgenics for abiotic stress in legumes of the semi-arid tropics. Japan International Research Center for Agricultural Sciences No. 23: 61-73.

Silim S.N. & Saxena M.C. (1993). Adaptation of spring sown chickpea to the Mediterranean basin II. Factors influencing yield under drought. Field Crops Research 34(2) 137: 146.

Sinclair T.R. & Ludlow M.M. (1986). Influence of soil water supply on the plant water balance of four tropical grain legumes. Australian Journal of Plant Physiol. 13: 329-341.

Singh A.K. & Singh R.A. (1999a). Effect of salinity on photosynthetic pigments in chickpea (*Cicer arietinum* L.) leaves. Indian Journal of Plant Physiol. 4(1): 49-51.

Singh A.K. & Singh R.A. (1999b). Effect of salt stress on chickpea germination. Journal of Research (BAU) 11(2): 201-204.

Singh A.K. & Singh R.A. (1999c). Effect of salinity on sodium, potassium and proline content of chickpea genotypes. Indian Journal of Plant Physiol. 4(2): 111-113.

Singh A.K., Singh R.A. & Sharma S.G. (2001a). Influence of salinity on activity of hydrolytic and oxidative enzymes in chickpea seedlings. Indian Journal of Plant Physiol. 6(1): 84-86.

Singh A.K., Singh R.A. & Sharma S.G. (2001b). Salt stress induced changes in certain organic metabolites during seedling growth of chickpea. Legume Research 24 (1): 11-15.

Singh K.B., Malhotra R.S. & Saxena M.C. (1993). Relationship between cold severity and yield loss in chickpea (*Cicer arietinum* L.). Journal of Agronomy and Crop Science 170(2): 121-127.

Singh K.B., Omar M., Saxena M.C. & Johansen C. (1996). Registration of FLIP 87-59C, a drought resistant chickpea germplasm line. Crop Science 36: 472.

Singh K.P. (1990). Evaluation of osmotic drought tolerance in chickpea (*Cicer arietinum* L.) varieties during germination and seedling growth. Indian Agriculturist 34(2): 103-106.

Singh R.A. & Kumar Prabhat (1992). Effect of water stress and hardening on metabolic changes in chickpea (*Cicer arietinum* L.). Indian J. Plant Physiol. XXXV(3): 252-257.

Soussi M., Khadri M., Lluch C. & Ocana A. (2001). Carbon metabolism and bacteriod respiration in nodules of chickpea (*Cicer arietinum* L.) plants grown under saline conditions. Plant Biosystems 135(2): 157-164.

Srinivasan A., Saxena N.P. & Johansen C. (1999). Cold tolerance during early reproductive growth of chickpea (*Cicer arietinum* L.): Genetic variation in gamete development and function. Field Crops Research 60(3): 209-222.

Srinivasan A., Takedu H. & Senboku T. (1996). Heat tolerance in food legumes as evaluated by cell membrane thermostability and chlorophyll fluorescence techniques. Euphytica 88: 35-45.

Subbarao G.V., Johansen C., Slinkard A.E., Rao R.C.N., Saxena N.P. & Chauhan Y.S. (1995). Strategies for improving drought resistance in grain legumes. Critical Review Plant Science 14: 469-523.

Taiz L. & Zeiger E. (1998). Stress physiology. In: Plant physiology by L. Taiz and E. Zeiger. Sinauer Inc. Pub., Sunderland, Massatusates, pp. 725-757.

Toker C. & Cagiran I. (1996). Winter sowing chickpea (*Cicer arietinum* L.) and breeding approaches for cold tolerance. Akdeniz university Journal of Faculty of Agriculture (Turkey), 9(1): 123-137.

Turner N.C. (1986). Crop water deficits : a decade of progress. Adv. Agronomy 39: 1-51.

Turner N.C. (1991). Measurement and influence of environmental and plant factors on stomatal conductance in the field. Agric. Forest Meteorol. 54: 137-154.

Turner N.C. & Begg J.E. (1981). Plant Water relations and adaptation to stress. Plant Soil, 58: 97-131.

Turner N.C. & Henson I.E. (1989). Comparative water relations and gas exchange of wheat and lupins in the field. In: Structural and functional Responses to Environmental Stresses: Water shortage (Eds. K.H. Kreeb, H. Richter & T.M. Hinckley), The Hague, SPB Academic Publ., pp. 293-304.

Turner N.C. & Jones M.M. (1980). Turgor maintenance by osmotic adjustment : A review and evaluation. In: Adaptation of Plants to Water and High Temperature Stress (Eds. N.C. Turner and P.J. Kramer), New York, N.Y., John Wiley Publ., pp. 87-103.

Uberoi S.K., Gupta A.K., Narinder Kaur & Kaur N. (2001). A comparison of amylase activites in kabuli chickpea (*Cicer arietinum* L.) varieties showing differential response to sodium chloride stress during seedling growth. Journal of Plant Biology 28(1): 71-77.

Vierling E. (1991). The roles of heat shock proteins in plants. Annu. Rev. Plant Physiol. Plant Mol. Biol. 42: 579-620.

Wery J., Turc O., Lecoeur J., Singh K.B. & Saxena M.C. (1993). Mechanisms of resistance to cold, heat and drought in cool season legumes, with special reference to chickpea and pea. In: Breeding for stress tolerance in cool season food legumes (Ed. K.B. Singh), John Wiley & Sons Ltd., Chichester, U.K., pp. 271-291.

White J.W. & Castillo J.A. (1990). Studies at CIAT on mechanisms of drought tolerance in beans. In: Research on Drought Tolerance in Common Bean (Eds. J.W. White, G. Hoogenboom, F. Ibarra and S.P. Singh), Cali, Centro International de Agriculture Tropical, pp. 146-151.

Williams J.P., Khan M.U., Mitchell K. & Johnson G. (1988). The effect of temperature on the level and biosynthesis of unsaturated fatty acids in diacylglycerols of *Brassica napus* leaves. Plant Physiol. 87: 904-910.

Winicov I. (2000). Alfin 1 transcription factor over expression enhances plant root growth under normal and saline conditions and improves salt tolerance in alfalfa. Planta 210 (3): 416-422.

Wright G.C., Hubick K.T. & Farquhar G.D. (1991). Physiological analysis of peanut cultivar response to timing and duration of drought stress. Australian Journal of Agricultural Research, 42: 453-470.

Zenoff A.M., Hilal M., Galo M. & Moreno H. (1994). Changes in roots lipid composition and inhibition of the extrusion of protons during salt stress in two genotypes of soybean resistant or susceptible to stress: Varietal differences. Plant and Cell Physiology 35 (5): 729-735.

Zhang J. & Zhang X. (1994). Can early wilting of old leaves account for much of the ABA accumulation in flooded pea plants ? J. Exp.Bot. 45: 1335-1342.

Zhang J., Schurr U. & Davies W.J. (1987). Control of stomatal behavior by abscisic acid which apparently originates in the roots. J.Exp.Bot. 38: 1174-1181.

CHAPTER 7

Stress Caused by Nagaon Paper Mill Effluents on Growth, Development and Yield of Rice (*Oryza sativa* L. var. Mahsuri) Plant

Sarat Kumar Dutta

Abstract

It was found that colour, odour and pH of the water samples collected from the experimental sites give within the acceptable range. Other physico-chemical properties high value were obtained. These values indicate that water of the rice field was polluted by the effluent of the paper mill. The height of rice plant, its length of shoots were very much reduced in plants growing in polluted areas, but the number of roots and its length were increased. The dry matter production increased with age in both control and polluted plants, however, at different rates. The percentage of root-shoot ratio were increased at early stage of growth and decreased at maturity. The study revealed that nitrogen, phosphorus and potassium contents of rice plants from polluted areas show significantly difference at their growth and productivity from that of non-polluted rice plants. Increase of NPK contents in grain and straw were accompanied by decreased plant length and grain yield. It was found that chlorophyll content and leaf area could not be directly correlated in both polluted and non-polluted rice plants.

The study has revealed that effluents particularly at high concentrations inhibit germination and growth of seedlings. Further, it has been seen that rice seeds collected from effluent affected areas are less viable and even the visible seeds show delayed germination in comparison to the ones collected from control areas. The yield components of rice plant such as total number of flower per

Address for correspondence: Department of Botany, Nowgong College, Nagaon-782001. Assam, India.

panicle, number of fertilised flower per panicle, average length of ear per plant, number of grain per ear, test weight, volume of 1000 grains and moisture percentage of fresh seed in polluted plants were significantly reduced when compared to plant grown in non-polluted areas.

INTRODUCTION

Among the major industries in India, pulp and paper is one of those that contribute to water pollution to a significant level. Indiscriminate discharge of industrial effluents into natural water bodies is posing problems to cultivated fields. One of the central sector industries i.e. M/S Hindustan Paper Corporation Limited (HPC), Nagaon Paper Mill, Jagiroad, Morigaon, Assam is causing problem in the Jagiroad area due to the discharge of effluents into the nearby Elenga Beel system. The treated effluent is disposed of through a closed pipeline to Elenga Beel, a natural storm water drain, which after travelling a distance of about 25 km downstream joins the River Kollong. The beel covers command area of 2500 hectares of land for which no irrigational facilities exist at present. The effluent discharged from the Nagaon Paper Mill (NPM) is directly open to the nearby paddy fields. In summer, the effluent quite adversely affect the rice cultivation. According to the local farmers, the land which once produced good crops does not produce expected quantity of crops in the effluent affected areas now.

Rice (*Oryza sativa* L.) an important cereal crop, requires proper water management as well as nutrient supply. Paper mill effluent contains organic as well as inorganic ingredients, which to a certain extent may affect the growth, development and yield components of rice.

Plants sensitive to particular pollutant show visible symptons like chlorosis, necrosis and growth retardation (Jocobsen and Hill, 1970; Pandey and Shrivastava, 1980). These morphological manifestations result due to physiological disturbances caused by pollutants. So physiological studies are of paramount importance in detecting effect of pollution within the plant.

There has been several reports relating to damage caused by particular water pollutant on the plant life. Published experimental results are confirmed principally, in the effluents which adversely affect on vegetation directly or indirectly. Most of the research has centred around the direct effect of effluents on leaves, twigs, flower as well as yield component and indirect effects from effluent accummulation in the soil.

Dhaneswar *et al*, (1982) showed that disposal of waste water of Ghur Paper Mill at Kankinara in West Bengal, for agricultural purpose involves requirements of huge primary treatment to meet ISI standards.

According to Kumar *et al*, (1990), the effluent of chemical factories in lower concentration, however, enhance plant growth. So, the recycling of effluent for irrigation may not only solve the disposal problem, but also serve as an additional source of liquid fertiliser in dilute form.

Gomathi and Oblisami, (1992) reported that the pulp mill wastewater can be used for irrigation by sufficiently diluting without causing any severe problem.

Bansai *et al* (1992) observed that wheat grain yield decreased significantly with the use of industrial wastewater.

Biologically treated wastewater from pulp and paper mills can be used for growing crops with no adverse effect on plant growth, crop yield and even soil conditions. Soil also acts as a good absorber for the colour present in the effluent. Raghubeer (1994) has reported that the soluble compounds present in the wastewater help plant growth and soil conditioning.

In the present study, an attempt has been made to assess the different parameters viz. plant height, number of roots and shoots, and its dry weight, plant biomass, presence of inorganic substance N.P.K. etc., leaf area and chlorophyll content, number of flowers, fertilised flower, length of ear per plant, number of grain per ear, test weight, volume of seeds and percentage of moisture content of seeds collected from field affected by the effluent of Nagaon Paper Mill, Jagiroad, Assam.

MATERIAL AND METHODS

The water samples in and around of Nagaon Paper Mill, Jagiroad, Assam, were collected from different stations at regular intervals of 15 days from the rice fields affected by the discharge of paper mill effluent. The paper mill effluent affected areas were termed as 'Polluted' (P) and the areas which were out of reach of the effluent were termed as 'Control' (C) or Non-polluted (NP) as the case may be. A total of 5 sampling stations from polluted and non-polluted areas were selected for the study.

Generally, the water samples were collected during the morning hours. Usually grab samples were collected from the beel as well as the effluent affected rice fields in 2.5 litre pre-cleaned polyethylene bottles and were labelled properly. The samples were taken to the laboratory immediately after collection and protected from direct sunlight during transportation. For determination of dissolved oxygen and BOD, samples were collected separately in BOD bottles. Before collection the containers were rinsed with the water samples to be collected.

The paddy fields selected for experimentation were located from areas within a distance of two km from the paper mill.

The technique for preservation of the samples was as described by Trivedy and Goel (1984).

The collected samples were then analysed by standard methods described by APHA, AWWA and WPCF (1975).

Samples of rice (*Oryza sativa* L. var Mahsuri) plants were collected for determination of the following parameters:

1. **Plant height:** The plant heights were measured after every 15 days from the day of transplanting to mature stage. The maximum heights of the plants were measured (from root tip to the leaf tip and as well as panicle tip as the case may be) separately in centimetre (cm).
2. **Number of shoot, its length and dry weight:** The number of stem was counted after every 15 days. The maximum length of shoot, from the bases of roots up to emergence of leaves were measured separately.

For the determination of dry weight of the shoots, the shoots were wrapped with brown paper and kept in an oven for 24 hours at 80°C.

3. **Number of roots, its length and dry weight:** The plants were carefully uprooted to avoid possible loss of root system. The roots were separated from soil by spray of water aided by hand manipulation. The entire undamaged root system of a single plant was counted to provide an estimate of their total number. The root length was measured by placing the wet roots in a flat glass-dish. The roots were straightened with forceps and measured by use of scale. The height of root was considered the root length.

   Root weight is the most commonly used parameter for studies of root growth in response to environment. To determine dry weight, the washed and cleaned roots were wrapped with brown paper and dried in an oven at 100°C for about 10 to 24 hours depending upon the amount of roots.

4. **Root-shoot ratio:** Root-shoot ratio was obtained by the equation:

$$\frac{\text{Root dry weight}}{\text{Shoot dry weight}}$$

(a) Ecological Characters

1. **Determination of Plant Biomass:** For determination of plant biomass of rice plant, the roots, stems, leaves and panicle of a randomly selected plant were separated and wrapped with brown paper then kept in an oven for 24 hours at 80°C (as described by Rao, 1993). The plants were then weighed to determine the biomass of the plant. This was repeated with plant collected after every 15 days and continued over a period of 135 days. This process was followed both in control and polluted plants.

2. **Determination of Primary Productivity (PP):** The primary productivity of rice plant was determined according to the method of Harvest (as described by Sen, 1978).

$$\text{Rate of production of productivity} = \frac{W_2 - W_1}{T_2 - T_1}$$

where, W_1 = Biomass (g) at time T_1. W_2 = Biomass (g) at time T_2.

(b) Physical Characters

1. **Determination of Nitrogen, Phosphorus and Potassium:** Nitrogen in grain and straw samples was estimated by Kjeldahl method (A.O.A.C, 1970) and for estimation of phosphorus and potassium, the samples were digested with di-acid extract, phosphorus was estimated by Chapmar and Pratt (1961) method and potassium with the help of flame photometric method (Jackson, 1958).

2. **Determination of Leaf Area (LA) and Leaf Chlorophyll:** The average leaf area was measured as suggested by Yoshida *et al* (1976).

   Five plots were selected from each of polluted and non-polluted areas. Rice leaves of similar age and size were separately collected in three stages of growth i.e., seedling stage (30 days after transplanting), flowering stage (90 days after transplanting), and mature stage (135 days from transplanting) from both polluted and non-polluted areas. In each collection only those leave were selected which have no symptoms like chlorosis, necrosis, etc. Chlorophyll contents (a, b and total a+b) of these leaves were then determined as per methods described by Arnon (1949).

3. **Yield Components:** The yield components of the rice plant were measured at the time of flowering of the plant, the number of flowers were counted in each plant at random. The number of fertilised flowers were also similarly counted. The number of grains were counted in each plant separately.

   The weight (test weight) of 1000 grains was measured.

   The volume of 1000 grains (c.c) was measured with the help of measuring cylinder.

   The moisture percentage (%) of fresh grains was calculated as:

$$\text{Moisture percentage} = \frac{\text{Fresh weight}}{\text{Dry weight}} \times 100$$

   For drying, the seeds were kept in an oven maintained at 80°C for 24 hours and subsequently cooled, the seeds were then weighed in previously weighed stoppered glass bottles.

4. **Germination Studies:** Rice (*Oryza sativa* L. var. Mahsuri) seeds were collected from both the polluted and non-polluted fields (beyond the reach of effluent) for studying germination behaviour. Germination studies were made in laboratory conditions. Sets of each fifty healthy seeds either from polluted or non-polluted fields were taken for separate treatments. They were surface-sterilised with 1% $HgCl_2$ and repeatedly washed with distilled water.

   The effluent collected from the paddy field was designed as 100% concentration. It was diluted to 10%, 30%, 50% and 70% (V/V) respectively. Distilled water was used as the control. Various sets of seeds were placed on circular filter paper (Whatsman No.1) covered with a thin layer of sterilised cotton wool in petridishes. Filter papers and cotton wool, were moistened by sprinkling with distilled water in the petridishes containing the control set. Other sets were moistened by respective concentrations of effluent viz. 10%, 30%, 50%, 70%, 90% and 100% respectively. Five replicates of each set were maintained in a germination-chamber at a temperature 28°C ± 3°C. Similar procedure was followed in both types of seeds. Protusion of the radicle through the seed coat was considered the criteria for germination. The germination behaviour was recorded every 24 hours from the initial start and thereafter it was followed after every 12 hours in case of seeds collected from non-polluted fields. In case of seeds collected from polluted fields the germination percentage was taken after 84 hours as described earlier followed by its counting after every 24 hours.

The percentage germination was calculated after 48 hour in case of control seeds as almost all seeds were seen to germinate during this period. However, in case of seeds collected from polluted fields, the germination of the seeds was studied over a period of 168 hours as some of these seeds showed delayed germination.

Percent inhibition in seed germination was computed as per the method described by Rao *et al*, (1979).

RESULTS AND DISCUSSION

The results obtained by analysing water samples and effluents collected from different sampling points of the beel and paddy fields are presented in Table 7.1.

Table 7.1: Physico-characteristics of water collected from lagoon (treatment effluent) polluted and non-polluted paddy fields as affected by the effluent of Nagaon Paper Mill, Jagiroad

S. No.	*Parameters*	*Value of treated effluent*	*Value of polluted fields*	*Value of non-polluted fields*
1.	Colour and odour	Dark, brown with foul smelling	Brown with dis-agreeable odour	Slightly yellowish practically no odour
2.	pH	8.7	7.6	7.2
3.	Suspended solids	935 mg/L	300 mg/L	78 mg/L
4.	Total dissolved solids	1480 mg/L	1390 mg/L	830mg/L
5.	Turbidity	148 NTU	106 NTU	28 NTU
6.	Dissolved Oxygen (DO)	0.0 mg/L	0.5 mg/L	3.6 mg/L
7.	Bio-chemical Oxygen Demand (BOD)	425 mg/L	1.38 mg/L	22 mg/L
8.	Chemical Oxygen Demand (COD)	845 mg/L	470 mg/L	61 mg/L
9.	Total alkalinity (as Na_2O)	628 mg/L	260 mg/L	93 mg/L

1. **Colour:** The whole paper mill effluent whether collected from the beel or from paddy field exhibited light brown to dark brown colour. The natural colour of the water was changed due to contamination by paper mill effluents, hence, the water appeared to be undesirably dark brown. The water of the samples collected from non-polluted fields also showed a pale yellowish colour.
2. **Odour:** A foul and disagreeable odour was to emit in the watercourse. The odour of charring wood and chemicals used in the paper mill might contribute to odour of the effluents.
3. **pH:** In the present study, the pH values were within the acceptable range. There seems to be a trend of gradual increase of pH as a result of increasing quantities of effluent. According to Desai *et al*, (1990) the tolerance limits of pH for irrigation varied between 6 and 9. The pH values of the paddy fields affected by the effluent of the paper mill were within the permissible limits with respect to its use in agriculture. Hence, it might be agreed that pH of the paddy fields effluents, although higher than the adjoining non-affected fields, might not be the sole factor for affecting the rice plants adversely and there might be some other factor more responsible to cause serious effects. This has been supported by the finding of Som *et al* (1994).

4. **Turbidity:** In the present study, the turbidity values were exceptionally higher than the permissible limits. It was found to be the highest at 0.5 km downstream (140 NTU) and lowest in 2.5 km downstream (85 NTU) in the samples collected from effluent affected paddy fields. These values were manifold higher than the permissible limits. In non-affected areas, it was around permissible limit (28 NTU). Hence, very high turbidity values recorded all along the water stretch would certainly account for the ill effect on the paddy fields in general, and the standing paddy plants in particular. It might be concluded that high turbidity of the effluents might pollute the paddy fields by being very toxic to the paddy plants.
5. **Total Suspended Solids:** On the scrutiny of the results the total suspended solids (TSS) were seen to vary in different plants and in different time periods. It ranged from 200 mg/L to 80 mg/L as the case may be, but the effluents collected from the lagoon showed the minimum (935 mg/L) and water collected from non-polluted rice fields showed minimum (78 mg/L)

   The TSS imparted brown colouration to water as well as reduced the transparency to nearby nil. Suspended solids, therefore, retarded or obstructed light penetration and thereby reduced photosynthetic activity of plants. This has also been suggested by Clausen (1973).
6. **Total Dissolved Solids:** It was seen that the amount of total dissolved solids (TDS) varied inversely with the increasing distance of the experimental area from the point of the origin of effluent. The TDS were found to be high (1480 mg/L) at lagoon which gradually decreases towards the downstream of the beel as well as affected paddy fields.

   High concentration of dissolved solids might adversely affect plants and also cause higher salinity (Trivedy and Goel, 1984). Therefore, there was every reason to believe that high TDS values might be detrimental to paddy plants particularly in the areas affected with effluents. Venkateswarlu (1986) has well documented the bad effects of high concentration of TDS and low dissolved oxygen on plants.
7. **Biochemical Oxygen Demand:** Water having biochemical oxygen demand (BOD) levels less than 20 mg/L was considered of good quality with high rating (Botton *et al*, 1977). According to Ross (1977) water having BOD less than 20 mg/L was considered the highest rating while BOD level higher than 50 mg/L was considered zero rating. Handa (1980) suggested that BOD level of 20 mg/L could be accepted as the maximum permissible limits for unpolluted waters.

   The results of the present experiment indicated that the non-polluted rice fields had BOD values almost in the permissible range, while the pollutant fields had BOD values extremely higher than the prescribed value for plant tolerance.

   When the contentions of Ross (1977), Handa (1980) were accepted and the results of the present experiments were considered, the increase in the BOD level in the present experiments clearly suggested that the effluen caused a high degree of pollution. The high concentration of BOD in the effluent at lagoon as well as in the fields polluted the paddy plants. This has been earlier indicated by Rao (1976) and Campbell (1978).

8. **Chemical Oxygen Demand:** Chemical oxygen demand (COD) values were usually greater than the BOD values, especially the when the wastewater contained a significant amount of biologically non-degradable organic matter. When the results of BOD and COD values were compared (Table 7.1) especially in the effluents of the beel rice field and non-polluted areas, it was seen that COD values were proportionately higher than the BOD values. However, the increase of beel (845 mg/L) and lowest in the non-polluted areas (61 mg/L). The high value of COD over BOD was the expression of the amount of dissolved oxygen which made the effluent unsuitable for irrigation to cultivated fields. This idea has also been supported by Agrawal *et al* (1994) and Subramani *et al* (1995).
9. **Dissolved Oxygen:** When paper mill effluent directly mixed with the water of paddy field at the time of transplanting of rice seedlings, the concentration of dissolved oxygen (DO) was very low or near to zero.

   It was evident from the results that the little DO available in the effluent affected rice fields probably helped the growth of the plants. This was in accordance with the observations of Madhappan (1993) and Verma and Verma (1995).

   Absence of DO damage the plant cells, hence, the paddy plants could not grow. It emerged, therefore, that DO at near zero level in water of the paddy fields mixed with paper mill effluent was harmful to the growth of the plants. This agrument was supported by the findings of Hariom *et al* (1994).
10. **Total Alkalinity:** The present study had revealed that water of the entire experimental area was alkaline, being more towards the source of effluent and less away from the source.

Maximum alkalinity (628 mg/L) had been found in the effluent and 260 mg/L in effluent affected rice fields. These results clearly suggested that the rice field water was highly polluted and also toxic to life of the paddy plants. Abebisi (1980) suggested that water was having 60 mg/L alkalinity added hardness to water. The results obtained in the present experiments clearly suggested that the effluent as well as the effluent affected rice field waters were considerably hard and, hence, toxic to rice plants. Further it was seen that less hard water i.e., less than 60 mg/L alkalinity value was beneficial to rice plants.

The plants of the control and polluted sites were significantly different in height; the maximum height being 138.25 cm and 105 cm, respectively. The reduction in seedlings might be due to greater accumulation of Ca, Mg and solid materials present in the effluents, as suggested by Israelsen and Hansen (1962) and Oblisami (1979) and Indhirabai *et al* (1989).

The maximum shoot length was observed 105.6 cm (control) and 82.3 cm in (polluted) plants at the time of maturity. The height of plants reduced considerably due to reduction internodal elongation in the polluted plants in comparison to control plants. Indhirabai *et al* (1989), also reported similar findings.

The data revealed that the growth of root was faster from transplanting to panicle-initiation stage in both polluted and control rice plant. But the root number, length and weight in effluent affected rice plant were significantly higher than non-polluted (control) plants. It may be argued that the effluent affected soil might have sufficient space available in the form of large pores, hence, the roots of these plants could grow freely than those of control plants where soil could be more compact. Comparatively vigorous root growth of rice plants in polluted soil over those of control areas may be accounted for more aeration of polluted soil resulting in enhanced uptake of nutrients and water. That the quality of soil in root growth was responsible for aeration and uptake nutrients and water had been supported by Van Noordwijk *et al* (1993).

The root number and length shows highest in 90 DAS and 30 DAS in case of control plant and 90 DAS and 60 DAS in polluted plant respectively. But the percentage of reduction of root length shows from vegetative stage to ripening stage. The results were in conformity with the observation of Sorte *et al*, 1992, Peter and Runkles, 1967. Growth of roots was faster from transplanting to panicle initiation stage of rice. However, at subsequent stages the root number and length were decreased in both types of plants mainly due to decay in older mass with an addition to little root mass. Reduction in root length was significantly prominent in floral initiation stage (90 DAS) in both types of plants. The increase in root length particularly in the polluted plants could possibly be attributed to the lowest soil mechanical resistance due to continuous submergence in the effluents. Roots in this case could grow without any waviness as reported by Reddy and Kuladavielu (1992). However, higher resistance offered by the non-polluted soil in comparison to polluted soil, was the possible reason for shorter roots. Similar results were reported by O'Toole (1982).

A drastic reduction in dry matter accumulation in shoot was evident at vegetative stage while the differences were minimised in later stages of polluted rice plants reduced rice matter production (DMP) in shoot significantly at all stages and the reduction was highest (68.36%) at vegetative stage and 55.35% at maturity (Table 7.3).

The dry matter production of roots increased in vegetative stage and decreased in ripening stage in polluted plants. The DMP in polluted plants showed maximum up to vegetative stage, which was even greater than that of the polluted plants after 60 DAS (0.7076g) (Table -3). At panicle initiation stage (75 DAS), the total dry matter of roots began to decline which continued up to plants maturity. This was the case in both polluted and control plants. The decrease of DMP in roots may be due to decay in old root mass and slow increase of active roots. This had been confirmed by the Reddy and Kuladavietu (1992).

The most striking effect of pollution on rice plants was exhibited in root-shoot ratio (548.45%) (Table 7.3), being the highest during the vegetative stage up to 90 DAS which declined subsequently. This pattern was found to be present in both polluted and control plants. However, there existed some difference between the ratio in both types of plants.

From the above findings it was concluded that the physiology of root-shoot ratio indicates that rice plant affected by the paper mill effluent enhances the vegetative growth but retards the reproductive growth.

Results presented in Tables 7.1 and 7.2 show higher rate of increase in the biomass and Net Primary Production (NPP) accompanied with age of the plant in the non-polluted plants. However in polluted plants, the biomass and NPP are reduced to certain extent.

The maximum biomass was recorded at the age of 135 days i.e. maturity of the plant. It is 7.9443 g/plant/day in non-polluted and 3.1521 g/plant/day in polluted plants respectively. The maximum biomass values were recorded at the age of 15 days in non-polluted and polluted

Table 7.2: Plant height, shoot length, root number and root length of rice plants collected from Polluted (P) and Control (C) rice fields at various stages of their growth (Average value of five different plants).

Days	Plant Height		Reduction Induction	Shoot length		Reduction Induction	Root Number		Reduction Induction	Root length		Reduction Induction
	C	P	%	C	P	%	C	P	%	C	P	%
15	43.2	44.5	+3.0	16.8	16.3	+2.97	25	46	+84	14.1	22.6	+60.28
30	45.4	50.2	+10.5	17.7	17.2	+2.82	25	62	+148	20.5	23.2	+13.17
45	70.8	58.5	–17.37	34.1	24.1	–29.32	30	61	+103	17.4	22.9	+31.60
60	98.5	61.2	–37.86	36.5	26.1	–28.49	51	62	+21.5	20.2	23.5	+16.33
75	100.4	55.9	–44.41	43.8	24.3	–44.52	70	68	–2.85	12.6	17.1	+35.71
90	102.5	59.7	–41.75	44.6	25.2	–43.49	82	72	–12.19	12.3	18.5	+50.40
105	128.5	90.5	–29.57	101.3	70.8	+30.10	76	57	–25	11.5	17.5	+52.17
120	137.4	101.1	–28.03	104.2	81.1	–22.16	92	52	–35.08	16.5	19.2	+16.36
135	138.2	102.5	–25.53	105.6	82.3	–22.06	60	43	–28.33	14.4	15.6	+8.33
Mean SEm ±		0.3385			0.0882		0.9329			0.0809		
(Age x C Vs P) CD (5%)		0.9712			0.253		2.676			0.2321		

plants (Table 7.1). NPP was not similar with that of plant biomass, as it was maximum in 75 days old plant i.e., 0.1449 g/plant/day in non-polluted plants and 0.0503 g/plant/day in polluted plants at seedling stage i.e. 30 days after transplantation.

Biomass is a manifestation of net production. Biomass production, a primary value of productivity was increased almost linearly up as the crop growth advanced. The rate of biomass production in cereal crops was proportional to the intercepted solar radiation, as reported by Biscoe and Gallagher (1977). It was observed that crop plants which intercepted higher amount of solar radiation-conversion efficiency had more dry matter accumulation (Gallagher and Biscoe, 1978). The results obtained in the present study, were similar to the findings of above-mentioned workers, but at the panicle initiation stage the dry matter production was found to be diminished. This might be accounted for the reduced solar radiation during flowering to ripening in wet season also decreased current photosynthesis, resulting in decreased dry matter production. This contention was supported by the study of Tripathy and Nayak (1995). Similar reports had been forwarded by Venketeswaralu (1977) and Thangaraj and Sivasubramanian (1990).

The growth and yield in the effluent plants is a complex process influenced by quantities of micro-nutrients, heavy metals decomposition products and salt toxicity as well as effect from soil structure, porosity and aeration, as reported by Dolar *et al* (1972). Reduction in growth by industrial effluents has also been noted in agricultural plants (Thabaraj *et al*, 1964; Sahai, *et al*, 1983) and wheat (Bansal *et al*, 1992). This inhibitory effect of growth and yield, may be due to higher concentration of effluent contents, high levels of total dissolved solids like Na, Cl and sulphate, etc. which possibly inhabit uptake of other elements like K, Ca, P and Mg by the plants, as proposed Ramesh Kumar *et al* (1990).

It was found that N and K increased, whereas P content decreased considerably in the polluted plants (Table 7.3 and 7.4). The decrease in phosphorus content might be due to the presence of oxygen phosphates co-precipitate with the complex insoluble oxides of iron, manganese, aluminium etc. consequently, at higher redox potential (oxidising conditions). The net tendency of phosphorus in waters is to get precipitated and loss of the sediments as reported by Trivedy and Goel (1984).

Table 7.3: Showing dry wts. of shoot, roots and root-shoot ratio of rice plants collected from polluted (P) and control (C) rice fields at various stages of their growth (average value of five different plots)

DAS	Dry wt. of Shoot		Reduction Induction	Dry wt. of Root		Reduction Induction	Root-Shoot Ratio		Reduction Induction
	C	P	%	C	P	%	C	P	%
15	0.4305	0.1362	–68.36	0.0782	0.1604	+105.11	0.1816	1.1776	+548.45
30	0.8246	0.4124	–49.98	0.1331	0.3522	+164.61	0.1614	0.8540	+429.12
45	1.0935	0.6963	–36.32	0.1976	0.3883	+96.50	0.1807	0.5576	+208.57
60	1.1192	0.6649	–0.59	0.2034	0.6	+247.88	0.1817	1.0642	+485.69
75	2.2995	0.7267	–63.39	0.6316	0.7149	+13.10	0.2746	0.9837	+158.23
90	2.9685	0.7091	–79.11	0.9569	0.8695	–9.13	0.3223	1.2262	+280.45
105	3.5796	1.6893	–52.80	0.8382	0.2457	70.68	0.2341	0.1454	–37.88
120	2.9716	1.3325	–55.15	0.5696	0.1885	–66.90	0.1916	0.1414	–26.20
135	2.8406	1.10167	–61.03	0.7783	0.1937	–75.11	0.2739	0.1750	–36.10

Table 7.4: Showing biomass (dry wt. in g/plant/day) production by rice plants from Polluted (P) and Non-Polluted (NP) fields at various stages of their growth (average value of five different plots)

Days	Dry Wt. (g/plant/day)									
	Root		Stem		Leaf		Panicle		Total	
	NP	P	NP	P	NP	P	NP	P	NP	P
15	0.0752	0.1504	0.4305	0.1352	0.2895	0.1194	–	–	0.7752	0.4150
30	0.1331	0.3522	0.5246	0.4124	0.5731	0.4062	–	–	1.5308	1.1728
45	0.1976	0.3883	1.0935	0.5963	0.9628	0.4856	–	–	2.2539	1.5702
60	0.2034	0.7076	1.1192	0.6649	1.0707	0.4149	–	–	2.3935	1.7874
75	0.6313	0.7149	2.2995	0.7267	1.6390	0.5600	–	–	4.5701	2.0016
90	0.9569	0.8695	2.9685	0.7091	1.0969	0.4625	0.3634	–	5.3857	2.0411
105	0.8382	0.2457	3.5796	1.6893	1.0020	0.2605	0.7596	0.2931	5.1794	2.4886
120	0.5696	0.1885	2.9716	1.3325	0.8570	0.2283	3.3126	0.8192	7.7108	2.5685
135	0.7783	0.1937	2.8408	1.1067	0.4410	0.2104	3.8844	1.6413	7.7944	3.1521

SEm (Non-Polluted Vs Polluted)			
±	0.0014	0.0069	0.0105
CD at 5%	0.0040	0.0197	0.0301
SEm (Age x Non-Polluted Vs Polluted)			
±	0.0043	0.0206	0.0314
CD at 5%	0.0123	0.0591	0.090

The presence of organic materials in the effluents and the paper mill added nitrogen and phosphorus in the form of urea and single super phosphate at the rate of 100 BOD : 5 N : 1 P as nutrient (Barthakur *et al*, 1989), might be responsible for the increase of available nutrients in soil. Such type of increased level of nutrients in soil was reported by Igbounamba (1972), Patra and Behera (1974) and Kadu *et al* (1991).

From the results it is seen that polluted plants show maximum amount of total chlorophyll, particularly in flowering and mature stages (2.84 mg/g FW and 0.95 mg/g FW respectively). It is possible that the effluent does not affect on the leaf surface as a result of which light falls on the surface without any disturbance. However, in the seedling stage the non-polluted plants exhibit maximum amount of total chlorophyll.

Each of chlorophyll 'a' and chlorophyll 'b' increases in seedling stage and abruptly decreases with the flowering of plants. This is equally true in both polluted and non-polluted plants. This decline in chlorophyll content continues, starting from the flowering stage to the mature stage of the plants. The pattern of chlorophyll contents (a and b) during the studied stages are similar in both polluted and non-polluted plants. However, the quantities are different in both the cases. The total chlorophyll content of polluted plants are higher than that of the non-polluted plants, during flowering and mature stage only. This increase of chlorophyll content in polluted plants is due to rise in nutrients status of the effluents.

Comparatively chlorophyll 'a' is found to be more than chlorophyll 'b' in both polluted and non-polluted plants.

The variations in the total chlorophyll contents in the different experimental sites may be attributed possibly due to edaphic factors, as well as due to different concentration of the effluents, that chlorophyll contents of some other plants vary in different plots. This report has been corroborated by (Wright and Smith, 1983; Singh and Patil, 1986; Ashraf and Khan, 1990 and Estill *et al*, 1991).

The non-polluted plants show maximum leaf area and grain yield in comparison to those of polluted plants. The results have revealed that rice plants with higher leaf area contribute to

Table 7.5: Net Primary Production (NPP) (g/plant/day) plants (root, stem, leaf and panicle) collected from Polluted (P) and Non-Polluted (NP) fields at various stages of growth (average value of five different plots)

Days	NPP (Root)		NPP (Stem)		NPP (Leaf)		NPP (Panicle)		NPP (Total)	
	NP	P	NP	P	NP	P	NP	P	NP	P
15	0.0057	0.0106	0.0267	0.0090	0.0179	0.0079	–	–	0.0518	0.0275
30	0.0036	0.0127	0.0262	0.0184	0.0269	0.0192	–	–	0.0567	0.0503
45	0.0043	0.0024	0.0179	0.0189	0.0193	0.0051	–	–	0.0415	0.0264
60	0.0003	0.00312	0.0017	–0.0020	0.0071	–0.0047	–	–	0.0091	0.0145
75	0.0285	0.0004	0.0786	0.0041	0.0378	0.0096	–	–	0.1449	0.0141
90	0.0216	0.0103	0.0446	–0.0011	–0.0361	–0.005	0.0242	–	0.0543	0.0027
105	–0.0079	–0.0415	0.0407	0.0653	–0.0063	–0.0134	0.064	0.0195	0.0529	0.0299
120	–0.0179	–0.0038	–0.0405	–0.0237	–0.0096	–0.0021	0.1702	0.0350	0.1022	0.0054
135	0.0193	0.003	–0.0087	–0.0150	–0.0277	–0.0011	0.0381	0.0548	0.156	0.0390

Wright G.S., Smith R.C.G. and Morgan J.M. (1983). Differences between two grain sorghum genetyts in adaption to drought stress. III. Physiological responses. Aust. J. Agric, Res. 34: 1.

Yoshida S. Douglasa., A. Forno. JAMESH. Cock Kwanchai, A. and Gomez. (1976). Measurement of leaf area and leaf area index, Laboratory Manual for Physiological Studies of Rice. 69-71.

CHAPTER 8

Advances in Physiology of Vegetable Crops: A Brief Background

S.K. Arora[1] and Rakesh Mehra[2]

Abstract

The present production of vegetables in India is estimated to be 84 million tons from an area of over 6.2 million ha against a requirement of 100 million tons, having 2.6 per cent annual growth. Thus, in past three decades, India has made a quantum jump in vegetable production securing the second position in the world. However, this amount is much less than our requirement and serves per caput intake of 210 g of vegetables against the requirement of 285 g per caput per day for balanced diet. There is a need to achieve the target of 100 million tons for the supply of 250 g vegetables per day per caput to our estimated population of 1 billion by 2000 A.D. However, our demand by 2020 A.D. will be around 250 million tons. More than fifty vegetables are being grown in the country. Hybrid varieties have been developed in 15 vegetable crops. The public sector has developed 25 hybrids in 11 vegetable crops which are now under cultivation. Private companies have developed hybrid varieties and producing/importing large quantities of F_1 hybrid seeds. Thus, there is immediate need to develop F_1 hybrid cultivation technology and generate vast infrastructure for the production of F_1 hybrid seeds to meet the requirement of farmers. The future thrust is to reflect hybrid seed production technology in the national vegetable research and seed production programmes so that vegetable production is increased as per our minimum requirements. This can be further strengthened by having regional and international collaboration. However, several production constraints result in low yields per hectare and insignificant share in the exports. There are still distinct gaps which need to be plugged. Technologies for cultivation of

Address for correspondence: Department of Vegetable Science, CCS Haryana Agricultural University, Hisar - 125 004 (Haryana)

[1]Professor (Vegetable Crops)

[2]Asstt. Scientist (Spice Crops)

various vegetables under different agro-climatic situations have been standardised to bring about marked increase in productivity. Seed production technology for vegetables has been standardised to make the country self-reliant in the production of seeds.

Thus, it can be summarised that the research in the area in India remained almost neglected in the first half of the century and gained momentum thereafter to acquire the present status having moderate infrastructure facilities and research priorities for vegetables in different institutions. Still there is a wide gap between production and availability of vegetables in our country. The strategies given may serve as guidelines for our endeavours to bridge this gap.

INTRODUCTION

Research on vegetable crops was started by ICAR with the sanctioning of a nucleus "Plant Introduction Scheme" at the IARI, New Delhi in 1947-48. Simultaneously, the ICAR also started such ad-hoc schemes in different states like Punjab, Uttar Pradesh, West Bengal, Maharashtra, Himachal Pradesh, Jammu & Kashmir and Tamil Nadu.

Prior to World War II, entire seeds of temperate vegetables were imported from European countries. When during World War II, supply of vegetable seeds from Europe suffered a serious setback, the Government of India felt the need for production of these seeds within the country. As a result of this, the Government of India established a Vegetable Breeding Station at Katrain in Kullu Valley, Himachal Pradesh in 1949. This station was transferred to Indian Agricultural Research Institute in 1955 to undertake research on improvement of temperate vegetable crops, standardisation of their seed production technology and to produce seeds of identified improved varieties of temperate vegetables. This research station is carrying out intensive research on all aspects of temperate vegetables including improvements, production technology and their seed production.

Further, a Division of Horticulture was created at the Indian Agricultural Research Institute, New Delhi in 1956-57. However, vegetable research was organised on sound footing with the establishment of the Indian Institute of Horticultural Research (IIHR) at Bangalore with a full-fledged Division of Vegetable Crops in 1968. The vegetable research in India received a further boost when a separate Division of Vegetable Crops and Floriculture was started at IARI, New Delhi in 1970. Later the floriculture work was separated and the division was then named as Division of Vegetable Crops, which was assigned the research work exclusively on different vegetable crops. In the Sixth Five Year Plan, IIHR, Bangalore started two Central Horticultural Experiment Stations, one each at Ranchi (Bihar) and Godhra (Gujarat) to intensify research on vegetable crops. Establishment of Agricultural Universities and All India Coordinated Vegetable Improvement Project gave further fillip to vegetable research in India. Ever since the start of vegetable research in the country in the post-independence period, efforts have been made to establish a sound research infrastructure to meet the requirements of vegetable research. This includes research by Central Institutes, 26 State Agricultural Universities, Coordinated Projects and ad-hoc schemes (Chadha, 1998).

India is fortunate to have a variety of agro-climatic conditions, which enable us to produce both tropical and temperate vegetables. Since independence, vegetable production of our country has increased tremendously and thus India is the second largest vegetable producer next to China with an estimated production of about 71 million tons from an area of over 6.2 million hectares against a requirement of 100 million tons. With this production, the availability of vegetables is 210 g per caput per day which is less than the dietary requirement of 300 g per caput per day.

India's population is growing at the rate of 1.8 per cent and it is expected to be over 1.2 billion by the year 2020. Population growth and urbanisation are creating increased demand of food. There is growing concern over malnutrition, especially in rural areas. To meet the full dietary needs of the common man, to eliminate malnutrition, deficiency diseases and to relieve over stress on cereals, there is greater need for enhanced vegetable production.

PRESENT SCENARIO OF VEGETABLE PRODUCTION IN THE COUNTRY

According to the statewise distribution of area under various vegetable crops, the maximum area lies in U.P. (12.6 lakh ha) followed by Tamil Nadu (8.8 lakh ha), Bihar (8.4 lakh ha) and West Bengal (4.5 lakh ha). U.P. contributes maximum share in country's vegetable production while Bihar and Orissa occupy second and third place, respectively.

VARIATION IN PRODUCTIVITY OF VEGETABLE CROPS AMONG STATES

There are variations among different states of the country with regard to productivity of vegetable crops. Karnataka recorded the highest productivity of 23.49 t/ha in the country, while Punjab ranked second. In India one of the major reasons of low production of vegetables is steep variation in productivity of vegetables among states. Andhra Pradesh, Gujarat, Karnataka, Kerala, Madhya Pradesh, Meghalaya, Punjab, Taml Nadu, U.P. and Delhi are having productivity more than 10 t/ha while there are certain states, where productivity is even less than 5 t/ha. Our country has witnessed 2.6 per cent increase in vegetable production per year in last 10 years (1983-94), which was at par with the region's growth rate (2.7%) and higher than the world growth rate (1.81%).

LOW PRODUCTIVITY OF THE VEGETABLES IN THE COUNTRY

Although India is the second largest vegetable-producing country in the world, yet productivity of different vegetables is comparatively low as compared to other developed countries. The average productivity of tomato in our country is 157.10 q/ha while its productivity in USA is 587.62 q/ha, in Greece 498.6 q/ha, in Italy 465.55 q/ha and in Spain 466.29 q/ha. The poor productivity of different vegetables in our country is due to several reasons. In developed countries, maximum production of vegetables is under controlled conditions and production does not suffer from severe weather onslaughts, while in our country area under controlled conditions till date is negligible. Similarly in developed countries, hybrids occupy maximum area and advanced production and protection technology have been standardised to realise the maximum productivity of vegetable crops. In India, area under hybrid vegetables has increased considerably in last few years, however, maximum share in production is contributed by open-pollinated varieties with the application of farmer's own thinking and their traditional vision. During the last 50 years, concerted efforts have been made in different directions to boost the vegetable production of the country. These newly developed aspects of vegetable production are mentioned below:

Nursery Management in Vegetable Crops

Young seedlings are raised and nurtured in the nursery beds before planting them into the main field. For raising good crop, it is utmost essential that seedlings should be healthy, vigorous and disease free. Seedlings are susceptible to a number of diseases due to its delicate, succulent and highly tender nature. Further sky-rocketing cost of hybrid seeds warrants attention of the grower to produce the seedlings under protected conditions, because every hybrid seed has its accountability owing to its high cost which is 20 to 50 times higher than open pollinated seeds. Raising the nursery under polyhouse/greenhouse and sowing the individual seed of hybrid in potting plug with artificial culture media is well established practice in developed countries. However, in India most of the nurseries are raised under open conditions and it is one of the main reasons of setback to development of hybrid programme during rainy seasons in northern India. Therefore, it is essential to standardise the nursery raising programme on scientific lines.

Soil Solarisation

Recently soil solarisation has proved as cheapest and eco-friendly approach for soil disinfection of nursery beds. Soil mulching is done with transparent polyethylene of 25-100 cm thickness during the period of high temperature and solar radiation. Soil should be kept moist before mulching to increase the latent heat and thermal sensitivity of resting structures of soil- borne pathogens, harmful pests and weeds. Following this simple practice, population of soil- borne pathogen, nematodes, pests and weeds can be reduced to a substantial level.

Raising Nursery Under Polyhouse/Agronet

Nursery raising of vegetables has become a specialised job with increasing susceptibility of vegetables to biotic and abiotic stresses. The major nursery pests and diseases are white fly, cut worms, borers, nematodes, damping off, fungal and bacterial leaf spots. Besides, high and low temperature, excess rains, drought, hot and cold winds affect nursery raising. It has now become possible to raise the seedlings in low tunnels and low cost polyhouses. Production of early cauliflowers, tomato, capsicum, chillies often poses problems in high temperature and high rainfall areas. Use of transparent plastic sheet as low tunnel provides ideal conditions for successful raising of seedling than conventional methods. The time taken for seed germination and seedling growth is reduced under low tunnels compared to open environment in tomato, brinjal and chilli crops sown in winter months for spring transplanting. The utility of low cost polyhouses for raising of seedlings of cucurbitaceous vegetables in winter months to harvest early summer crops has been very successful and commercially adopted by farmers in Punjab, Haryana and U.P. The seeds sown in polybags show early germination and make faster growth under protected cover. When outside conditions are favourable, seedlings are transplanted along with earth ball in the open field in early spring, the crops flower and set fruits in very short times. This technique is capable of giving highly remunerative yields in northern plains. Besides, the seedlings of tomato, brinjal and chilli also get ready for planting in spring season in plains. The use of shade and agronets have been found very effective in raising of seedlings during high temperature. Shade reduces temperature by 5 to 10°C and protect the seedlings from sun scorching. Agronet protects the seedlings from insect-pests and reduces the vector-borne viruses infestation and damage by other insects.

Improved Vegetable Varieties

Lot of progress has been made in the development of high yielding open pollinated varieties, disease and pest resistant varieties and hybrid varieties. Till date 160 improved varieties in 19 vegetable crops have been identified and recommended for cultivation in various agro-climatic conditions based on multi location and multidisciplinary testing of vegetable varieties. These varieties are given in Table 8.1.

Breeding for Disease Resistant Varieties

Breeding disease resistant varieties is a novel option to avoid the excessive use of chemicals and to reduce the grower's burden on the cost of production (Kalloo, 1988). In past, concentrated efforts were made to develop Fusarium wilt, bacterial wilt, early blight, root knot nematode, tomato leaf curl virus, tomato spotted wilt virus resistant varieties in tomato. Against Fusarium wilt a number of varieties like Columbia, Walter, Floradade, HS-110, Homestead and Sel-28 have been found resistant. BT-1 and BT-10, BWR-l, BWR-5, LE-79 are resistant to bacterial wilt. These varieties can be grown successfully in Karnataka, Kerala, Orissa and other bacterial wilt infested areas. Floradade variety has been found resistant to early blight which is a serious problem in many parts of the country. NT-3, NT-8, NT-12 and Sel-120 have been developed as root knot nematode resistant varieties. Varieties H-24, H-36, H-86, H-88 are resistant to tomato leaf curl virus which have been developed utilizing *L. hirsutum* f. glabratum as source of resistance (Table 8.2). In brinjal, which is highly susceptible to bacterial wilt in certain parts of the country, BB-7, BWR-12 and SM-6-7 have been found resistant to bacterial wilt. There is immediate need to develop a variety resistant to Phomopsis and shoot and fruit borer in egg plant. Chilli is susceptible to anthracnose, Fusarium wilt, die back, bacterial wilt, bacterial leaf spot, bacterial blight and leaf curl etc. Punjab Lal, Hisar Shakti and Hisar Vijay are the varieties resistant to TMV, CMV and leaf curl virus and moderately resistant to fruit rot and die back have been developed.

In peas, JP-4, JP-83, FC-l, PRS-4 and NDVP-4 have been recommended as powdery mildew resistant varieties. JM-l, JM-2 and Kala Nagin have been found to be resistant to Fusarium wilt. In many parts of the country pea is susceptible to rust, however, there is no definite source of resistance to this disease. Attempts have been made to develop varieties resistant to multiple diseases and pests in peas and JP-179 resistant to powdery mildew and bruchus have been developed. Cowpea is susceptible to *Pseudocerospora, Anthracnose, Macrophomina,* wilt, cowpea mosaic virus and golden mosaic virus. However, there is no systematic disease resistant breeding programme in the country. Only the sources of resistance to *Pseudocercospora,* cowpea

Table 8.1: Improved vegetable varieties

Vegetable		*Variety/ies*
Amaranthus	:	Badi Chaulai, Co. 1, Co. 2, Co. 3
Brinjal (Long)	:	Pusa Purple Long, Pusa Purple Cluster, BB-13, PH-4, S-4, PBR-129-5, Azad Kranti, Arka Sheel, Arka Shirish, ARU-1, ARU-2C, KT-4, Hisar Pragati, NDB-25, BB-26, Punjab Barsati, Sel-4, DBSR-31, Punjab Chamkila, Punjba Sada Bahar Baigan
Brinjal (Round)	:	T3, Sel-1, PBR-92-2, K-202-9, Hisar Shyamal, Pusa Upkar, Pusa Uttam, KS-224, DBR-8, Punjab Bhar, Jamuni Gola, Punjab Neelam, Arka Navneet
Brinjal (Oblong)	:	Hisar Jamuni, BH-1, BH-2
Brinjal (Green)	:	Arka Kusumakar
Brinjal (Small round)	:	Aruna, DBSR-44, Punjab Moti, Punjab Bindu, PLR-1
Cabbage	:	Sel-8, Pride of India, Golden Acre
Carrot	:	Hisar Gairic, Chantenay, Pusa Kesar, Nantes, Pusa Yamdagni, Danvers, Imperator, Selection-233, Selection-21, Sel-5, Pusa Meghali
Capsicum	:	California Wonder, Yolo Wonder, King of North, Early Giant, Vinedale, Bullnose, Chinese Giant, Ruby King, Punjab-27
Cauliflower	:	Pusa Early Synthetic, Pusa Synthetic, Punjab Kunwari, Pusa Deepali, Early Kunwari, 327-14-8-3, 351-4-1, Improved Japanese, Synthetic-1, EC-12012, Sel-1, Pusa Snowball-2, 235-S, K1, 114-S-1, 6-1-2-1, Ealry Synthetic, Punjab Giant-26, Punjab Giant-35, Pusa Snowball-1
Chilli	:	G-4, G-5, JCA-154, K-1, K-2, Kalianpur Selection, NP46-A, Pant C-1, Pant C-2, Pusa Jwala, Pusa Sadabahar, Sindhur, X-197, Hisar Shakti, Hisar Vijay, Sel-1., JCA-283, Phula-5, J-218, X-235, Musalwadi, LCA-206 B, Punjab Lal, Punjab Surukh, Punjab Guchhadar
Cluster bean	:	Pardeshi, Pusa Mausami, Pusa Navbahar, Pusa Sadabahar, S-279
Cowpea	:	Pusa Barsati, Pusa Dofasli, Pusa Komal, Pusa Phalguni, L-1552, Deepaliwali, Sel 61-B, Sel 2-1, Sel 263
Cucurbits	:	
(a) Ash gourd	:	Co. 1, Co. 2
(b) Bitter gourd	:	Arkaharit, Coimbatore Long, Pusa Do Mausmi, VK-1, Priya
(c) Bottle gourd	:	Pusa Naveen, Pusa Summer Prolific Long, Pusa Summer Prolific Round
(d) Cucumber	:	Balam Khira, Japanese Long Green, Poinsette, Priya
(e) Muskmelon	:	Arka Jeet, Arka Rajhans, Durgapura Madhu, Hara Madhu, Edisto, Pusa Sarbati, S1 445, Sel-2, Punjab Sunehri, Punjab Hybrid, Punjab Rasila, MH-10
(f) Pumpkin	:	Arka Suryamukhi, Arka Chandan, Pusa Biswas
(g) Ridge gourd	:	Co 1, Pusa Nasdar, Satputia, HRG-14
(h) Snake gourd	:	Co 1
(i) Sponge gourd	:	Pusa Chikni, Sel-99, Pusa Supriya
(j) Summer squash	:	Pusa Early Prolific, Early Yellow Prolific, Pusa Alankar
(k) Wanga	:	Punjab Wanga-1
(l) Watermelon	:	Arka Jyoti, Arka Manik, Durgapur Kesar, Durgapura Meetha, Sugar Baby, Shipper, Congo, Improved Shipper, AHW-19, AHW-65 (Mateera)

(Table 8.1 Contd.)

(Table 8.1 Contd.)

Fenugreek	:	Pusa Early Bunching, Kasuri, Methi No. 14, Methi No. 47, Hisar Methi-57
French Bean	:	VL-Boni-1, Arka Komal, UPF-191
Garlic	:	Fawari, G-41, G-1, G-50, Punjab Garlic-1 Rajalle Gaddi
Lablab Bean	:	Hisar Kirit, HD-60, T-1, Hebbel Avarez, Hebbel Avare 4, IIHR Sel-1, IIHR Sel-2
Letuce	:	Punjab Lettuce-1
Okra	:	Harbhajan, Lucknow Dwarf, Pb 7, Parbhani Kranti, Pusa Makhmali, Punjab no. 13, Pusa Sawani, varsha Uphar, Punjab-8, Hisr Unnat, Punjab Padmini
Onion	:	Pusa ratnar, Pusa Red, Hisar-2, Punjab Selection Line-181, Line 106, S-4, S-131, N-257-9-1, N-2-4-1, Arka Kalyan, Line 102, Arka Niketan, Agrifound Dark Red, VL-3, Agrifound Light Red, Punjab Red Round, Punjab-48, Punjab Naroya, Punjab White, PBR-5, Early Grano, Punjab-48, Punjab Selection, Pusa White Flat, Pusa White Round, Ratna Red
Pea	:	Asauji, Bonneville, Early Badger, Early December, Early Superb, Meteor, Perfection New Line, Sylvia, Arkel, GC-195, GC-141, IP-3, P-88, PM-2, VL-3, Lincoln, VL-7, VL-06, PH-1, Mata Ageta-6, Punjab-87, Punjab-88, Har Bona, Mithi Phali, NDVP-8
Potato	:	Kufri, Alankar, K. Bahar, K. Badshah, K. Chamatkar, K. Deva, K. Chandramukhi, K. Himalini, K. Jeevan, K. Jyoti, K. Kisarn, K Kuber, K. Kumar, K. Kundan, K. Lalilma, K. Lauvkar, K. Mathu, K. Neela, K. Red, K. Safed, K. Sheetman, K. Sindhuri, K. Sutlej, K. Sherpa, K. Swarna
Radish	:	Jaunpuri Giant, Japanese White, Kalyanpur T-1, Punjab Safed, Punjab Pasand, Pusa Chetki, Pusa Desi, Pusa Himani, Pusa Reshmi, White Icicle, HR-1, Punjab Ageti
Spinach beet/ Beet leaf (Palak)	:	All Green, Banerjee's Giant, HS-23, Jobner Green, Palak No. 51, Pusa Harit, Pusa Jyoti, Pusa Palak, Punjab Green
Sprouting Broccoli	:	Punjab Broccoli-1
Sweet potato	:	H-41, H-42, H-260, Pusa Red, Pusa Safed, V-35, Verde
Tomato	:	Best of All, HS-102, HS-101, HS-110, Marglobe, Punjab Chhuhara, Pusa 120, Pusa Early Dwarf, Pusa Ruby, Pusa Sheetal, Roma, S-12, Sel 152, Sioux, Sweet-72, Pusa Uphar, Hisar Arun, BT-12, Hisar Lalima, Hisar Lalit, Sel-120, T-1, T-3, KS-2, AC-238, Arka Vikas, Arka Saurabh, Pant T-3, CO-3, Punjab Kesri, LaBonita, Sel-7, Sel-1-6-4, Punjab Tropic, Keck Ruth, Keck Ruth Ageti, Punjab NR-7, Utkal Kumari (BT 10)
Turnip	:	Purple Top Milan, Purple Top White Globe, Pusa Chandrima, Pusa Kanchan, Pusa Swarnima, Pusa Sweit, L-1
Winged Bean	:	Colombo, GRWB-11, IIHR Selection, Karikari, Puttalam, Tyt-1, 3, 14, WBC-2, UPS 62

(Source: "Vegetable Crops" ed. By T.K. Bose, M.G. Som and J. Kabir; 1993; Kalloo G., 1993; Anonymous, 1999; Pareek and Samadia, 1999)

mosaic virus and golden mosaic virus have been identified which are required to be utilised to develop commercially viable resistant variety.

Okra is highly susceptible to yellow vein mosaic virus, Fusarium wilt, *Microphomina*, *Cercospora* and powdery mildew. Varieties P-7, Parbhani Kranti, Arka Anamika, Arka Abhay, Varsha Uphar have been developed as potential resistant varieties utilising *Abelmoschus manihot* as donor parent (Table 8.3). These varieties are widely being grown in the country at commercial scale. Sources of resistance to *Fusarium* wilt, *Microphomina* and mildew and insect pests have been identified in *Abelmoschus* caillei. There is a need to develop multiple disease and pest resistant, i.e. YVMV, microphomina, *Cercospora* blight, jassid, fruit and shoot borer and root'; knot nematode resistant variety in okra utilising wild taxa which are easily accessible in the country.

In cururbits, muskmelon is highly susceptible to powdery mildew and downy mildew. Source of resistance to both of diseases are available. Arka Rajhans is resistant to powdery mildew. DMDR-l and DMDR-2 varieties have been developed which are resistant to downy mildew and cucumber green mottle mosaic virus.

In watermelon, Arka Manik has been developed which is resistant to anthracnose, downy mildew and powdery mildew and this is one of the ruling variety of watermelon widely grown in the country. In cucumber Poinsette is resistant to downy mildew, powdery mildew, anthracnose and angular leaf spot. Bottlegourd is susceptible to downy mildew and summer squash is susceptible to virus diseases and there is need to undertake research programme to develop varieties resistant to common diseases in a number of cucurbitaceous crops.

Onion and garlic are an important bulb crops. Onion is highly susceptible to purple blotch, *Stemphyllium* blight and basal rot caused by *Fusarium oxysporum* sp. cepae. Only one genotype IIHR 56-1 has been identified as resistant to purple blotch which can be used to develop new commercial purple blotch resistant variety in onion (Table 8.2).

Increasing the Area Under Improved Variety/Hybrids

Vegetable production of our country is still dominated by the locally available genotypes or open pollinated varieties. This is mainly due to farmers' ignorance and poor extension activities. Hybrid varieties in India are of recent origin. Hybrid technology has been most successful in tomato resulting in large area coverage. Karnataka, Maharashtra, Gujarat and some parts of Andhra Pradesh were first to adopt hybrid tomato varieties because of longer growing period. Hybrid cabbage is next in area coverage to tomato. The entire cabbage production in southern part of Maharashtra and West Bengal is under F_1 hybrids. In capsicum, brinjal, cucumber and watermelon area under F_1 hybrids is small. With awareness of advantages of cultivation of F_1 hybrids, the area is bound to extend. Data presented in Table 8.4 shows the extent of cultivation of F_1 hybrids, open pollinated and unidentified varieties in India.

A well planned strategy is needed to cover all the vegetable-growing areas under noted identified varieties/hybrids to boost the vegetable production in the country. It is estimated that under eggplant approximately 32.20 per cent, cauliflower 46.71 per cent, chillies 60 per cent and under tomato 18.49 per cent area are under unidentified or local varieties. Here, great scope exists to replace the local cultivars with the improved cultivars having high quality and other preference of the region (Kirti Singh, 1998).

Availability of Quality Seeds and Hybrid Seed Production Technology

Over the last two decades, India has made significant progress in making available the quality seeds both of temperate and tropical vegetables. Many joint ventures for the production and distribution of vegetable seeds have recently come up as a result of the open seed policy leading to global recognition of Indian seeds industry. However, lack of availability of quality seeds at the right time still continues to be one of the major constraints in increasing the vegetable

Table 8.2: Vegetable varieties resistant to diseases and insect pests

Crop	Disease/insect-pests	Varieties
Brinjal	Bacteiral wilt	BB-7, BB-26 (Utkal Keshar), BWR-12, Pant Rituraj, Pusa Purple Cluster, SM-6-7, SM-6-6, BB-44 (Utkal Madhuri)
	Phomopsis blight	Pant Bhairav, Pant Samrat
	Shoot and fruit borer and Jassids	Pant Samrat
	Fusaraium wilt	BB-26, BB-44
	Nematode wilt	BB-44
Chilli	Leaf curl	Hisar Shakti, Hisar vijay, Punjab Lal
	Powdery mildew	-do-
	Mosaic	-do-
	Bacterial wilt	Ujwala
Tomato	Bacterial wilt	BWR-5, BT 10, BT-1, LE-79-5
	Verticillium wilt and Fusarium wilt	Pant Bahar
	Root kot nematode	Sel-120, Hisar Lalit
	Leaf curl	H-24, H36
Capsicum	Bacterial wilt	Arka Gaurav
Cabbage	Black Rot	Pusa Mukta
Cauliflower	Black Rot	Pusa Shubhra, Pusa Snowball K-1
Watermelon	Anthracnose	Arka Manik
	Powdery mildew	-do-
	Donwny mildew	-do-
Muskmelon	Downy mildew	Punjab Rasila
Cowpea	Bacterial blight	Pusa Komal
Pea	Powdery Mildew	FC-1, JP-83, JP-4, PRS-4, PM-2, NSM-2, NDVP-2
Okra	Yellow vein mosaic virus	Arka Anamika, Arka Abhay, Sel-2, Pusa A-4, Parbhani Kranti, Varsha Uphar, Hisar Unnat
Cucumber	Powdery mildew	Poinsette
	Downy mildew	-do-
Potato	Early blight	Kufri Pukraj
	Late blight	Kufri Pukraj, Kufri Chipsona, Kufri Chipsona 2
	Tolerant to frost	Kufri Chipsona 2
Onion	Purple blotch	IIHR-56-1

(Source: Kallo, G., 1995; Narendra Singh and Verma, V.K., 1998; Bhatnagar, P.S., 1998; Arora, S.K. 1998; Gopala Krishnan, 1996)

Table 8.3: IIHR varieties/hybrids with biotic and abiotic stress resistance

Vegetable crop	*Stress*	*Hybrid*
Tomato	Tolerance to fruit cracking	Arka Vishal (F1 hybrid)
	Resistant to nematode	Arka Vardan
	Resistance to bacterial wilt	Arka Abha and Arka Alok Arka Shresta (BRH 1) and Arka Abhijit (BRH-2) (Both are hybrids)
	Resistance to powdery mildew	Arka Ashish
	Resistance to moisture stress	Arka Vikas
	Reistance to moisture stress	Arka Meghali
Brinjal	Resistant to bacteiral wilt	Arka Nidhi, Arka Keshav and Arka Neelkanth
Watermelon	Resistance to anthracnose, powdery mildew and downy mildew	Arka Manik
Roundmelon	Tolerance to fruit-fly	Arka Tinda
Pumpkin	Resstiance to fruit-fly	Arka Suryamukhi
Okra	Resistance to yellow vein mosaic virus	Arka Abhay and Arka Anamika
Onion	Tolerance to purple blotch	Arka Kalyan
Dolichos	Photoinsensitive	Arka Jay and Arka Vijay
Chilli	Tolerance to moisture stress	Arka Lohit
Cucumber	Tolerance to powdery mildew	Swarna Poorna
Pea	Resistance to rust and powdery mildew	Arka Ajit

(*Source:* Yada, I.S., 1996; Yadav I.S., 1998)

Table 8.4: The estimated share of hybrid, open pollinated and local or unidentified varieties in total cropped area in India (1997-98).

Crop	*Total Area (ha)*	*Area under F_1 hybrid (ha)*	*Share of hybrid (%)*	*Share of open pollinated (%)**	*Share of unidentified varieties (%)**
Egg plant	479095	85300	17.80	50.0	32.2
Cabbage	242140	76000	31.39	68.61	-
Cauliflower	304156	10000	3.29	50.00	46.71
Chillies	573529	14000	2.44	37.56	60.00
Gourds	409270	10000	2.44	20.00	77.56
Melons	169350	67800	4.02	25.98	70.00
Okra	371665	20000	5.38	80.00	14.62
Tomato	485520	153000	31.51	50.00	18.49

*Source: Seed Association of India

*Estimated figures

production and productivity especially when dealing with as many as 180 recommended varieties. No doubt India has achieved self-sufficiency in seeds of temperate vegetables and has even started exporting but still much work is required to be done. Seed production jumped from 1.8 million tons to 7 million tons in 1991-92 within a short span of time. According to an estimate, value of hybrid seed sold in 1994-95 was Rs. 730 million. Seeds of F_1 hybrids in vegetables in India are primarily produced in the private sector (Table 8.8). Seed companies primarily concentrated in South (Karnataka, Andhra Pradesh) and West (Maharashtra) and public institutions like NSC, SFCI have been concentrating on seed production mainly of high volume, low value items like okra and pea. While most of the seeds of cole crops are imported, seeds of other vegetables are produced in India mostly in open except capsicum in which case glasshouse facility is required. Most private companies have trained women in hand emasculation and hand pollination technique. In Punjab, PAU, Ludhiana has trained farmers in utilisation of genic male sterility for muskmelon seeds production.

F_1 hybrid seed, mostly of cabbage and capsicum, are imported from Japan, Korea, Holland, Denmark and USA. In recent years, Gherkin seeds have been imported from Netherlands. The import component in hybrid seed contributes the major share in the cost of hybrid seeds, is not substitute for ever, and can only serve as a catalyst in development of indigenous hybrid seed production technology. The cost effective technology for growing F_1 hybrid variety has to be developed by the use of (a) indigenous parent (b) indigenously produced hybrid seed (c) disease resistant parents to reduce the cost of plant protection cover. The Public Sector Seed Industry has taken commercial hybrid seed production technology (Agrawal, 1997; (Table 8.5 to Table 8.7).

Production Technology for Rare/Minor Vegetable Crops

In our country more than 70 types of vegetables are grown but maximum emphasis has been given on important vegetables like tomato, brinjal, chilli, cauliflower, cabbage, pea and few important cucurbits. Besides, a number of minor vegetables are grown in different parts of the country. Though presently the area under these minor vegetables is very limited, it is expected that in near future there will be great demand of these vegetables. Even today, these vegetables are fetching very high price in multistar hotels and other tourist restaurants. In Himachal Pradesh, cultivation of asparagus and broccoli is increasing very fast. Among all the vegetables in cole family, broccoli is rated as more remunerative. Recently, two improved varieties of broccoli, namely, Palam Samridhi (Green headed) and DOG PPB. 1 (Purple headed) have also been developed for cultivation. Demand for supply of seed of these vegetables in now coming from various states.

Like asparagus and broccoli there are a number of other minor vegetables such as Faba bean, Summer bean, Lima bean, Winged bean, Velvet bean, Chive, Leak, Welsh onion, Brussels' sprouts, Chinese cabbage, *Momordica cochinchinensis, Coccinia indica, C. cordifolia, Melorthia heterophylla,* Curry leaf, drumstick, agathi, waterleaf, Indian spinach, parsnip, celery, lettuce, globe artichoke, sweet com, baby corn etc. Looking to the importance of these crops, it is desired that there should be well chalked out plan to cover these vegetables under various research projects to generate the information on various aspects so that commercial cultivation of these minor vegetables can be popularised.

Table 8.5: IARI released vegetable hybrids for all-over India

	Released synthetics and F1 hybrids			
Crop	*Hybrid*	*Yield (in tonnes/ha)*		*Centre showing potential yield*
		Realized	*Potential*	
Eggplant (Brinjal)	Pusa Hybrid-5	51	90	Pantnagar
	Pusa Hybrid-6	42	62	Kalianpur
	Pusa Hybrid-9	50	148	Jalna
Tomato	Pusa Hybrid-2	55	65	Aurangabad
	Pusa Hybrid-1	32	42	Delhi
	Pusa Hybrid-4	65	112	Rahuri
Capsicum	Pusa Deepti	24	50	Katrain
Cauliflower	Pusa Synthetic	19	24	Ludhiana
	Pusa Early Synthetic	10	12	Coimbatore
	Pusa Hybrid-2	23	35	Delhi
Muskmelon	Pusa Rasraj	17	30	Ludhiana
Bottlegourd	Pusa Meghdoot	19	26	Delhi
	Pusa Manjari	16	25	Delhi
Summer Squash	Pusa Alankar	45	54	Delhi
Cucumber	Pusa Sanyong	45	50	Delhi
Bittergourd	Pusa Hybrid	22	40	Delhi

Source: Narendra Singh, 1997; Siro P1I, P.S., (1999).

Table 8.6: Vegetable hybrids identified/released by Public Sector

Brinjal (Egg plant)	:	Arka Navneet, Pusa Hybrid-5, Pusa Hybrid-6, Pusa Hybrid-9, NDBH-1, NDBH-6, ABH-1, ABH-2, ABG-2 Phule Hybrid-5
Tomato	:	Pusa Hybrid-1, Pusa Hybrid-2, Pusa Hybrid-4, FMH-1, Arka vardhan Pusa Sambadh (KT-4), TH-802
Cauliflower	:	Pusa Hybrid-2
Cabbage	:	Pusa Synthetic
Muskmelon	:	Pusa Rasraj, Punjab Hybrid-1
Watermelon	:	Arka Jyoti
Bottle gourd	:	Pusa Mnjari, Pusa Meghdoot, Pusa Hybrid-3
Pumpkin	:	Pusa Hybrid-1
Cucumber	:	Pusa Sanyog
Summer squash	:	Pusa Alankar
Capsicum	:	Pusa Deepti
Chilil	:	CH-1

(*Source:* Narendra Singh and V.K. Verma, 1998)

Table 8.7: Vegetable hybrid identified/released for cultivation through AICVIP (1990-1997)

	Brinjal		*Tomato*		*Capsicum*
1.	Arka Navneet	1.	ARTH-4	1.	KT1
2.	Pusa Hybrid-6	2.	MTH-6,		
3.	Pusa Hybrid-5	3.	ARTH-3		Cabbage
4.	ARBH-201	4.	Arka Vardhan	1.	Pusa Synthetic
5.	NDBH-.1	5.	Pusa Hybrid-2	2.	Sri Ganesh Gol
6.	ABH-1	6.	KT-4 (Pusa Sambandh)	3.	NATH-401
7.	MHB-10	7.	NA-601	4.	BSS-32
8.	MBH-39	8.	FMH-1		Cauliflower
9.	NDBH-6	9.	NA-501	1.	Pusa Hybrid-2
10.	ABH-2	10.	DTH-4		Muskmelon
				1.	M-3 hybrid
					Watermelon
				1.	Arka Jyoti
					Carrot
				1.	Hybrid-1

Table 8.8: List of popular hybrids of private sector

Tomato	:	Vaishali, Karnataka Hybrid, Lerica and NS-815
Brinjal	:	Suphal and HOE-404
Chilli	:	Agri, NS-1101 and HOE-808
Capsicum	:	Indira, Bharat, Lerio, Early Bonty, NS-33 and Hira
Cabbage	:	Cabbage No.8, Questo, NS-22, Hari Rani Gol, Krishna, Vishesh
Cauliflower	:	Snow Crown, BSS-18, NS-60 and NS-60 and NS-90
Bottle gourd	:	Barad and Kavery
Watermelon	:	Madhur, Milan, Na-101 and MHW-4
Muskmelon	:	Abhjeet, MCH-5 and MCH-6
Okra	:	Varsha, Vijay, Vishal

(Source: Kalloo *et al.,* 1998 National Symposium ISVS, Souvenir, Silver Jubilee)

Table 8.9: List of promising hybrids, an achievement of "promotion of hybrid project"

Tomato	:	Avinash-2, Hemlata, TH-2312, Ratna, RHTH-13, DTH-6, ARTH-3
Brinjal (Egg plant)	:	NDBH-8, ARBH-201, PDBH-13, IIHR-500A, X Arka Keshav and IIHR-104 X IIHR-491
Chilli	:	BSS-188, BSS-141, HOE-818, HOE-888, MHP-59, ARCH-236, Kiran
Cabbage	:	K-I, CBH-127, Kt CBH-95 and Kt CBH-139; Golden Acre, Pride of India
Cucumber	:	CH-13, PCUCH-I, DH-9. 18 and 17 (developed from gynoeceous' lines), KH-I
Onion	:	MS-65 X PBR-243 and MS-65 X PBR-245, Hybrid-63, Hybrid-35

(Source: Kalloo *et al.,* 1998. National Symposium, ISVS, Souvenir, Silver Jubilee)

Table 8.10: Hybrid seed marketed in vegetable crops (1995-96)

Crop	Seed marketed (kg)	Source
Tomato	21000	Country
Brinjal	8250	Country
Chilli	1700	Imported
Cabbage	18000	Imported
Cauliflower	4000	Country
Gourds	16300	-do-
Melons	8700	-do-

(Source: Seed Association of India)

Chemical Weed Control in Vegetable Crops

Weeds cause substantial damage to crop plants and reduce the yield considerably. Grasses and broad leaved weeds frequently harbour insect-pest diseases which may later attack the crop. Several herbicides have been tested to control the weeds in a number of vegetable crops. Among herbicides Stomp, Goal, Basalin, Sencor, Lasso and Tribunil have been found very effective in several vegetable crops. These herbicides are widely being used by the farmers to control the weeds in vegetables.

Growth Regulators in Vegetable Crops

Plant Growth Regulators (PGRs) are organic compounds, other than nutrients, which in small amounts promote, inhibit or otherwise modify any plant's physiological processes. Nowadays plant growth regulators are being used to increase the production of a number of vegetable crops (Table 8.12). Application of GA_3 @ 325 ppm induces staminate flowers in gynoecious lines of muskmelon and cucumber, thus are used for maintaining such lines for hybrid seed production. Cycocel is effective to induce leaf curl virus tolerance in tomato. Ethephon sprayed on the plant at 2-4 true leaf stage increased female and hermaphrodite flowers in several cucurbits thus increased the yield. In tomato, chilli and pepper, spray of ethephon at ripening stage enhances ripening. Parachlorophenoxy acetic acid at 50 to 100 ppm sprayed at the stage of flowering increased fruit set at high and low temperature conditions in tomato (Panditor *et al.*, 1979).

Integrated Plant Nutrient System for Vegetable Production

In most of the Indian soils, nitrogen is deficient but phosphorus and potash are available in sufficient quantities. However, in areas deficient in phosphorus and potassium as indicated by soil testing, about 60-100 kg/ha P_2O_5 and 50-120 kg nitrogen/ha are recommended for vegetable crops, depending on the requirement of individual crop. The use of micro-nutrients in vegetable production is gradually becoming important. Although a part of the requirement is met by application of manure, compost or poultry manure, deficiencies in certain areas of the country are still common, and use of micro-nutrient fertilisers in such regions has been found

Table 8.11: Selected herbicides for control of weeds in important vegetable crops

Sr. No.	Herbicides	Active ingredient (kg/ha)	Crops	State of application	Control of weeds
1.	Pendimethalin	1.0	Brinjal	1-10 days after transplanting	Broad leaf weeds & Grasses
		0.5	Garden pea	Pre-emergence	-do-
2.	Oxybluorfen	0.25	Tomato	Pre-plant spray	Broad leaf weeds & Grasses
3.	Fluchlonalin	1.0	Tomato	Pre-plant spray	Broad leaf weeds
		2.0	Onion	Pre-plant spray	Broad leaf weeds
		2.5	Okra	4 days before sowing	Broad leaf weeds
4.	Metribuzin	0.3-1.1	Tomato & pepper	Pre-plant soil spray	Broad leaf weeds
5.	Alachlor	1.5	Okra	Pre-emergence	Grasses and broad leaf weeds
6.	Methabenzthi-ozuron	1.87	Garden pea	Pre-emergence spray	Broad leaf weeds
7.	Decthal	6.7-11.2	Brinjal	Pre-emergence after earthing	Annuals primarily Grasses
		6.7-11.2	Cauliflower	Immediately after sowing and also as per plant	Annuals primarily Grasses
		6.7-11.2	Onion	Immediately after seed sowing or transplanting	Annuals primarily Grasses
		6.7-11.2	Radish	Immediately after seed sowing	Annuals primarily Grasses

beneficial. Sprays of 2 per cent sulphate and borax solution have been found to increase the yield of tomato.

In simple terms Integrated Plant Nutrient System (IPNS) is a balance use of chemical fertiliser, organic manure, such as the form yard manure (FYM), crop residues, animal and human wastes, and biofertilizers. The IPNS aims at sustainable productivity with minimum deleterious effects of chemical fertilisers on soil health and environment.

In vegetable production, role of organic manure is much more as compared to other cereal crops. It has been proved experimentally that vegetables produced by the application of organic manure possess good firmness and have long storage life. Organic amendments in the form of oil cakes and neem, mahua, GNC and Castor suppress the population of nematodes and other harmful pathogens like *Fusarium* and *Rhizoctonia.* Beside, press mud and digested sludge are being used to raise other crops.

The micro-nutrients, namely, boron, manganese, iron, molybdenum and copper play an important role in vegetable production. Boron is effective in controlling the malady of fruit cracking in tomato and quality produce in terms of total soluble solids, ascorbic acid and flesh appearance.

Table 8.12: List of growth regulators along with their important uses in vegetable crops

Sr. No.	PGR	Concentration (mg/l)	Method of application	Crops	Attributes affected
1.	Gibberllic acid (GA_3)	10-20	Foliar spray	Watermelon, Tomato	Sex expression, Good yield
		40-100	Seed treatment or foliar spray	Okra, Tomato, Brinjal	Seed germination, fruit set and yield
		1000	Foliear spray	Cucumber	Maintaining Gynoecism
2.	Cycocel (CCC)	500	Seed soaking	Okra	Salt Tolerance
		500	Root dipping and foliar spray	Tomato	Control leaf curl virus, fruiting and yield
		250	Root dipping of transplants	Tomato	Salt tolerance
		1%	-do-	Onion	-do-
3.	Ethephon	100-500	Foliar spray	Cucurbits, Tomato	Flowering fruiting, sex expression and yield were improved
		1000	Pre-harvest	Tomato	Fruit ripening
		2000	Post-harvest	Tomato chillies	Fruit ripening
4.	Parachlorophe noxy acetic acid (PCPA)	50-100	Foliar spray at full bloom	Tomato	Fruit set at low and high temperature conditions
5.	Planofix (NAA)	10	Foliar spray	Chilli	Reduces flower and fruit drop and improved fruit set and yield
6.	Malic hydrazide (MH)	50-150	Foliar spray at 2 and 4 true leaf stage	Bottle gourd Round melon	Fruit set and yield

Recently, application of bio-fertilisers in vegetable crops has been found very effective. Bio-fertilisers harper atmospheric nitrogen with the help of specialized soil micro-organisms. Such nitrogen fixing micro-organisms are either free living in soil and symbiotic with plants and directly or indirectly contribute towards the nitrogen nutrition of the plants. In onion, inoculation with *Azospirillium brasiliens* sp. Nov. and VAM at nursery stage and in the main field recorded an additional yield of 22.93 t/ha and saved 25 per cent recommended doses of NPK fertilisers. Thus production of vegetables through organic farming has high demands in export and some farmers have started organic farming by trial and error methods (Pandey and Singh, 1998).

Water Management

Vegetables in general are heavy users of moisture and draw a large share of irrigation resources. The need for high yielding and short duration varieties for water use efficiency is much more than low yielding ones. Drip or trickle irrigation has proved useful in growing tomato and okra in Maharashtra, with a saving of 56 and 69 per cent, respectively. At the Tamil

Nadu Agricultural University, the comparative effect of furrow and drip irrigation were tested on tomato, okra, brinjal, beet root, radish, chillies and sweet potato. The germination and growth were greater in drip method than the furrow method. The weed infestation was less in the drip method compared to the furrow method. In drip method, water reaches only at the root zone to the required depth, thus, plants use nutrients.

Increase in Production Period of Some Crops

On the basis of the temperature requirement for curd irritation and development, cauliflower varieties are available nowadays which can supply curds in the market throughout the year in the northern plains of the country (Narendra Singh, 1998). Similarly radish is cultivated and used around the year. It is a winter season crop, however, in north Indian plains it is grown as late Kharif and Rabi vegetable. Generally, most of the tropical varieties start bolting without forming marketable roots in the later part of December or January. So, it is not practicable to grow one variety of any group (temperate, sub-tropical). Thus, varieties suited to various conditions/periods have been developed. Therefore, it is very important to select the right variety for sowing in a particular season. Nowadays, tomato can be grown almost throughout the year in the country. This is due to the development of heat tolerant, cold tolerant and disease resistant varieties. In northern plains tomato is grown continuously from June to March covering most period of year. Low temperature setting varieties such as Ostenkinskiz, Cold Set and Pusa Sheetal can be grown in severe winter conditions. Similarly Hot Set, HS-102, Saladette can set fruit in summer. Spray of Parachlorophenoxy Acetic Acid at 50 and 100 ppm at flowering stage sets fruits during summer and winter conditions. During rainy season due to severe incidence of leaf curl tomatoes are not being grown in many parts of the country. However, the development of tomato leaf curl resistant varieties H-24 and H-36 have led to the production of tomato under such conditions.

Due to development of Kharif onion, two crops of onion are being taken in northern plains of India. Kharif onion is the growing of onion in Kharif season (off-season), i.e. from June to November. There is scarcity of onion from November to March and during such period *Kharif*

Table 8.13: Water requirements and yield of vegetabJe crops

Vegetable	*Water used (cm)*		*Water saving (%)*	*Yield increased (%)*
	Drip	*Control*		
Tomato	10.76	49.20	78.0	43.00
Okra	8.60	53.53	83.0	13.00
Beetroot	17.73	85:76	79.0	4.00
Radish	10.81	46.41	76.0	3.00
Chilli	41.77	109.71	61.0	20.75
Brinjal	24.47	69.18	64.0	0.81
Sweet potato	25.20	63.14	60.0	38.0

(*Source:* "Production Tecfutology of Vegetable Crops" by Shanmugavelu, K.G. 1989).

onion proves to be a boon to the farmers. This not only increases the total onion production but also provides fresh onion to the consumers at reasonable price and good return to the farmers. Apart from this, it avoids the storage losses of Rabi onion, because if there is no Kharif onion we have to store the Rabi onion from November to March. During this period most of the onion starts sprouting, shrinking and rotting. The varieties which are most suitable for cultivation in Kharif season are N-53, Agrifound Dark Red, Arka Kalyan and Baswant-780. These varieties require mild climate without extreme of low and high temperatures. The seeds are first treated with Thirum 2-3 g/kg of seeds and sown in the nursery from the last week of May to first week of June. Nursery beds must be treated with Captaf @ 2g/I of water before sowing. Seedlings are transplanted from first week of July to August or when they are 7-8 weeks old. Depending upon the rains, 8-10 irrigations are enough. Irrigation should be stopped 15-20 days before obtaining the maturity to improve the keeping quality of bulb.

Integrated Pest Management

Generally, vegetables are attracted by a number of insect pests. Due to plant canopy and succulency, vegetables suffer maximum from pest damage. In vegetables more than 40 per cent yield loss is caused due to pest attack. As a result, growers face heavy loss due to degradation of quality and marketability. Unlike cereals, the green pods, fruits and leaves are mostly used as vegetables, which are more prone to retain the pesticide residue. Even although vegetables covers only 2.5 per cent of total cropped area, 17.16 per cent of total insecticides are consumed in vegetable cultivation. The persistency of toxic chemicals in consumable parts limits the choice of pesticides and make vegetable pest management more sensitive and challenging. Residual toxicity is of particular consequence in produce for export. Shipment from the country are rejected because of residual toxicity above a maximum residue level (MRL). According to the chairman of the spices exports, chilli farming of Andhra Pradesh has to adopt IPM technology to reduce the use of pesticides to improve exports, as consignments have been rejected due to pesticides residues. Under the IPM, we assess exhibiting pest situation and utilise all the available pest control methods in a compatible manner to keep the pest population below economic threshold level and above maintenance threshold level of natural enemies.

In recent past, throughout the world efforts have been made to control the vegetable pest by use of natural enemies, parasite predators besides host specific insect viruses and other entomo-pathogenic micro-organisms and to a considerable extent success has been achieved. In tomato, fruit borer can be successfully controlled by use of *Trichogramma* and Ha NPV (Nuclear Polyhedfosis Virus). Okra shoot and fruit borer can be checked by *Trichogramma brassliensis.* Diamondback moth of cabbage can be controlled successfully by *Bacillus thuringiensis.* No doubt IPM technology has emerged an effective and ecofriendly approach to combat the pest problem in vegetable crops. But farmer's ignorance about this approach and apprehension that natural enemies do not totally suppress the pest population and low populations of pest is enough to cause heavy loss. Further lack of availability of these bio-agents and their multiplication is another problem in adoption. To make the IPM technology more popular among the growers, an intensive effort on research and extension is needed.

Pollination Management in Vegetable Crops

An ecofriendly and relatively less known practice in developing countries of enhanced vegetable yield is pollination management. This method of bringing the pollinators to the crop in bloom has been widely used in many countries. A honeybee colony with around 20000 bees in one colony of *Apis indica* in capable of locating accurately and foraging on vast areas. Today in the west, bee keepers rent their hives to agriculturist in the flowering season. Use of bees in cross pollination is the cheapest, ecofriendly and self-sustainable method to boost the yield to meet the growing demand for the food. In Israel, bumblebee has been found more effective pollinator than honeybee. As like honeybee and bumblebee, carpenterbees visit only some plants but are excellent pollinators (Pandy and Singh, 1998).

Inter-generic Grafting in Vegetable Production

In vegetable production, there are several problems of soil-borne diseases and pests and nematodes and it is strenuous to incorporate resistance to such stresses in the varieties. Therefore, intergeneric grafting has been proved very effective in a number of vegetable crops. Cucumber grafted on pumpkin gives resistance against *Fusarium* spp. and grafting on figleaf gourd provides low temperature resistance. Watermelons are generally grafted on bottlegourd and rarely on pumpkin. Melons are grafted on the same species but are very rarely on white gourd and pumpkin. Interspecific grafting is very common in brinjal plant also. Scarlet brinjal and *Solanum torvum* are popular root stocks for egg plant to give resistance mainly against *Fusarium* and bacterial wilt, respectively. In grafting technology, Japan has taken lead and it is growing 32 per cent tomato, 50 per cent brinjal, 93 per cent watermelon and 72 per cent cucumber as grafted plants (Paroda, and Kalloo, 1995).

Pruning and Training of Vegetable Crops

Pruning away some branches within that tangled mass of vegetation inhibits the growth of certain diseases, producing micro-organisms and the combination of light and air circulation promotes more rapid drying. Removing crossing limb also prevents wound created by rubbing. Proper pruning and training have been found to improve earliness and yield in several cucurbits. In muskmelon the female/hermaphrodite flowers are born on the side branches arising from higher nodes. If the pruning of the side branches is done up to 7th node, the growth and vigour of the plant improves and flowering is early and ultimately yield is improved. It has been observed that fruits arising from 9-12th nodes are more sweet in taste. In watermelon the female flowers are born on the main branches. Hence, female flowers and yield can be increased by nipping the main shoot from 5-6th node and allowing 3-4 well developed main branches to develop. By this technique number, yield and sweetness of fruits have been improved substantially. In round melon also pruning to four primary branches has resulted in larger number of fruits and higher yield because of the same reason.

Training is also very important practice to train the tomato, cucumber and other beans and cucurbits. Training of indeterminate tomato provides ideal conditions to express the full potential of growth and fruiting. Vegetable growers of MP and Maharashtra are cultivating cucurbits mainly on bower system of training. This system of training does not only enhance the vegetative growth and yield but also give quality fruits. Bottlegourd grown on this system bears uniform size and colour of ftuits which fetch high price in the market.

Vegetable-based Cropping System

Vegetables are short or medium duration crops. They vacate fields early or in a short time. Small and marginal farmers can even cultivate vegetable crops round the year. Vegetables have cropping intensity of 400 per cent. In North India, two crops each of radish, turnip, beet root, cauliflower, knol-khol and potato are quite possible during rabi season. Long duration vegetable crops like brinjal, tomato, beans, pea, okra, cluster bean and cucurbits though begin yielding early and continue for longer period. Inter-cropping of vegetables in sugarcane, maize, gobhi, saron and sraya is possible. Inter-cropping of palak, onion and radish with tomato, cauliflower and cabbage is more profitable. Relay cropping of vegetables as filler crop in a normal cropping is best for the small and marginal farmers. Some of the vegetable crop rotations which have been found remunerative are:

- Maize—Potato—Onion
- Okra—Potato—Tomato
- Early cauliflower—Potato (early)—Round melon
- Tomato—Potato—Watermelon
- Cowpea—Radish—Potato—Pumpkin
- Kharif onion (green)—Potato—Chilli
- Buttergourd—Potato—Bottlegourd
- Luffa—Early Potato—Garden pea—Round melon

Use of Mulches in Vegetable Production

Mulches are covering material to the soil surface. They are mainly used to reduce or increase the temperature. High temperatures affect seed germination, vegetative growth, flowering, pollination, fruit set and yield. Mulch also improves the quality of vegetables by avoiding their direct contact with soil.

Organic mulches can reduce the soil temperature up to some extent. In potato, mulching hastened and improved the sprouting, plant growth and yield in early crop. It reduces the rotting of seed tuber at high temperature. In tomato during summer, mulching gives higher yield. Organic mulches have been used successfully in cucurbits, chilli, brinjal, and early cauliflower for obtaining comparatively high yields. Mulches protect the roots of the plants from heat, check evaporation, and conserve soil moisture, reduce weed infestation and modify the soil and air micro-climate.

In general, plastics are used to increase the soil temperature specially during winter. Polyethylene sheets (clear and black) are used as mulches. Under the transparent film, the soil temperature rises by 2-10°C during day time but the retention of the heat energy is poor during night. However, black film shows better retention. Covering of nursery bed with the transparent plastic sheet of 300 or 400 gauge for 30 to 45 days during summer months has been found to effectively control the soil pathogens. Soil solarisation prevents the seeds and seedling rot caused by *Verticillium, Rhizoctonia, Fusarium, Sclerotium* etc. and thereby improves seedling growth of brinjal, chilli, cauliflower, cabbage, onion etc. Plastic sheets used immediately after seed sowing are helpful in early germination by increasing soil temperature and conserve the moisture.

River Bed Cultivation of Vegetable Crops

The river bed cultivation is a very old practice of growing vegetables on the bank or basin of rivers. There are several advantages of river bed cultivation viz. early yield, ease in irrigation, high net return per unit area and additional crop. Mostly cucurbits are grown in river beds. Besides, cucurbits, tomatoes are successfully cultivated due to its deep root system. Cucurbits are sown in the month of November to December in trenches/channels. The trenches are dug in North-West direction to manage the availability of higher temperature. The depth of channels is 45-90 cm depending on height of water table. Generally 60 cm to 90 cm is the height of water table in river beds. Width of channels is kept about 45 cm and length depends upon area and location. Sometimes circular pits of about 35 to 45 cm in diameter are prepared having a depth of about 90 cm. The trenches are filled with compost/FYM, soil and N, P, K requirements of the crops. These trenches are left for a few days and if possible pre-sowing irrigation is done to settle down the various layers of materials in the trenches. Distance between trenches is kept 2-2.5 m depending upon the variety and seed of the crop to be raised. Seed sowing for early crop is done in 1st or 2nd week of November at a spacing of 45-60 cm in the trench at a depth of 3 to 4 cm. Two seeds are generally sown at one place. If the temperature is very low, pre-sprouted seeds are sown for smooth germination. Mulching is done to enhance seed germination. In North-West India, the temperature in December to January goes down to 1-2°C, the protection is provided by thatch screen made of locally available material. This protection serves as wind breaker to save the crop from chilling winds and checks the sand drifting.

Biotechnology

In recent past, biotechnology has emerged as a powerful tool for alleviating hunger, disease, overcoming energy shortage and improving self-sufficiency of developing countries. It is useful in many areas like *in vitro* multiplication of elite planting material genome mapping and marker assisted selection and in development of transgenic plants. Genetic engineering of vegetable Brassicas for insect and fungal resistance is imperative to achieve an eco-friendly approach towards integrated pest management. Transgenic plants carrying Bt genes for insect resistance are about to be marketed in the USA. No doubt this branch of science has immense potential and can play an important role in understanding the numerous and mysterious complex phenomenon in plants and animals which will prove a boon to the welfare of the mankind. However, a point of caution is essential at each and every step. Good efforts in this direction are underway both by public and private sector.

Post-harvest Management of Vegetable Crops

Green vegetables due to high moisture content are inherently more vulnerable especially under tropical conditions such as ours. Owing to lack of production, planning and climatic factors, gluts occur frequently leading to distress sale, poor prices to growers and wastage. The post-harvest losses estimated at 20-40 per cent of the total fruit and vegetable production go as waste, costing more than Rs. 30, 000 million annually. Thus income of the growers and per caputa availability of produce come down.

Research in post-harvest has probed into screening of vegetable varieties and hybrids for prolonged shelf-life and processing character, standardisation of different storage systems including controlled atmosphere system for different vegetable crops, improvement of packaging and transportation system suited to Indian conditions; harnessing solar energy for drying as well as storage utilisation of wastes for the development of economically viable products and initiating post-harvest research specific to improve export potential.

Export Oriented Vegetable Farming

Vegetables are now being considered one of the most potential commodities for export. The value of refresh, canned and dehydrated vegetable exports has been to the tune of Rs. 363.44 crores during 1996-97. The Agricultural and Processed Food Product Export Development (APEDA) has identified traditional vegetables like onion, potato, bittergourd and chilli and non-traditional vegetables like asparagus, celery, sweet pepper, paprika, sweet com, baby com, green peas, French bean, cherry, tomato and gherkin having good export potential. Onion accounts for 70 per cent of total foreign exchange earning among fresh vegetables. Among other vegetables, 60 per cent share goes to okra, 20 per cent to green chilli and 20 per cent bitter-gourd, Frenchbean, capsicum and other mixed vegetables. Among the traditional category of vegetables, onion, okra, bittergourd, green chillies etc. meet the requirement of Indian expatriates in South-east Asia and Gulf countries and to some extent in the U.K. The non-traditional items including vegetables like cabbage, cauliflower, lettuce, capsicum etc. go to E.C. countries, where growing conditions are not ideal during November to March. India's varied agro-climatic conditions are major asset in terms of natural resources, which can be utilised in accelerating the vegetable production.

REFERENCES

Agrawal P.K. (1997). Hybrid Vegetable Seed Requirements-supply and scope in India. In summer school on Hybrid Seed Productions Technology of Vegetables, April, 7-26, 1997.

Anonymous (1972-93). Annual Reports, All India Coordinated Vegetable Improvement Project, (AICVIP), New Delhi, Varanasi.

Anonymous (1990-97). Annual Reports, All India Coordinated Vegetable Improvement Project, (AICVrp) New Delhi, Varanasi.

Anonymous (1999). Vegetable Research at a Glance Souvenir of the Department of Vegetable Crops, Punjab Agricultural University, Ludhiana-141 004 (XVIII Group meeting of All India Coordinated Vegetable Improvement Project October, 11-14, 1999).

Arora S.K. (1999). Present Scenario of Vegetable Production in India. Published in *"Recent Advances in Bio-sciences for Sustainable Food Security"* Eds. R.C. Sihag and other, AAREM, CCS Haryana Agricultural University, Hisar - 125 004.

Bhatnagar P.S. (1998). Notified Varieties of Horticultural Crops. ICAR News (Oct.-Dec. 1998). page 15.

Bose T.K., Som M.G. and Kabir J. (1993). *"Vegetable Crops"* Naya Prakash, Kalyani, West Bengal.

Chadha K.L. (1998). *"History of Vegetable Research in India"*. Silver Jubilee National Symposium of Indian Society of Vegetable Science held at Varanasi (U.P.), December 12-14, 1998.

Gopala Krishnan T.R. (1996). Ujwala, a Bacterial Wilt Resistant Chilli Variety. ICAR News (Oct.-Dec., 1996). Pp-23.

Kalloo G. (1988). *"Vegetable Breeding"* Vol. III C.R.C. Press, Fla, U.S.A.

Kalloo G. (1995). Status of Vegetable Production in India with special reference to Vegetable Hybrid Technology, Vegetable Research In India, PDVR, Varanasi. Page 125-152.

Kalloo G., Banerjee M.K. Sanjeet Kumar and Chander Prakash (1998). Hybrid Vegetable Technology In India: All Over, View. Silver Jubilee National Symposium of ISVS Souvenir, December 12-14, 1998 at Varanasi (U.P.).

Kirti Singh (1998). "Challenges in Vegetable Production in Twenty First Century." National Symposium on Emerging Scenario in Vegetable Research and Development. Dec. 12-14, 1998. Page 1-11.

Narendra Singh (1997). IARI released vegetable hybrids for all over India. ICAR News (July-Sept., 1997) pp. 14.

Narendra Singh (1998). Cauliflower for all seasons. ICAR News (April-June, 1998) pp.15.

Narendra Singh and Verma V.K. (1998). *"Recent Trends in Vegetable Production Technology"*. ISVS Souvenir: Silver Jubilee National Symposium, December 12-14, 1998, Varanasi (U.P.), India.

Pandey A.K. and Singh K.P. (1998). Vegetable Production Technology Towards Productivity Enhancement. ISVS Souvenir: Silver Jubilee National Symposium, December 12-14, 1998, Varanasi (U.P.), India.

Pandita M.L., Arora S.K. and Kirti Singh (1979). Effect of plant regulators on the fruit set, early and total yield of tomato var. HS-1 0 1 (*L. esculentum* Mill.) during summer season. *Haryana J. Hart. Sci.* 8: 112-16.

Pareek O.P. and Samadia D.K. (1999). Watermelon (Mateera) varieties developed by NRCAH. ICAR News (April-June, 1999).

Paroda R.S. and Kalloo G. (1995). Vegetable Research with special reference to hybrid technology in the Asia and Pacific Region.

Shanmugavelu K.G. (1989). *"Production Technology of Vegetable Crops"* A Book Chapter.

Sirohi P.S. (1999). Bitter gourd Pusa Hybrid-1, First Commercial F_1 Hybrid. ICAR News (April-June, 1999) pp. 14.

Y adav I.S. (1996). Horticultural varieties/Hybrids with biotic and abiotic stress resistance. ICAR News (Oct.-Dec., 1996) pp. 21.

Yadav I.S. (1998). Bacterial-wilt resistant tomato hybrids—A success story, ICAR News (July-Sept., 1998) pp. 14.

CHAPTER 9

Protected Cultivation of Vegetables in North Indian Conditions

S.K. Arora, A.K. Bhatia, S.P.S. Yadav and Rakesh Mehra

INTRODUCTION

India is the largest producer of vegetable crops next to China. Despite, its vegetable production in the country is much less than the requirement if balanced diet is provided to every individual. The present production of 93.92 million tons (13.5% of total world production of vegetables) is to be raised to 196.5 million tons by 2020. There are different ways and means to achieve this target, e.g., bringing additional area under vegetable crops, using hybrid seeds and use of improved agro techniques. Another potential approach is perfection and promotion of protected cultivation of vegetables. Production of vegetables under protected conditions involves protection of production stages of vegetables mainly from adverse environmental conditions such as temperatures, hail, scorching sun, heavy rains and snow. Protected conditions can be used as conventional devices to computerised environment controlled greenhouses.

Protected conditions for vegetable production are created locally by using different types of structures. These structures are designated as per climate modification requirement of the area. For different set of climatic conditions, plant protection structures are designed. Besides temperature, humidity, wind velocity and soil conditions etc. play major role in the design of protection structures for growing vegetable crops which has been discussed in this chapter.

PLANT PROTECTION STRUCTURE

Greenhouse, low tunnels, shading net houses, anti-hail nets, bird protection nets, trench (pit type greenhouse), mulch, wind break, floating crop cover etc. are some of the common plant protection structures used for raising vegetable crops or their nurseries.

Address for correspondence: Department of Vegetable Science, CCS Haryana Agricultural University, Hisar-125004, India

OPPORTUNITIES IN INDIA

There is a great need throughout India for effective greenhouse facility to boost vegetable production. The need exists in vegetable research institutes and laboratories to standardise greenhouse production technologies. There are potentials and opportunities in various agroclimatic zones in India for commercial use of greenhouses in vegetable production. In temperate climates, vegetable farmers can substantially increase their income if they can start early vegetable production under protected conditions. Vegetables raised in open fields would reach market later than greenhouse crops. Besides in hilly areas, tropical vegetables can be grown in greenhouses.

Vegetable nursery production in greenhouses enables taking early crop and better management of nurseries. In arid regions such as Rajasthan and adjoining states, greenhouses supported with cooling device would make vegetable production possible during extreme summer. In tropical climate, greenhouses can provide opportunities of raising his value vegetable crops throughout the year. Greenhouse vegetable production provides appropriate opportunities for production of vegetables under organic system. Production of vegetables in greenhouses in long run works out to be cheaper in terms of per unit cost of production.

In the present scenario of perpetual demand of vegetables and shrinking: land holding drastically, protected cultivation is the best alternative and drudgeryless approach or using land and other resources more efficiently. In protected environment (greenhouse, glasshouse or polyhouse), the natural environment is modified to suitable conditions for optimum plant growth which ultimately provides quality vegetables. The greenhouses are usually covered structures of plastic film which allow the solar radiation to pass through it but traps the thermal radiation emitted by plants inside. The CO_2 released by the plants at night is also trapped inside, which increases the rate of photosynthesis at day time. The evaporation from the soil and plants also raises the humidity inside. The ultra-mode devices are also used for controlling the environment inside.

Growing vegetables under protected environment is a fascinating art. It is expensive. In India, the structure of the greenhouse depends on the economic condition of the growers/ farmers. Vegetable cultivation in low- and medium-cost greenhouse is a technical reality in India and is the need of the hour. Such production system has not only extended the growing season of vegetables and their availability but also encouraged to grow vegetables suitable for export. The seed production of vegetables under protected environment is also a major step to increase vegetables production in India. Priorities must be focused to raise vegetables throughout the year by using greenhouse technology to compete the foreign markets.

DEFINITION OF A GREENHOUSE

A greenhouse is a built-up enclosed space, covered with a transparent or translucent material for the propagation, growing or storage of living plants under partially controlled environmental conditions. Glasshouse, polyhouses, rain shelters and shadehouse are all different names given to greenhouses depending on the type of covering material used and the utility of the facility. In simple words, greenhouse is a general term for the framed or inflated structure covered with transparent/translucent material large enough to grow crops under partial or fully controlled environmental conditions such as temperature, solar radiation, rainfall, humidity etc. to get optimum growth and productivity round the year. When the greenhouse is constructed with glass as the covering material is known as a glasshouse, whereas, a greenhouse using polythene as the covering material is termed as a polyhouse.

A plant grows best when exposed to an environment that is optimal for that particular plant species and its stage of growth. The plant's environment can be specified by five basic factors namely, light, temperature, humidity, carbon dioxide and nutrients. A greenhouse, due to its closed boundaries, permits the control over anyone or more of these factors, which is not possible in open field conditions.

ADVANTAGES OF THE GREENHOUSE

The major benefits of greenhouses are given below:

1. Higher production per unit area.
2. It is possible to multiply the plant material throughout the year, which is not possible under open conditions.
3. Plant propagation is possible in problematic areas and extreme conditions.
4. Vegetable crops can be grown under adverse climatic conditions when it is not possible to grow under open fields.
5. Certain vegetable crops can be grown round the year in a particular place.
6. Greenhouses provide an excellent opportunity to produce vegetables for exports.
7. Productivity is manifold in greenhouses in comparison to growing vegetables in open field.
8. Greenhouses are ideally suited for raising nurseries of vegetable crops both sexually and asexually.
9. Management of insects, pests, weeds etc. is easier in greenhouses.
10. Greenhouses are ideally suited to farmers having small holdings.
11. Organic farming of vegetables is easier in greenhouses.
12. Greenhouses are ideally suited for production of genetically engineered and micropropagated vegetable varieties and hybrids.

DISADVANTAGES OF THE GREENHOUSE

1. It is very costly because it involves very high initial expenditure.
2. Greenhouses can be managed properly only by skilled personnel
3. Intensive and difficult cultural operations.

CLASSIFICATION OF THE GREENHOUSES

Greenhouses could be broadly classified into the following three groups:

1. Low cost greenhouses
2. Medium cost greenhouses

3. High cost greenhouses
   (i) **Low cost greenhouse:** As the name indicates, this type of greenhouse involves less initial cost because they are naturally ventilated. Natural ventilation refers to providing sufficient open area in greenhouse structure so that ambient air by itself enters into the greenhouse after displacing an equal amount of greenhouse air.
   (ii) **Medium cost greenhouse:** The initial cost for the establishment of such type of greenhouses is more than the low cost greenhouses, since partial control of the environmental factors is necessary.
   (iii) **High cost greenhouse:** Such type of greenhouses involve very high initial cost of establishment because the conditions inside the greenhouse are to be fully controlled. These are most efficient with respect to plant propagation particularly under extreme climatic conditions. However, preference should be given to low cost greenhouses wherever, capital is a scarce input.

INDIAN GREENHOUSES: Following types of greenhouses are in use in the country.

1. Single span 24 m × 5 or 6 or 9 m.
2. Four span 42 m × 20 m
3. Saw tooth type 36 m × 24 m
4. Mud brick greenhouse 25 m × 4 m Leh design
5. IPCL design of wooden framed greenhouse
6. Jorhat design of bamboo framed greenhouse
7. Other localised types

PRESENT STATUS: Commercial greenhouses with climate controlled devices are very few in the country. Solar greenhouses comprising glass and polyethylene houses are becoming increasingly popular both in temperate and tropical regions. In early sixties, Field Research Laboratory (FRL) of DRDO at Leh attempted solar greenhouse vegetable production research and made an outstanding contribution to the extent that almost every rural family in Leh valley possesses a polyhouse these days. In other parts of the country, state agricultural universities and research institutes started research on the use of solar greenhouses since late 1980's onward. Indian Petro Chemical Corporation Ltd. (IPCL) boosted the greenhouse research and application for raising vegetables by providing Ultra Violet (UV) stabilised cladding film and aluminium polyhouse structures. Several private seed production agencies have promoted greenhouse production of vegetables. Solar greenhouses are being used for raising vegetable nursery, protecting imported germplasm, raising off-season vegetable crops and their seeds. In comparison to other countries, India has very little area under greenhouses as is evident from the following statistics:

Solar greenhouses, mainly polyhouses, are becoming popular in Ladakh region of Jammu and Kashmir and almost 70 per cent of total area under polyhouses in India is in Ladakh.

PRINCIPLE: As mentioned in the preceding section, a greenhouse is covered with a transparent or a translucent material such as glass or plastic film. Depending on the cover transparency, a major portion of sunlight is admitted which is absorbed by the crop, soil and other objects present in the greenhouse. These objects in the greenhouse, in turn, emit long

Table 9.1: Approximate area (ha) under greenhouses

Country	Area	Country	Area
Holland	9600	Spain	25000
Italy	18500	USA	4000
Canada	400	Israel	1500
Turkey	10000	China	48000
Japan	54000	South Korea	21000
India	100		

wave thermal radiation. But the greenhouse cover materials have low transmitivity for the long wave thermal radiation. Consequently, the solar energy trapped in the greenhouse causes a rise in the temperature of the inner space. This phenomenon is known as "greenhouse effect" in relation to warming up of the earth.

The rise in temperature of greenhouse space due to the above effect is utilised to grow crops in unfavourable cold climates. However, this natural and unavoidable warming requires greenhouse cooling during summers to maintain temperatures favourable to plant growth. The closed boundaries of a greenhouse permit design of systems for controlling, to an extent, the environmental variables like temperature, humidity, light, photoperiod, carbon dioxide, nutrients etc.

CONSIDERATIONS FOR ESTABLISHMENT AND MANAGEMENT OF THE GREENHOUSES

A. Type of plant material to be propagated.
B. Agro-climatic conditions of the area.
C. Requirement/demand of the planting material in the vicinity.

1. **Planning of a greenhouse:** The greenhouse should be so designed that it could withstand the load of covering material, intensity of the rain, wind velocity and the cultural activities inside the greenhouse. Keeping in view the above factors, the design of a greenhouse depends upon the type of construction material, control system used and the type of plant material to be propagated. In addition to being strong enough, it should admit enough sunlight for the plants. At the same time, it should require minimum energy for maintaining the micro-climate. Local conditions and availability of materials should be taken into consideration while planning a greenhouse.
2. **Selection of site:** It is one of the most important factors for a successful greenhouse. A good site with perfect drainage and good quality underground water should be selected. Site should be away from high buildings but should be close to a passing by electric line and if possible a metalled road.

3. **Orientation of a greenhouse:** Orientation can be in any direction when they are single span, but in case of multi-span only north-south direction should be selected, in order to avoid continuous shading of a part of the greenhouse.
4. **Components of a greenhouse:** A low cost greenhouse has only two components viz., the frame and cladding material but in medium and high cost greenhouses there is one more component, i.e., environmental control system.

Structural Design of a Greenhouse

It is determined by certain factors like rain, wind, storm, snow, fire, frost, dead load (load of materials used in construction) and live weight (weight superimposed on the greenhouse including handing baskets, shelves, ropes etc.).

Construction of greenhouse

Low cost greenhouses depending upon their size and durability are generally constructed from mild steel or galvanised iron pipes since these are less expensive than aluminium angles. Size of the greenhouse may vary from 100-500 square metre. A polyhouse of 100 square metre is sufficient and economical for plant propagation. The length, breadth and height of the greenhouse depend upon local situation and the purpose for which it is constructed.

Covering material

For covering of the greenhouse 200 micron (800 guage) polythene film or any other ultraviolet stabilised material can be used for this purpose. It is advisable that the cladding material should be only UV stabilised film since UV stabiliser provides protection against the ultraviolet rays of sun. This material would cost about Rs. 40 per m^2 as against Rs. 50-60 per m^2 for fiberglass sheets. One kg of UV film of 200 micron thickness covers about 5.3 m^2 area. Hence, the total quantity can be calculated accordingly.

Environmental control system

In a low cost, naturally ventilated greenhouse, sufficient ventilation (60-70%) is provided with 40-60 mesh net all round the greenhouse. But nowadays, greenhouses are well equipped with appropriate heating (with electric heaters during winter) and cooling (evaporative cooling using fan and pad system along with exhaust fans) systems. Exact specifications of these systems vary with the size of the structure and prevailing climatic conditions of the area. Provision of proper ventilation is an equally important consideration.

MINOR PLASTICULTURE APPLICATIONS

Low Tunnels

Low tunnels are miniature form of greenhouses to protect the plants from rains, winds, low temperature, frost and other vagaries of weather. However, under low tunnels, artificial controlling of conditions is not possible. The low tunnels are very simple structures requiring very limited skills to maintain, are easy to construct and offers multitudinal advantages. It has been proven that with low tunnels seedling can be raised very quickly. Sometimes the advancement of the period is by more than one month giving the farmer extra one month for his main crop. It is, therefore, felt that the application needs to be supported as it will attract the small and marginal farmers who may not be in a position to take up greenhouses.

For construction of low tunnels film of 100 micron would be sufficient. The cost of a 100 micron thick UV film would be about Rs. 10/- per m^2.

Shade Net Houses

Shading nets are used to provide relief to the plants from the scorching sunlight, high winds, direct rainfall as well as insects and pests. Shading nets are being used for raising nursery, indoor plants, hardening of tissue cultured platelets and growing of vegetables also. The shading nets are being used almost throughout the country.

Anti-Hail Nets

These nets are used to protect the crop from hails in hail-prone areas of states like Himachal Pradesh, Jammu & Kashmir, Uttar Pradesh etc. for protecting crops like apple, litchi etc. in high altitude areas. The anti-hail nets are normally provided for each individual tree.

Soil Trench

This structure is underground solar greenhouse. Solar trench of 5′ width and 3′ depth with different lengths depending on the availability of land is made for raising vegetables in extreme cold conditions. UV stabilised polyethylene film is spread over them. Regidex film has been found having longer life hence cheaper than LDPE. These structures harness soil and sun heat for the growth of vegetables. Such structures have been found quite useful for cold desert regions of the country for raising early vegetables and for growing leafy and other vegetables during winter when ambient temperature is subzero for considerable period. These structures are becoming poplar in Ladakh region and are likely to green this frozen barren desert in winter months. This technology of trench cultivation of vegetable crops in cold deserts has been perfected by FRL, Leh through long experimentation. In addition to above, locally designed and improvised structures made of different materials are used for raising vegetables under protected conditions. River bed cultivation of cucurbits is mainly done by providing protection to young seedlings using different material and techniques. This protected cultivation of vegetables is practised in the country since ages.

In general vegetable farming involves variety of protection structures. Their standard design, cladding and other materials, vegetable production technology has not been researched systematically and intensively in the country.

Plastic Mulch

Use of mulch in vegetable farming helps in many ways. Mulch is a material spread in the field to cut off direct sun to soil. This can be done by using organic materials or inorganic material like LDPE plastics. Mulch facilitates early production of certain vegetables, reduces water requirement and weed population. Plastic film is generally used in vegetable production as mulch. The film provides greenhouse effect to the root zone of vegetables which facilitates better nutrient and water absorption leading to early and high yield.

Besides early crop, weed growth and water requirement is reduced in vegetable production. Plastic mulch helps raising cucurbits at high altitudes. Commercial cultivation of cucurbits using plastic mulch has successfully been demonstrated by FRL in Leh valley. Cultivation of cucurbits in Ladakh region till recent years was considered impossible. Crops like tomato, cabbage and cauliflower yielded 20-25 days early and 15-20 per cent more under black polyethylene mulch at FRL, Leh. This technology needs to be popularised by making available photodegradable plastic film.

GREENHOUSE OPERATIONS

Following are the most important greenhouse operations

Control of Temperature, Humidity and Light

In general, the temperature inside the greenhouse will always be somewhat higher than the outside environment depending on the climate, location, season and the degree of ventilation. Therefore, reducing the temperature especially during summer is of great importance. Temperature can be reduced by ventilation, roof shading, with the help of both shades and evaporative cooling using fan and pad system. It is possible to reduce the temperature to some extent by increasing the humidity by providing additional gadgets like misting nozzles. Temperature can be increased by heating through electric heaters, reducing the degree of ventilation and reduced speed of exhaust fan. Humidity can be regulated by regulating cooling, the speed of exhaust fans and the degree of ventilation. It can also be increased by increasing the number of misting nozzles. Provision should be made for penetration of enough light into the greenhouse.

Nutrition and Irrigation

In general, the water requirement in the greenhouse is lower than the under open conditions because of reduced rate of evapotranspiration. In the modern greenhouses, irrigation and fertilisation is done through micro-irrigation systems like drip irrigation or micro-sprinkler system etc. which not only economise water use but also help in getting better results. Fertiliser requirement in the greenhouse is higher than under conventional cultivation. Fertilisation increases the fertiliser use efficiency also. Fertilisers are mixed in nutrient tank along with irrigation water.

Control of Pests and Diseases

Incidence of pests and diseases is generally lesser in the greenhouse as compared to open cultivation as the greenhouse area is covered. However, effective prophylactic and control measures should be adopted as and when incidence of insect-pest or disease is noticed.

VEGETABLE PRODUCTION

Nursery Raising

Vegetable nursery raising under protected conditions is becoming popular throughout the country especially in hilly regions. Management of vegetable nursery in protected structures is easier and early nursery can be raised. Needless to emphasise that this practice eliminates dangers of destruction of nurseries by hailstorms, rains, protection against biotic and abiotic stresses becomes easier. As a result, cultivation of vegetables like cabbage, cauliflower, knol-khol, tomato and onion has become commercial in cold desert region of the country such as Ladakh. Of late, nursery raising of cucurbits in poly pouch under greenhouses or protected structures has become common in most of the areas. This practice has helped river bed cultivation of cucurbits. Raising nursery in containers (plastic pouch, earthen trays etc.) under protected structures makes it possible to transport them to long distances facilitating transplanting at will.

Cold desert/temperate region vegetable production: A sizeable area in Jammu & Kashmir and Himachal Pradesh comprising Ladakh and Lahaul Spiti is cold desert. Besides, boosting vegetable cultivation during summer it has now become possible to grow vegetables under subzero ambient conditions of long winter months using solar greenhouses. Vegetables like spinach, vegetable mustard, fenugreek, lettuce, coriander, mint, radish, turnip, knol-khol, green onions and potato can be successfully grown under different protected conditions during winter months. In Leh, out of several plant protection structures, soil trench has been found suitable for vegetable production during winter months. The soil trench yields more and its cost per sq. m is affordable by farmers.

Production of Off-season Vegetable Crops

Use of polyhouses for production of off-season vegetable crops, during summer months in plain areas of the country is picking up. Crops like tomato, capsicum and cucurbits are raised by providing cooling devices in greenhouses. Greenhouses are becoming popular among the vegetable growers for growing early as well as late vegetable crops and raising their seeds particularly hybrid seeds. DRDO has successfully demonstrated production of certain vegetable crops like tomato, cucumber, brinjal and capsicum in greenhouses in UP hills and Antarctica using hydroponics technology. In temperate regions of UP hills, vegetable production technology in greenhouses has been developed by DRDO in Pithorgarh and Chamoli districts (Auli, Badrinath and Harsil). Polyhouses have been found suitable for vegetable production during winter months when it snows heavily making open field cultivation impossible. Leafy vegetables, peas, root crops are grown in polyhouses. During summer months cultivation of cucurbits in high altitudes in polyhouse has become possible. Crops like cucumber, bottlegourd, bittergourd can be commercially grown. DRDO is using local service units in above areas as demonstration centres for cultivation of vegetables under protected conditions. Besides, this technology is becoming handy to supply fresh vegetables to troops in landlocked high altitude areas during winter months.

Furthermore, in protected structure such as green/polyhouse, vegetable farming for domestic consumption, raising off-season nurseries, plant propagation, vegetable seed production can be carried out in adverse conditions. The cost on construction of a polyhouse in a 1000 m^2 comes to Rs. 7 lakh. The net income comes to Rs. 10/m^2/season in cucumber. Under Indian conditions, hi-tech greenhouse or polyhouse are highly expensive, requiring qualified operators, maintenance, care and precautions. So, the low and medium cost greenhouse have wide scope in production of off-season quality vegetables and their seeds.

Crop Yield Improvement through Greenhouses

Vegetables are an important source of mineral and vitamins for mankind. Balance human diet must include, on an average, about 285 g of vegetables everyday. The production of vegetable crops in 2002-03 was 93.92 million tonnes and projection for 2002-03 and 2007-08 is 146.50 and 175.20 million tons, respectively. The present production and consumption of vegetables in the country are very inadequate, being about one-fourth to one-third of the requirement. There are different ways and means to increase the vegetable production by bringing additional area under vegetable crops, using hybrid seeds and use of improved agro-techniques. Another potential approach is perfection and promotion of greenhouse cultivation of vegetables.

Cultivation of vegetable under protected condition assumes better quality, uniformity and ensuring availability throughout the year. Under low temperature and harsh climatic conditions, production of quality vegetables can be assured. In winter season, 4-6°C higher temperature can be maintained inside the polyhouse without any heating unit which provides better condition for plant growth. During summer (March-June), desert coolers and sprinklers are used to reduce the temperature by 5-10°C and desired relative humidity under controlled condition. The fruiting duration under controlled condition is also prolonged with better quality fruits. Looking to the various advantages of the greenhouses in vegetable production, experiment conducted at CCS HAU, Hisar showed that the greenhouse vegetable cultivation increased the crop yield and there is the possibility to cultivate vegetables all round the year. In capsicum cv. California Wonder and hybrid Alankar have performed best under greenhouse conditions. Tomato hybrids, namely, HYB-99K-126, INDAM-2101, NP-5002, Naveen and an Israeli cultivar R-144 performed better. Results showed that cucumber variety Poinsette, Rani, Phule Prachi and a gynoecious line, namely, Kasan produced maximum fruit yield. Among muskmelon hybrids Punjab Hybrid No.1 and Deepti produced maximum fruit yield with highest T.S.S. content. Economic analysis has indicated that the greenhouse vegetable cultivation can be profitable if the produce is available when the open field grown produce is not available.

CONSTRAINTS IN PROTECTED VEGETABLE PRODUCTION

As stated earlier, protected vegetable production in the country, though very ancient, is in its infancy. Very little efforts have been made to exploit vast potentials of protected vegetable production. This technology involves joint working of engineers, physiologists, physicists, horticulturists and environmentalists. Problems requiring concerted efforts are:

- Standardisation of greenhouse and other structures design for different agro-climatic regions of the country.

- Cladding materials are very costly. Intensive research to make cladding material affordable by average vegetable grower is required. Plastic films available in the country are not only costly but have short life. UV stabilised LDPE is not easily available.
- Lack of professional greenhouse manufacturers leading to use of improvised structures.
- Instruments for environment control in the greenhouse are not easily available.
- Vegetable production technologies for different crops under different types of protected structures for different agro-climatic zones of the country are not worked out and documented.
- Lack of major research programmes on protected vegetable farming.
- Lack of awareness among farmers pertaining to potentials of protected vegetable production.
- Import of greenhouses which are not suitable for Indian conditions.

Due attention to above problems will certainly lead to popularisation of protected vegetable production in the country.

CHAPTER 10

System of Rice Intensification: A Physiological Perspective

Anitha S. and Jose Mathew

Rice is the most important staple food in Asia contributing to 35 to 80 per cent of the total calorie uptake (IRRI, 1997) of the people. Irrigated lowland rice is considered the most important agricultural ecosystem in Asia and the current and future food security of majority of the Asian population depends on rice. But it is alarming to note that the sustainability of such a critical system is seriously threatened largely due to the declining water availability. The reasons for this are diverse but mostly location-specific. It predominantly includes decreasing quality and quantity of water and increased competition for water from urban and industrial users (Postel, 1997). To meet the water crisis head-on, it is required to grow rice with less water. It may be remembered that traditionally it requires from 3,000 to 5,000 liters of freshwater to produce one kg of rice. It emphasises the need for developing new approaches to reduce water requirement of rice significantly. The need of the hour is to produce more rice with less water. System of Rice Intensification (SRI) is a new concept in this direction for growing more with less water and simultaneously modifying the physiology of the rice plant.

The System of Rice Intensification or *le Systéme de Riziculture Intensive* — known as SRI is a methodology for increasing the productivity of irrigated rice by changing the management of plants, soil, water and nutrients. This system encourages greater root growth and soil microbial diversity contributing to improved soil and plant health. Fr. de Laulanié and his colleagues who developed SRI in Madagascar have never considered SRI as a technology, i.e., a fixed set of practices to be applied with relatively invariant results. Rather, they consider it a methodology that is based on empirical observations and inductive principles to help rice plants to become most productive (de Laulanié, 1993). The methodology capitalises on genetic potential that has always existed in rice but which has been inhibited by several common management practices

Address for correspondence: Department of Agronomy, College of Horticulture, Kerala Agricultural University, 680 656 Thrissur, Kerala such as flooded paddies, transplanting of mature seedlings, close spacing, and use of chemical fertilisers.

PHYSIOLOGICAL FEATURES OF RICE GROWN UNDER CONVENTIONAL SYSTEM

The practice of submerging paddy fields came to existence to check weed growth. Flooding makes weed control easier or even unnecessary, thereby saving labour. It is true that rice grows more abundantly in standing water than it does under upland conditions. But this does not make continuous flooding the best practice.

Submerging paddy fields continuously from the date of planting till harvest has so many disadvantages. Several observable physiological features of rice grown under continuously flooded condition suggest that these are not ideal circumstances for the best rice plant performance.

One important consequence of growing rice in continuously flooded paddies is that aerobic microbes, including fungi, cannot grow in inundated soil. The rice plants cannot benefit from the services of mycorrhizal associations, which are increasingly recognised as important intermediaries in the nutrition of most terrestrial plants (Habte and Osario, 2001). The microbial population will be affected due to the absence of oxygen essential for their survival in the soil during submergence. This in turn affects the conversion of soil organic matter into humus, the ultimate source of plant nutrition. Since seedlings are densely planted in submerged conditions, roots do not grow deeper and wider, thus cannot absorb nutrients from deeper zones. A month after transplanting, about 75 per cent of the rice roots growing in saturated soils are concentrated in the upper 6 cm of soil as they remain near the surface to obtain whatever dissolved oxygen they can get from irrigation water (Kirk and Solivas, 1997). Such truncated root system can access nutrients from only a limited volume of soil, having to rely mostly on nutrients provided through fertilisers.

When rice plants are grown under continuously flooded conditions, much of the root cortex disintegrates to form aerenchyma (air pockets) (Kirk and Bouldin, 1991 as quoted by Uphoff and Randriamiharisoa, 2002). This process occurs both in varieties bred for irrigated and upland cultivation. However, neither irrigated nor upland varieties form aerenchyma when they are grown in well drained soil (Puard *et al*, 1989 as quoted by Uphoff and Randriamiharisoa, 2002). The difference between these two categories of rice is that the former are able to create larger and more regular aerenchyma that enable roots to continue functioning and to survive longer in flooded soil. Formation of aerenchyma appears to be a sub-optimising rather than an ideal adaptation to hypoxic conditions. Rice plants spend lot of its energy to develop aerenchyma tissue in its roots under continuous inundation.

Roots of rice plants that are growing in continuously saturated soil do not get sufficient oxygen. By the time of flowering, about 75 per cent of rice root tips get degenerated in continuously saturated soils, whereas there is little or no degradation of roots growing in soil that is well drained (Kar *et al*, 1974).

Paddy has the capacity of doubling its tillers 13 times during its vegetative growth phase if they are provided with a conducive situation. Rice seedlings are conventionally raised in a thickly sown nursery for a period of 4 to 5 weeks. When seedlings are raised in flooded seedbeds and then plunged into flooded fields, usually after some time of exposure to the sun and air, they suffer from hypoxia as well as trauma to their roots. In conventional system transplanting is done densely with 5 to 10 seedlings per clump. Further transplanting is done like "J" making the roots point upwards. This causes trauma as the roots need 12 to 14 days from transplanting to get the roots established to start growing. In such a situation, 2 to 10 stages of tillering are lost (Reddy, 2002). Another issue is that dense planting does not provide enough light and air for the growing plants. They are increasingly susceptible to pests and diseases.

SYSTEM OF RICE INTENSIFICATION (SRI): BASIC CONCEPT

The SRI is centred on making the best use of the existing genetic potential of rice by breaking many of the conventional 'rules' of management. It is based on the insights that rice has the potential to produce more tillers and grains than now observed and that early transplanting and optimal growth conditions (spacing, humidity, biologically active and healthy soil, and aerobic soil conditions during the vegetative phase) can fulfill this potential (Uphoff and Fernandes, 2002). SRI changes some key practices and thereby, through synergistic effects of the new practices, increases the productivity of land, labour and water.

More tillers per plant, more panicles per tiller, more grains per fertile tiller and larger grains are contributory factors for increased yield. The success of SRI is based on the synergetic development of both the tillers and roots. With a more vigorous root growth, plants can become fuller and taller, and get better access to the nutrients and water they need to produce tillers and grains. With more aboveground growth to carry out photosynthesis, more energy is available for root growth. The stronger the plants, the more resistant they are to attacks by pests and diseases. There is synergy among various agronomic practices used under SRI that makes their use together more beneficial than using any of these practices alone. To have a positive correlation between tillering and grain filling, there needs to be reinforcement between tillering and root growth, with each process supporting the other: root growth supports tillering and vice versa. The emergence of tillers and root is biologically linked in all grass species such as rice, with both being components of the basic unit of plant growth in grasses, the phytomer, which emerges from the meristematic tissue (Nemoto *et al*, 1995).

Exploiting Tillering Potential in Rice

One of the "tricks" in SRI is that it capitalises on an in-built pattern of development in rice that had been identified many years ago T. Katayama, a by the Japanese researcher (Katayama, 1951 as quoted by Rabenanadrasana, 1999). He studied the patterns of growth and development in rice, wheat and barley and found that these plants put out tillers in a regular sequential pattern. In rice, each tiller produces another tiller two phyllochrons later (a phyllochron is an interval of plant growth, usually about 5 days but it can be longer or shorter depending on temperature and soil conditions). When soil and other conditions are favourable, the rice plant can go through as many as 12 phyllochrons (or more) before it moves from the vegetative growth phase to the reproductive phase (marked by panicle formation and flowering). The number of tillers can increase exponentially with as many as 84 or more forming on a single plant. The full potential for tillering can be exploited by the following management practices:

Transplantation of young seedlings

Transplanting after the fourth phyllochron begins (after 12 to 18 days) sets back the growth momentum of rice plants so that their full potential for producing tillers, roots and grains is not achieved. If the 10th through the 12th phyllochrons of growth do not materialise because of late transplanting, 75 per cent of rice plants' potential for tillering is lost. If transplantation is done poorly, the first two tillers get damaged in such case not more than 16 tillers can grow. Under SRI, seedlings should be transplanted at 8 to 12 days old and certainly less than 15 days old. Transplant seedlings with just two small leaves when the seed sac is still attached to the root (indicating that the plants have not yet started their fourth phyllochron of growth). This preserves the plants' potential for massive tillering and root growth that is lost if transplanting is done after the start of the fourth phyllochron (de Laulanie, 1993). Tillering and root growth potential are preserved and plant growth is not set back by 7-14 days as usually happens with standard transplanting methods (Kirk and Solivas, 1997). With SRI, plants typically have 50-80 tillers compared with 5-20 for conventional ones.

Plant carefully and quickly

Transplanting should be done carefully and quickly to have minimum trauma to the roots. Transplanting should be done very soon after removing the seedlings from the seedbed and within 15 to 30 minutes after the tiny plants have been gently uprooted. Seedlings are planted fairly shallow (1 to 2 cm deep) and care is taken so that the root tip is not inverted upward by thrusting the plant straight downward into the soil. An inverted tip delays resumption of plant growth until the plant is once again oriented downward. The tiny roots should be placed horizontally in the soil so that the tip of the root can easily resume its downward growth. In this way the leaves of the seedlings will not become yellow and the plant will start to grow again within a few hours. The plants need time to adjust to their new environment before the first tiller starts to grow.

Transplanting of single seedlings

If more than two seedlings are transplanted together in a clump, competition among their roots limits tillering to five per plant at the most. To enhance the development of roots and tillers and minimise competition between plants, transplanting should be done singly in SRI, only one per hill instead of 3 to 4 together in a hill. This is to avoid root competition that inhibits vigorous root growth. This reduces the seed requirement from 50-100 kg ha^{-1} in conventional system to about 5 to 10 kg ha^{-1} under SRI method.

Wide spacing

The close planting common in traditional rice cultivation could discourage tillering. Under SRI, seedlings in a square pattern with wide spacing (25 × 25 cm apart in a grid pattern rather than rows). Spacing can go up to 33 × 33 cm or 40 × 40 cm, even 50 × 50 cm with soils rich in biological terms. Such wide spacing reduce competition among plants' roots, encourage greater root and canopy growth and gives them more room to grow, a critical factor in SRI success. This facilitates mechanical weeding, as well as reducing seed use from 100 kg ha^{-1} to about 5 to 10 kg ha^{-1}. Wider spaced plants develop a different architecture, with more room for roots and tillers. Better root system means reduced lodging.

Exploiting the Full Potential for Root Growth

Inundating rice fields helps to suppress weeds and reduce the amount of labour needed. During reproductive phase, when plants go through flowering, panicle initiation, grain filling and maturation, maintaining 1 to 2 cm of water on rice fields has a beneficial effect. But during the preceding growth phase, rice plants grow better in unsaturated soil. This is because when there is no standing water and there is air in the soil, the roots can acquire oxygen much more easily through the *aerenchyma* (air pockets) in the root cells. Lack of oxygen in the root zone leads to soil acidification that causes the destruction of *aerenchyma* and hampers nutrient uptake, assimilation and plant growth. The nitrogen cycle in the soil is disturbed as well, and all kinds of toxicity will develop. The full potential for root growth can be exploited by:

Intermittent irrigation

Alternative wetting and drying of the field modifies the growing environment of rice: improves soil structure, gets more oxygen into the root zone, and enhances active soil life. As the soil dries, air replaces water and when it rains or irrigation is applied, this air is pushed downwards. Periodic water stress and the availability of oxygen facilitate root growth, and the volume of soil penetrated by the roots increases. In rice production, an effective drainage system to evacuate excessive rainfall and irrigation water from the field is as important as the irrigation system itself. At the beginning of tillering, there is still not much vegetative growth and the plant only requires a small amount of water. When the root system has been developed, 3 or 4 days of superficial dryness should not cause alarm even if some cracks develop in the field. During growth, irrigation will only be needed if rainfall is inadequate and then should be applied in moderate amounts and at favourable times preferably at night. In this way, irrigation requirements can be reduced by up to 50 per cent.

Under SRI paddy fields are not flooded during vegetative phase. Keep the soil moist but well drained, thereby avoiding root hypoxia. During the vegetative growth period, either (a) *light irrigation* is provided, keeping the soil moist but well aerated, with short periods (2 to 5 days) of drying; or (b) soil is *alternately flooded and dried* for 2 to 3 days at a time. In neither case the soil is kept continuously saturated. Once the plants begin their reproductive phase, after panicle initiation, just a thin layer of water is maintained on the fields (1 to 2 cm), which are then drained 10 to 20 days before harvest. Such management encourages more root growth. SRI requires only about half as much water as normally applied in irrigated rice. Best water management practices for a particular field depend on soil type, labour availability and other factors.

Early and frequent weeding

Late weeding not only reduces the yield, but also lose the opportunity to bring oxygen into their soil. Aeration of soil by weeding may be even more important in rice cultivation than the removal of weeds. When paddies are not kept flooded, weeding should start 10 days after transplanting and should be repeated 2 or 3 times at 10-day intervals and preferably 3 or 4 times before the canopy closes. Using a simple, inexpensive, mechanical push-weeder, often called a rotary hoe has the advantage of aerating the soil simultaneously eliminating the weeds. The weeds are left in the soil to decompose so their nutrients are not lost. Increased weeding is the most labour-intensive part of SRI, but it was found that an increase in yield associated with each additional weeding repays the added labour cost several times over.

Application of compost

Apply compost preferably some months before rice planting, e.g., on a preceding crop. Augmenting soil nutrient supplies preferably with compost and manure gives yield advantages. Possibly this can be attributed to the large volume of soil penetrated by roots and the high activity of soil life brought about by aerobic soil conditions and organic manures. Chemical fertiliser also gives good results, but not the best results for most varieties, since building up soil organic matter and microbial activity is part of the SRI strategy for increased productivity. At least initially, nutrient amendments are not necessary to achieve higher yields with the other SRI practices.

Seed and Seed Rate

It was found that all varieties traditional, improved and hybrid have given higher yield with SRI practices. Traditional varieties, which are considered low yielder, respond well to these SRI practices and increase the yield up to 6-10 kg ha^{-1}. Select healthy seed. SRI seeding rates are 5 to 10 kg ha^{-1}, compared to 50 to 100 kg ha^{-1} in conventional system.

Nursery Management

A Nursery area of 1 cent is required to produce seedling for acre. Prepare raised nursery bed and apply a layer of fine manure. Pre-sprouted seeds are sown on nursery bed sparsely. Cover with another layer of manure. Mulch with paddy straw and water the nursery carefully. Banana leaf sheath may be used for easy transport of seedlings.

Land Preparation

Land preparation is not different from standard practice. Land should be levelled properly so that water can be applied very evenly to get better results from applying small amount of water under the SRI practices. At the same time, a proper drainage system for the field should be established, i.e., at every 5-10 m distance from drainage channels so that alternate wetting and drying of the soil can be done effectively.

PHYSIOLOGICAL FEATURES OF RICE GROWN UNDER SRI

Physiological characteristics of rice plants under SRI cultivation show significant difference compared to rice plants under conventional system.

Plant Type with SRI

There was a significant difference in plant type with SRI. Individual plants were larger with SRI, hence, facilitating better light distribution (Tao *et al*, 2002). The base internode weight bearing ability (dry weight of the plant between 0-25 cm divided by the dry weight of the rest part of the plant) was 63 to 74 per cent with SRI compared to 33 to 39 per cent under conventional system. The plant height and total leaf area was higher with SRI. Plant leaf area was mainly distributed at a height between 50 to 100 cm with SRI and at 75 to 100 cm with conventional system. The plant type of SRI method was improved with better light distribution.

Increased Tillering

The plants under SRI produce 20 to 30 tillers per plant and in well-managed fields these reach 50 or even 70 tillers per plant (Uphoff and Randriamiharisoa, 2002). Also there is less senescence of leaves during the grain filling period, with the flag leaf remaining dark green until grain maturation occurs, probably reflecting the continuing functioning of plant roots. The leaf age at transplanting was young with SRI, thus tillers occurred earlier and at lower tillering nodes on the stem. Thiyagarajan *et al* (2002) reported that at active tillering stage, limited irrigation resulted in a significantly higher tiller density than conventional irrigation. Limited irrigation resulted in higher tiller density throughout the rest of the growth period. At the panicle initiation stage, there were significant differences in the number of tillers for seedling age, irrigation, and weeding and a significant interaction for seedling age with irrigation. With SRI the tillers were larger and stronger and single stem and sheath weight at heading was higher (Wang *et al*, 2002).

Root Growth with SRI

This is the key to SRI performance though it is unseen. Root pulling resistance with SRI plants, using the measurement test developed by IRRI is about 5 times greater per plant than for rice plants grown under conventional condition (Joelibarison, 1998 as quoted by Uphoff and Randriamiharisoa, 2002). The root growth is five times more than for rice plants grown under conventional conditions. Root volume increased from planting to the flowering stage. The increase in root volume from active tillering to panicle initiation was 110 per cent in young seedlings and 73 per cent in conventional seedlings (Thiyagarajan *et al*, 2002). Limited irrigation, weeding using a cono weeder and the application of green manure significantly increased the root volume. Root growth was greater in SRI plants. Roots of SRI plants extended 10 to 15 cm deeper than plants under conventional system. Root dry weight per plant was higher with SRI (Tao *et al*, 2002).

Root Physiological Activity

Root activity in SRI during each development stage was significantly higher than in conventional rice (Wang *et al*, 2002). The enhanced root activity during the entire growth period, especially during late growth stage was an important physiological characteristic in SRI rice plants. The amount of material exuded from the root when cut can be regarded as an index of root physiological activity and also as a signal of stem growth status (Tao *et al*, 2002). The material exuded by the root was less with SRI. This exudation increased as the stem and root system increased and reached a maximum value at panicle initiation stage then sharply decreases after heading stage. Note that this is not the same thing as root exudates, which are the photosynthate that is exuded by the root system into the rhizosphere and utilised by soil biota.

Physiological Activity in Leaf

Dry matter accumulation and distribution was greater in SRI plants. Tao *et al* (2002) reported that higher total plant dry weight and 10 per cent greater dry matter distribution with SRI plants. The total biomass accumulation at full heading was about 40 per cent of that for the total duration. Wang *et al* (2002) reported that with SRI, the absolute portioning rate of stored matter from vegetative organs was remarkably higher than with conventional method. The rate of dry matter and stored carbohydrate from leaves more than three times, and from stem and sheathes were about 1.4 to 1.7 times of those with conventional treatments. The total translocation of nitrogen from leaves, stem and sheathes was 66.9% higher, and apparent translocation percentage from leaves, stems and sheathes were distinctively higher than with conventional treatment. Higher translocation and conversion rates of stored matter from vegetative organs was of significant importance for enhanced grain filling and spike weight in SRI.

Wang *et al* (2002) reported obviously higher content of soluble sugar, non-protein nitrogen, proline and malondialdehyde (MDA) in leaves of SRI rice. Soluble sugar, non-protein nitrogen, MDA and proline contents in leaves of SRI rice were 54.5%, 24.6%, 32.7% and 11.7% higher at jointing stage and were 61.5%, 23.4%, 103.8% and 26.5% higher at heading stage than those of conventional rice. As main osmotic regulators in rice plant, accumulation of sugars and non-protein nitrogen would enhance rice ability for drought tolerance, which was another physiological characteristic of SRI rice. The higher content of MDA under SRI treatment implied that drought damage in SRI rice could be more serious than in conventional rice.

Yellowing of the leaf sheath (senescence) in the base stem appeared later with SRI than with conventional plants (Tao *et al*, 2002). They also noticed an increase in stomatal resistance rate after flowering and decrease in flag leaf photosynthesis rate with SRI.

More Grain Filling

With SRI practices, a positive correlation is found between the number of tillers per plant and the number of grains per panicle (Uphoff, 1999). But in conventionally grown rice, an inverse relation is observed between tillering and grain filling because due to closer planting roots get degenerated and this causes diminishing return to tillering. With SRI practices, keeping rice plant roots intact through different water management methods and less total application of water, they have found no evidence of a yield ceiling. Translocation rate and percentage of assimilate from leaves, stem and sheathes, conversion percentage of stored assimilate before heading and harvest index were all higher with SRI rice (Wang *et al*, 2002). With these methods the plant can produce panicles with 200-300 grains and some with as many as 400 or 500, even 600 grains (Uphoff and Randriamiharisoa, 2002). The SRI's combination of management practices gives expression to impressive genetic potential already existing within the rice plant.

RESEARCH RESULTS

Systematic evaluation of SRI was started only in 1994. The first evaluation outside of Madagascar was done at Nanjing Agricultural University in China in 1999, where yields of 9.2 to 10.5 t/ha were obtained, using only about half as much irrigation water as usual (Uphoff *et al*, 2002). The Indonesian Agency for Agricultural Research and Development got 9.5 t/ha in its wet-season trials in 2000 at its Sukamandi station. It is estimated that 20,000 farmers have now adapted the full SRI in Madagascar.

Over the last decade, it was reported that in Madagascar these methods can yield about 8 t ha^{-1}, with top yield is double or more, beyond what has been claimed to be the biological maximum for rice. SRI raises questions about present understanding of rice plant requirements and potentials. Experience in Madagascar over the past ten years has shown that the effects of these different management practices are not variety-dependent. So far, all varieties have responded positively to the recommended practices for plant, soil, water and nutrient management, though some high-yielding modern varieties have given the highest yields, 15 to 20 t ha^{-1} and even higher. Uphoff and Randriamiharisoa (2002) reported significantly higher yield to both traditional and high-yielding varieties under SRI management.

Uphoff and Fernandes (2002) opined that eight-day-old seedling planted singly per hill and application of compost under SRI method produced significantly higher yield compared to conventional method both in clay and loam soils of Madagascar. Ten days old seedling planted at 25 × 25 cm recorded higher yield compared to 21 days old seedling planted at 15 × 15 cm spacing (Makarim *et al*, 2002).

Thiyagarajan *et al* (2002) reported 56 per cent water savings without any significant effect on grain yield. The water productivity, defined as the grain yield divided by the total amount of water supplied was 60 per cent higher under SRI management practices in Tamil Nadu. They also observed that early and frequent weeding and incorporation of weeds and disturbance of the soil with a cono weeder significantly increased the grain yield to under SRI management compared to conventional method of weeding.

As many as 288 trial plots testing 96 combinations of variables for six factors, have shown both the *efficacy* of each practice and the *synergy* that results when SRI practices are used together (Uphoff and Randriamiharisoa, 2002). There is synergy among practices that makes their use together more beneficial than just using certain of the practices; but any of the practices should give some improvement in yield. Thiyagarajan *et al* (2002) reported that the significant yield increase because of weed incorporation with the cono weeder and the significant interaction effect for the age of seedlings and weed and nutrient management reveal the possibility for creating synergy among the agronomic practices.

Uphoff (2002) reported more than twofold yield increase in the entire where SRI is being practising. Despite very high yields, declines in yield have not been observed so far, even though the soils used have been described as some of the "poorest" in the world assessed in (inorganic) chemical terms (Johnson, 1994). Possibly the depletion of nutrients is being masked by increases in farmers' skills, which cause yields to move upward rather than let them decline. On the other hand, possibly soil fertility is being enhanced by the combination of plant-soil-water-nutrient-microbial interactions supported by SRI practices, making otherwise unavailable nutrients available through microbiological processes.

Broader Initiatives for Negros Development (BIND), a Philippine NGO-based Bacalod in Negros Occidental, has done a series of trials adapting SRI concepts and principles to unirrigated (rainfed) rice production. With upland rice, there is no transplanting. 3-4 seeds per hill were planted. At 10-12 days, the seedlings were thinned to 1 per hill. At that time, mulch was applied to control weeds, conserve soil moisture, lower soil temperature to encourage earthworm and other biotic activity, and possibly add nutrients to the soil. The average yield obtained, without irrigation but using mulching as the key innovation (together with wide spacing, single plants, and organic fertilization), was 7.2 t ha^{-1}, surpassing most yields from irrigated rice. The yield at 20 × 40 cm spacing was 7.7 t ha^{-1}. The calculated benefit: cost ratio was 2.6:1.

ADVANTAGES OF SRI

The most important benefit associated with SRI practices is an increase in total factor productivity. Other advantages reported by Uphoff and Fernandes (2002) included:

1. ***Higher yield***: Yield increase of 50-200 per cent, with yields 4-8 t ha^{-1} has been reported.
2. ***Increased returns to labour***: SRI requires more labour about 26 per cent in one Madagascar evaluation, 11 per cent in a Sri Lankan survey. However depending on the cost of labour, the value of increased production has been reported to increase the returns to labour by around 50 to several hundred per cent.
3. ***Water saving***: Since yields double or more with around 50 per cent of water, the productivity of water is greatly increased.
4. ***Improvement of soil quality*** and increased nutrient use efficiency.
5. ***Reduced requirement of seed***: This makes the use of improved and hybrid seeds much cheaper for farmers.
6. ***Less external inputs***: Low requirement of purchased inputs (fertiliser, seed and pesticides) and water and lower costs of production contribute to higher income for farmers.
7. ***Higher seed quality***: SRI methods make it possible to increase considerably the yields for traditional varieties grown organically for which higher prices can be obtained. Multiplication of breeder seed can be faster as many more grains can be produced from a single seed.
8. ***Equity Considerations***: SRI also has the benefit of being particularly accessible for farmers who have small landholdings and need to get the highest yields possible from their available land. Since poorer households have relatively more labour compared to land, SRI is one of the few agricultural innovations that have a bias in favour of equity.
9. ***Environmental and Sustainability Contributions of SRI:*** (Uphoff *et al*, 2002)
   a. *Reducing pressures to expand arable area*: By increasing the productivity of presently cultivated land, SRI can curtail the shifting cultivation of upland rice encroaching on remaining rainforests in several parts of the world. Which can be used for producing green manure or crops with higher value.
   b. *Reducing agricultural demands on the world's water supplies*: SRI methods can cut requirements for irrigated rice production by half to two-thirds. This water can be used for other purposes particularly in countries or places where water is scarce.

c. *Reducing the load of agrochemicals in soil and water*: Since rice plants are healthier and show increased tolerance to pest and disease problems, SRI methods makes it possible to reduce the use of agro chemicals.

d. *Soil that is not kept saturated has greater biodiversity.*

e. *Reducing methane gas emissions from rice paddies:* SRI methods reduce methane gas emission by not keeping them continuously flooded (Conway, 1997).

f. *Enhancing soil quality and long-term productivity:* Integration of plant, soil, water and nutrient management practices under SRI method contribute to the size, dynamism and diversity of soil microbial communities.

DISADVANTAGES OF SRI

The disadvantages of SRI according to Uphoff and Fernandes (2002) includes:

1. ***Requirement of good water control:*** In areas where there is no control or reliable access to water, the benefit from SRI practices will be less.
2. ***Requirement of more labour:*** SRI requires additional labour input ranging from 25 to 50 per cent, particularly for careful transplanting and for weeding. But it can be offset subsequently by increased productivity.
3. ***Drastic change of farming practices.***
4. ***Requirement of greater skill:*** Farmers are encouraged to experiment with the methods, evaluate the results and adopt the best ones. Recommendations for spacing, water management, age of seedling, etc. are not offered as universal. Only the principles are presented to understand, test to suit their own conditions.

CONCLUSION

SRI practices facilitates irrigated rice production with less water, under mostly aerobic soil conditions, the benefits thereof can be maximised by modifying the plant, soil and nutrient management practices. Physiological demand of water by the rice plant is indispensable for normal metabolic activities, but ecological water demand has certain flexibility. With the root zone kept moist under SRI cultivation, supplemental gas and heat were continuously provided into the soil on the basis of meeting physiological water demand, thus making the soil environment more suitable for rice root growth. This could be a major reason for enhanced root activity in rice plant of SRI. The reduced ecological water supply under SRI created a slight drought stress, as seen with increased content of soluble sugar, proline and malondialdehyde in leaves. This water stress had little impact on rice with strong root and vegetative growth. SRI significantly enhanced the dry matter in vegetative organs, partitioning percentage of stored carbohydrate and nitrogen, percentage of effective leaf area, single stem and sheath weight, percentage of productive tillers, grain yield and harvest index. Several of the principles and practices of SRI go against the present-day scientific and conventional principles of rice production. Hence thorough investigations on the physiological and agronomic perspectives are required before SRI gains widespread acceptance for general cultivation.

REFERENCES

Conway, Gordon (1997). *The Doubly Green Revolution: Food for All in the 21st Century*. Penguin Books, London.

de Laulanié H. (1993). System of rice intensification. *Tropicultura*(Brussels) 11(3): 110-114.

Habte M. and Osario N. W. (2001). Arbuscular mycorrhizas:producing and applying arbuscular mycorrhizal innoculam. College of Tropical Agriculture and Human Resources, University of Hawaii and Manoa, Hawaii, USA.

IRRI [International Rice Research Institute]. (1997). Rice almanac, Second edition, IRRI, Los Baños, Philippines, pp. 181.

Johnson Bruce K. (1994). Soil survey. In: *Final Report for the Agricultural Development Component of the Ranomafana National Park Project in Madagascar*, 5-12.

Kar S., Varade S. B., Subramanian T. K., Ghildyal B. P. (1974). Nature and pattern of rice root system under submerged and unsaturated conditions. Riso, Italy 23: 173-179.

Kirk G. J. D. and Solivas J. L. (1997). On the extent to which root properties and transport through the soil limit nitrogen uptake by lowland rice. *European Journal of Soil Science* 48: 613-621.

Makarim A. K., Balasubramanian V., Zaini Z., Syamsiah I., Diratmadja I. G. P. A., Handoko Arafah, Wardana I. P. and Gani A. (2002). System of Rice Intensification (SRI): evaluation of sedling age and selected components in Indonesia. *Proc. Int. workshop on Water wise rice production,* 8-11 April 2002, International Rice Research Institute, Los Banos, Philippines, pp. 129-139.

Nemoto K., Morita S. and Baba T. (1995). Shoot and root development in rice related to the phyllocron. *Crop Sci.*35: 24-29.

Postel S. (1997). Last Oasis. Facing water scarcity. Norton and Company, New York, USA, pp. 239.

Rabenandrasana J. (1999). Revolution in rice intensification in Madagascar. *ILEIA: Newsletter for Low External Input and Sustainable Agriculture*, 15: 3-4, 48-49.

Reddy N. L. (2002). SRI method of paddy cultivation. *LEISA- India* 4(3): 28-29.

Tao L., Wang X. and Min S. (2002). Physiological effects of SRI methods on the rice plant. *Proceedings of the International conference on Assessment of the System of Rice Intensification (SRI)* (eds. Uphoff, N., Fernandes, E., Long-Pin, Y., Jiming, P., Sebastine, R. and Rabenandrasana, J.), 1-4 April, 2002, Sanya, China, Ithaca, NY(USA); CIIFAD. pp. 132-136.

Thiyagarajan T. M., Velu V., Ramasamy S., Durgadevi D., Govindarajan K., Priyadarshini R., Sudhslakshmi C., Senthilkumar K., Nisha P. T., Gayathry G., Hengsdijk H. and Bindraban P. S. (2002). Effects of SRI practices on Hybrid rice performance in Tamil Nadu, India. *Proc. Int. workshop on Water wise rice production,* 8-11 April 2002, International Rice Research Institute Los Banos, Philippines, pp. 119-127.

Uphoff N. (1999). Agroecological implications of the system of rice intensification (SRI) in Madagascar. *Environment, Development and Sustainability*, 1: 3-4, 297-313.

Uphoff N. (2002). Opportunities for raising yields by changing management practices: The system of rice intensification in Madagascar. In: *Agroecological Innovations: Increasing Food Production with Participatory Development*, ed. N.Uphoff. London: Earthscan.

Uphoff N. and Fernandes E. (2002). System of Rice Intensification gains momentum. *LEISA- India* 4(3): 22-27.

Uphoff N. and Randriamiharisoa. (2002). Reducing water use in irrigated rice production with the Madagascar System of Rice Intensification (SRI). *Proc. Int. workshop on Water wise rice production,* 8-11, April 2002, International Rice Research Institute, Los Banos, Philippines, pp. 71-87.

Uphoff N., Yang K. S., Gypmantasiri P., Prinz K. and Kabir H. (2002). The System of Rice Intensification (SRI) and Its Rélevance for Food Security and Natural resource Management in Southeast Asia. *Proc. International Symposium Sustaining Food Security and Managing Natural Resources in Southeast Asia - Challenges for the 21st Century*, 8-11, January 2002, Chiang Mai, Thailand.

Wang S., Cao W., Jiang D., Dai T. and Zbu Y. (2002). Physiological characteristics of and high yield technique with SRI rice. *Proceedings of the International conference on Assessment of the System of Rice Intensification (SRI)* (eds. Uphoff, N., Fernandes, E., Long-Pin, Y., Jiming, P., Sebastine, R. and Rabenandrasana, J.), 1-4 April, 2002, Sanya, China, Ithaca, NY(USA); CIIFAD. pp. 116-124.

CHAPTER 11

Study of the Interactive Effect of ABA and Cytokinin on Mustard under Soil Water Stress Condition

Meera Srivastava

INTRODUCTION

Brassica juncea, czen & coss L. (Indian Mustard) is one of the important oilseed crops that is grown in our country, occupies nine per cent of total agricultural equivalent to a value of Rs. 1200 crore per year, and it is utilised by the people of North and North–Eastern regions of the country in variety of ways. Mustard oil is widely used for edible purpose and is highly prized for vegetable and fish frying, because of its peculiar taste. The seeds are also used as medicine and spices as component in the preparation of salad, sauces and pickles. Oil is as well used in the preparation of fractice (a rubber additive), in fat liquoring, leather chrome and tanning. Its oil cake is used as animal feed, which contains 35% crude protein, 30% nitrogen free extract, 25% digestible crude protein, 9% ash and nearly 15% of total digestible nutrients.

Growth regulators are known to influence the various facets of plant life so much so that the rate of action or influence of the one part of plant on the other is conditioned by these chemical substances. It is, therefore, pertinent to use these substances in modifying or in increasing the economic yield in such diverse situations as arid, semi-arid and areas with lack of substantial amount of rain water.

The situation with regard to pulses and oilseed is still worse in the sense that only 11-12 g of fat even after importing them from abroad is available per capita. This is so meager for about 60 g of fat is recommended by nutrition and health experts. In this sense we have to go a long way in increasing the productivity of oilseed. The problem of increasing the yield of oilseed is

Address for correspondence: Section of the Economic Botanist (*Rabi Cereals*) C.S.A. University of Agriculture & Technology, Kanpur –208 002 beset so many hazards particularly because the oilseed crops are mainly grown in marginal lands.

The objectives of this experiment was to study the interactive effect ABA and cytokinin in respect of relative water content (RWC), water potential, chlorophyll content, nitrate reductase activity, proline content and finally seed yield and its attributes.

MATERIALS AND METHODS

Mustard variety Varuna was taken for this experiment. The experiment was conducted in field at Research Farm Nawabganj of C.S. Azad University of Agriculture and Technology, Kanpur. There were ten treatment combinations comprising three levels of ABA (0, 25 and 50 ppm) and three levels of cytokinin (0, 10 and 20 ppm) with one water spray replicated four times. The observations were recorded at four stages (5 to 10 days after 1st spraying, 10 to 5 days before spraying, 5 to 10 days after 2nd spraying and at harvest). Spraying of growth regulators was done after one month and two months of sowing. This experiment was conducted in randomised block design (RBD).

Treatments and their combinations:

ABA0 ppm (AbA0)	Cytokinin 0 ppm CK0
AbA25 ppm (AbA25)	Cytokinin 10 ppm CK10
AbA30 ppm (AbA50)	Cytokinin 20 ppm CK20

1. ABA0 CK0
2. ABA0 CK10
3. ABA0 CK20
4. ABA25 CK0
5. ABA25 CK10
6. ABA25 CK20
7. ABA50 CK0
8. ABA50 CK10
9. ABA50 CK20
10. Water spray

Physiological and biochemical observations like relative water content (RWC%), water potential (bar), chlorophyll content, nitrate reductase activities, proline content and cytokinin content were taken after foliar spray of growth regulators. Yield observations were taken at the time of harvesting.

RESULT AND DISCUSSION

The object of this experiment was to study the relationship between exogenous application of ABA and cytokinin at the cytokinin content of the plants under rain-fed situations. ABA seems to play a substantial role in the stomatal closing and opening which ultimately helps plants in maximising photosynthesis and reducing transpiration rate. This is the fundamental property of plants (also animals) known as homeostasis. The results obtained in this experiment are discussed broadly with a view to find out how far these promote ultimate yield of the plants.

It was observed that after the first spray AbA25 increased RWC without cytokinin to the maximum. Further increase in its concentration reduced RWC. Cytokinin also increased to

RWC of the leaves, up to its highest concentration tried at this stage possibly due to increasing membrane permeability or by increasing the internal concentration of osmotically active solution as suggested by Kaufman and Ross (1970). But with AbA25 increasing dose of cytokinin showed numerical reduction, although with highest concentration of cytokinin there was a fall in the RWC but it increased with the increasing dose of cytokinin. The chlorophyll content did not behave in the same fashion. It did increase with the increasing dose of cytokinin even under the endogenous level of CK for cytokinin is known to promote chlorophyll synthesis but AbA reduced it and this supports the view of Pradhan and Basu (1982). This increase in the chlorophyll content and the water potential helped utilisation of proline in protein synthesis. With AbA, on the other hand, there was increase in proline content possible because of inhibition of protein synthesis as reported earlier by Trewavas (1972) and Cooke *et al* (1979) who showed that the increase in proline content could be either due to decrease in the rate of protein synthesis or by increase in protein degradation by AbA. This instance when AbA and cytokinin spread simultaneously the chlorophyll was reduced at the first stage to 3.9 in AbA50 and CK10 and CK20 thus showing no improvement in the chlorophyll content with application of cytokinin but with AbA25 spray of cytokinin help increase chlorophyll remarkably. In the first instance the degradation effect of AbA could not ameliorate with cytokinin but in the second instance i.e. with AbA25, cytokinin either prevented degradation or supported de-novo synthesis of chlorophyll over and above the loss in chlorophyll due to higher dose of AbA. At the second stage increased RWC and water potential of leaves due to increasing concentration of CK again promoted chlorophyll content and nitrate reductase activity but it decreased slightly proline content. At lowering RWC by higher dose of AbA resulted in marked increase in free proline content and decrease in chlorophyll content. At the final stage increase in RWC and water potential of leaves and flowering axis by AbA25 ppm only resulted into marked decrease in proline content but chlorophyll and endogenous cytokinin contents were also decreased by AbA25 at this stage. This support view of Alvim (1976) that AbA decreased endogenous cytokinin level. Enhanced water potential of leaves and flowering axis by cytokinin reflected into marked increase in chlorophyll and nitrate reductase activity, thus enhancing the photosynthesis in treated plant at the final stage.

Increased RWC by CK also reflected into marked reduction in the free proline content or it can be said that transamination of proline into protein was enhanced in cytokinin treated plants than in AbA treated plants. Oil content of mustard seeds remained unaffected by both the growth regulators. Higher doses of CK combined with increased RWC, water potential, and chlorophyll and nitrate reductase activity ultimately reflected into higher number of siliqua per plant. Seeds per siliqua and weight of seeds per siliqua was also increased by cytokinin for there photosynthetic rate could be increased due to higher chlorophyll content coupled with increased RWC in the leaves and as well as flowering axis. This ultimately resulted into higher seed yield per plant and area basis at the time of harvesting (supported by Prakash and Joshi 1973). Contrary to this with AbB, higher dose declined all the yield parameter and ultimately the yield at harvest this supports the earlier findings of Ray *et al* (1983). This may be due to reduced chlorophyll content and inhibition of protein synthesis as there was increase in free proline in the leaves particularly with the highest dose of AbA for even cytokinin could not help increase in yield parameter. In the lower dose of AbA, cytokinin promoted the yield parameter for reasons already discussed earlier.

Table 11.1: Interactive effect of AbA and cytokinin on relative water content water potential nitrate reductase activity proline content and cytokinin content of mustard leaves.

Treatment	Relative water content (%)			Water potential ψ (bar)		Nitrate reductase activity (μ Mole No2 g^{-1} F.W.)		Proline content (μ g proline 100 g^{-1} F.W.)			Cytokinin content (ppm)
	I	II	III	II	III	II	III	I	II	III	
Water spray	58.70	47.80	39.80	–14.60	–19.50	41.20	20.60	1133.7	1461.0	921.00	18 × 104
AbA0 × CK0	58.60	47.60	39.60	–14.80	–19.20	41.00	20.70	1125.7	1460.5	939.70	18×10^{-4}
AbA0 × CK10	66.60	56.40	50.10	–12.00	–16.50	45.20	23.60	866.2	1210.7	756.50	20×10^{-4}
AbA0 × CK20	68.60	52.00	50.60	–11.40	–15.40	46.60	23.90	810.0	1186.0	721.20	25×10^{-4}
AbA25 × CK0	70.60	56.40	50.20	–12.80	–15.90	40.90	20.60	1104.7	1310.5	845.00	25×10^{-4}
AbA25 × CK10	70.20	54.80	50.40	–12.80	–15.50	45.00	23.40	1013.7	1298.7	823.62	18×10^{-4}
AbA25 × CK20	69.90	55.10	50.60	–12.00	–14.30	46.60	23.80	1002.2	1196.0	810.00	19×10^{-4}
AbA50 × CK0	60.20	48.80	39.80	–14.10	–19.40	41.00	20.70	1251.2	1499.0	964.70	15×10^{-4}
AbA50 × CK10	61.60	49.20	39.60	–13.20	–19.00	45.00	23.20	1196.2	1485.5	957.50	16×10^{-4}
AbA50 × CK20	64.50	49.70	39.80	–12.80	–17.20	46.40	23.60	1144.5	1436.2	926.30	16×10^{-4}
SE±	0.12	0.10	1.10	0.12	0.17	0.10	0.16	17.4	250.5	15.10	
CD5%	0.24	0.21	2.20	0.25	0.35	0.25	0.34	35.7	42.1	31.00	

Table 11.2: Interactive effect of AbA and cytokinin on yield, yield attributes and oil content of mustard at harvest

Treatment	*Siliquae plant^{-1} (No.)*	*Length of siliquae plant^{-1} (cm)*	*Weight of siliquae plant^{-1} (g)*	*Seed siliquae plant^{-1} (No.)*	*1000 seed weight (g)*	*Seed yield/ plant (g)*	*Economic yield (kg ha^{-1})*	*Biological yield (Kg ha^{-1})*	*Harvest index (%)*	*Oil content (%)*
Water spray	786	4.00	72.2	8.0	3.3	24.10	1606	8812	18.2	40.10
AbA0 × CK0	740	3.90	71.1	8.5	3.4	26.20	1699	9399	18.0	40.10
AbA0 × CK10	900	5.20	91.2	11.4	3.5	32.10	1912	9722	19.5	40.40
AbA0 × CK20	1000	5.60	96.5	11.8	3.5	34.40	2015	9617	20.9	40.50
AbA25 × CK0	736	3.60	65.9	8.8	3.3	27.30	1612	9106	17.6	40.50
AbA25 × CK10	836	4.00	80.9	9.5	3.4	29.20	1711	9313	18.2	40.20
AbA25 × CK20	868	4.50	81.1	10.5	3.5	30.10	1806	9511	18.9	40.30
AbA50 × CK0	716	3.50	60.6	8.2	3.2	25.10	1503	8915	16.8	40.10
AbA50 × CK10	724	3.60	76.2	8.8	3.4	25.20	1513	8706	17.2	40.10
AbA50 × CK20	732	4.00	80.0	9.0	3.47	26.30	1723	9412	17.9	40.20
SE±	30.8	0.07	1.8	0.17	0.6	0.77	70.4	140.0	0.11	0.66
CD5 %	63.1	0.15	3.8	0.36	NS	1.58	140.8	300.0	0.23	NS

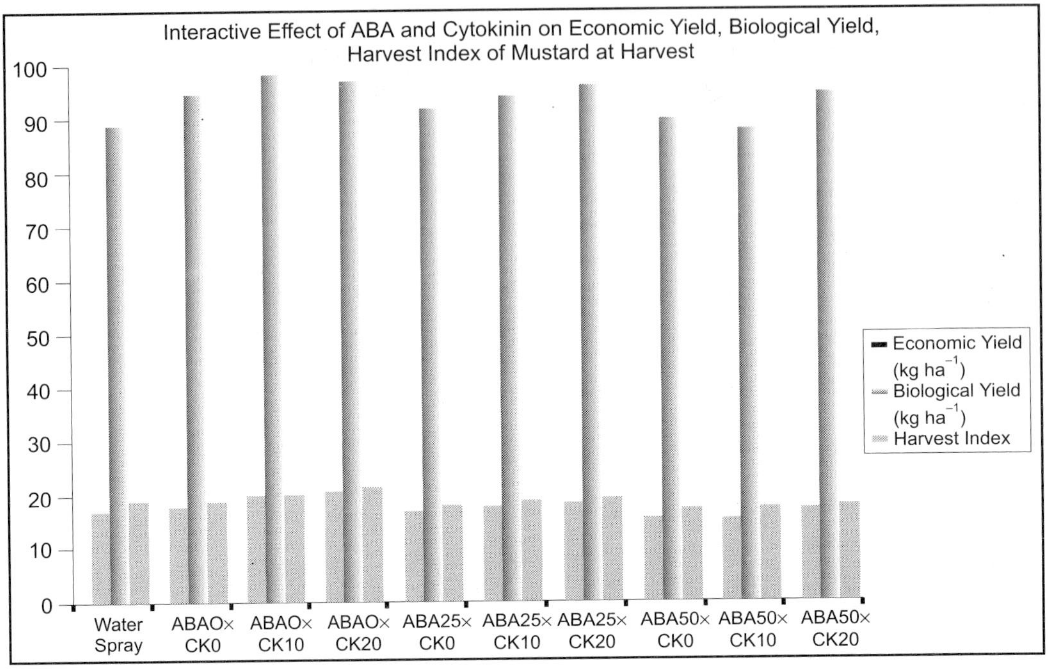
Interactive Effect of ABA and Cytokinin on Economic Yield, Biological Yield, Harvest Index of Mustard at Harvest
100
90
80
70
60
50
40
30
20
10
0
Water Spray
ABAO× CK0
ABAO× CK10
ABAO× CK20
ABA25× CK0
ABA25× CK10
ABA25× CK20
ABA50× CK0
ABA50× CK10
ABA50× CK20
Economic Yield (kg ha⁻¹)
Biological Yield (kg ha⁻¹)
Harvest Index

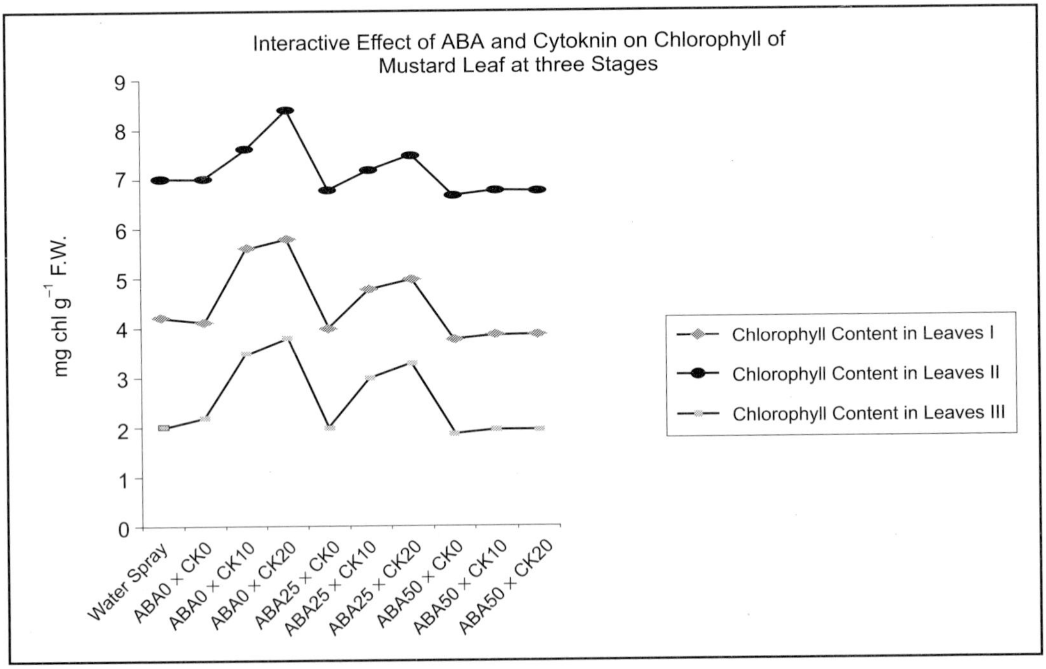
Interactive Effect of ABA and Cytoknin on Chlorophyll of Mustard Leaf at three Stages
mg chl g⁻¹ F.W.
9
8
7
6
5
4
3
2
1
0
Water Spray
ABA0 × CK0
ABA0 × CK10
ABA0 × CK20
ABA25 × CK0
ABA25 × CK10
ABA25 × CK20
ABA50 × CK0
ABA50 × CK10
ABA50 × CK20
Chlorophyll Content in Leaves I
Chlorophyll Content in Leaves II
Chlorophyll Content in Leaves III

Fig. 11.1:

In conclusion it could be said that cytokinin promotes yield probably because it acts as counter mole for AbA more so under water stress situation in which Indian mustard crop was grown in the experiment in question. In between AbA and cytokinin, AbA in higher concentration alone is harmful for mustard crop while cytokinin and combined with lower doses of AbA improved yield and quality of mustard crop under water stress situation.

REFERENCES

Alvim R.; Hewitt E.W. and Sounders R.F. (1976). Seasonal variation in the hormone content of willow. Changes in abscisic acid and cytokinin activity in Xylem sap. Plant Physiol. 57: 474-476.

Cooke R.J.; Oliver J. and Davis D.D. (1979). Stress and protein turnover in *Lemna minor.* Plant Physiol. 64: 1109-1113.

Kaufman M.R. and Ross K.J. (1970). Water potential temperature and kinetin effect on seed germination in soil and solute system. American J. Bot. 57: 413-418.

Pradhan J.P and Basu P.K. (1982). Antagonistic effect of abscisic acid rutin upon seed germination and seedling growth of *Tephrosia vogellii.* Indian J. Exp. Biol. 20 (6): 459-465.

Prakash V. and Joshi Y.C. (1973). Influence of foliar feeding of kinetin and potassium nitrate on grain setting of wheat. Agrokhimiya. 17 (3/4): 238-242.

Ray S.; Mondar W.A. and Chaudhuri (1983). Regulation of leaf senescence, grain filling and yield of rice *Oryza sativa* cul. Jaya by kinetin and abscisic acid Physiol. Plant. 59 (3): 343-346.

Trewavas A. (1972). Control of protein turnover ratio in *Lemna minor.* Plant Physiol. 49: 47-51.

CHAPTER 12

Effect of Different Photoquality on Germination and Early Seedling Growth of Three Varieties of *Vigna radiata* (L.) Wilczek

Arvind Kumar

Abstract

The present paper deals with the germinability and early seedling growth of three varieties of *Vigna radiata* (L.) wilczek Viz. SUNAINA, ML-65 and PANTMOONG-2 in response to continuous light, Dark, Red and Far red light. PANTMOONG-2 shows reduction of seedling length and fresh weight under Far red radiation. SUNAINA shows increased seedling length and fresh weight under Far red radiation.

INTRODUCTION

The germination of a seed is a physiological event comprising essentially the resumption of growth by the dormant embryo and its development into an independent seedling. The light is one of the most important environmental factors for germination of seeds. The seeds of many species do not germinate in the dark. They require a light stimulus. Such seeds are called positive photoblastic. Some seeds are negatively phostoblastic. Some seeds are also not photoblastic in response to phytochrome responses. Borthwick and his co-workers (1952) observed that there is a marked effect of Red and Far red radiation on germination. The Red region of spectrum is the most effective for the promotion of germination of the light requiring seeds. Far red region of spectrum has been reported to reverse the response induced by Red. The different plant varieties show different response in germination and their seedling growth in different photoquality.

Address for correspondence: Dept of Botany, T.N.B College, T.M. Bhagalpur University, Bhagalpur 812007 BIHAR, INDIA

METHODS

Effect of different photoquality on germination was evaluated in petridishes containing seeds with 60 watt incandescent electric bulb. Effects of Red and Far red light on germination of seeds were tested. One fold red and one fold blue cellophane papers were wrapped on petridishes to test the effect of Far red light on germination. Double fold of red cellophane papers were wrapped on petridishes to test the effect of Red light on germination. Complete darkness was created by wrapping petridishes into 4 wraps of carbon paper and by putting them in between the two folds of Black cloth. Seeds examined for effect of darkness, when once counted were discarded for next reading due to exposure to light during counting. Petridishes wrapped with double layers of colourless cellophane were treated as control.

RESULTS

The germination percentage in SUNAINA fluctuated from 100% in the Far red light to 90% in the Red light, 85% in the continuous dark and 80% in continuous light. ML-65 had 95% germination in continuous light. ML-65 had 95% germination in continuous dark, 85% in continuous light to 90% in Far red light and 75% in continuous dark.

SUNAINA had maximum seedling length and seedling fresh weight in Far Red light and minimum in Red light. ML-65 had maximum seedling length and fresh weight in red light and minimum in continuous dark. while PANTMOONG-2 has maximum seedling length and fresh weight in continuous dark and minimum in Far red light. Initial time lag was enhanced in Far red light in all the varieties but it was more in case of PANTMOONG-2.

DISCUSSION

In general the effect of varying light conditions (continuous light and continuous dark) did not show any variation either with regard to the treatment or at the level of variety. A minor fluctuation was apparent in SUNAINA with more length and fresh weight of seedlings in continuous dark. These results indicated that the varieties are non-light sensitive as regards the effect of Red and Far red radiations. Interestingly, the Far red radiation resulted into a reduction of seedling length and fresh weight in PANTMOONG-2 while in SUNAINA it increased the seedling length and fresh weight. Many workers have reported the enhancing effects of red radiation (Agarwal and Prakash, 1978; Valio and Jully, 1979). There are others (Sircar, 1970; Khare ; 1978; Widel, 1983) who have reported the inhibitory effect of red exposure. In all probabilities, the phytochrome level remained in well set condition during storage and incubation in these varieties. However, the inhibitory effect of Far red in PANTMOONG-2 is worth noting. Nwoke (1982) has given that events occurring during dark incubation involves a biochemical synthesis of some factors which could be rapidly utilised to stimulate germination and in terms of phytochrome action remaining correlated with Pr and Pfr balance. Heim and Schaefer (1984) are also of the opinion that Pr and Pfr are not the only light dependent factors controlling hypocotyl growth which is also evident from the present result. A great variation in the seedling growth and seedling fresh weight appears to be in consonance with those of Singhal *et al* (1984).

Table 12.1: Effect of continuous light, continuous dark and spectral composition on germination of SU, ML-65 and PM – 2 varieties of *Vigna radiata* (L.) Wilczek.

Treatment \ Attributes	Germination %		
	SU	ML-65	PM-2
Red	90	80	80
Far Red	100	80	90
Continuous light	80	85	95
Continuous dark	85	95	75

Table 12.2: Effect of continuous light, continuous dark and spectral composition on seedling growth of SU, ML-65 and PM-2 varieties of *Vigna radiata* (L.) Wilczek

Treatment \ Attributes	Seedling length after 96 hours (in cm)								
	SU			ML-65			PM-2		
	R	H	TOTAL	R	H	TOTAL	R	H	TOTAL
Red	2.83	2.62	5.45	4.08	5.00	9.08	2.37	2.45	4.82
Far Red	5.22	7.57	12.79	3.34	5.91	9.25	0.9	1.08	1.98
Continuous light	4.14	3.25	7.39	2.45	4.22	6.67	3.27	3.21	6.48
Continuous dark	3.05	6.37	9.42	2.68	4.11	6.79	2.83	4.01	6.84

R = Radicle H = Hypocotyl

Table 12.3: Effect of continuous light, continuous dark and spectral composition on seedling fresh weight of SU, ML-65 and PM-2 varieties of *Vigna radiata* (L.) Wilczek

Treatment \ Attributes	Seedling length after 96 hours (in cm)								
	SU			ML-65			PM-2		
	R	H	TOTAL	R	H	TOTAL	R	H	TOTAL
Red	27.77	116.11	143.88	29.28	149.28	178.56	10.5	100	110.5
Far Red	42.00	199.00	241.00	27.85	150.00	177.85	2.00	104	106.00
Continuous light	36.11	118.88	154.99	15.71	142.85	158.56	20.00	108	128.00
Continuous dark	25.00	225.00	250.00	13.33	133.33	146.66	11.66	131.11	142.77

R = Radicle H = Hypocotyl

REFERENCES

Carpenter William J. and Ostmark E.R (1988). Sensitivity seed germination of amaryllis to light and temperature. Hort science. 23 (6 Part I): 1002-1004.

Evenari M. (1956). Seed germination. In: radiation Biology. Vol.III MacGraw-Hill, Newyork, page 518.

Heim B. and Scheafer E. (1984). The effect of Red and Far red light in the high irradiance reaction of phytochrome (Hypocotyl grown in dark grown *Sinapsis alba*). Plant cell environ. 7 (i): 39-44.

Koller R.D (1972). Environmental control of seed germination. In: seed Biology. Academic Press, New york, Page 1-101.

Malik N. and Born W.H. Wanden (1987). Germination responses of *Galium spurium* L. to light. Weed. Research. 251-258.

Mayer A.M and Poljakoff – Mayber A. (1982). The germination of seeds. 3rd edition. Parganon press, Oxford, London.

CHAPTER 13

Absolute Growth and Growth Rate of Winter Crop Guava cv. Sardar of Different Maturities during the Ontogeny

K. Bhupinder*, K. Nirmaljit, S. Tejinder, K. Harinder and K. Gursharan

Abstract

The relative growth rate (6.975) increased rapidly during the early periods of the life cycle (90-120 days) whereas 50 per cent decrease in growth rate was observed in ripened stage of guava within 30 days of maturity. The mean weight of fresh guava increased from 92.38 to 146.80 g from unripened to ripened stage. The per cent rate of increase of fresh weight, was observed during increase in maturity stage was 41.2 per cent. The mean range of shape index was 0.95-1.07 cm showed spherical shape of the guava fruit of three maturities. The property of adherence to the stalk, decreased with maturity. Very little flavour was developed in unripened as compared to the half-ripened and ripened stage of guava which had little and good flavour respectively. The firmness decreased with increase in maturity at the rate of 54.8 per cent. The external colour of guava fruit changed from dark green to greenish yellow (144A-150A) within the increase in the maturity days i.e. unripened, half-ripened and ripened guava fruit. The unripened (90 days) guava fruit was hard on pressing and crisp on chewing as compared to half-ripened (120 days) fruit, was attained whereas smoothness and softness on chewing at ripened (150 days) stage of maturity. During relative rate of growth of guava, the changes in chemical composition range from 10-11.8°B, TSS; 3.6-4.1, pH; 11.30-29.30, Sugar : acid ratio; 198.86-230.81 mg/100 g, ascorbic acid and 6.8-8.8% total sugars increased whereas 0.60-0.30%, acidity; 0.88-0.68%, pectin and 27-12.2 lb, firmness decreased with increased in maturity stages.

Key words: Guava, absolute growth, growth rate, ontogeny, texture, colour.

*Address for correspondence: Deptt. of Food Science & Technology, Punjab Agricultural University, Ludhiana

INTRODUCTION

Amongst the fruits, guava is very popular with all types of consumers (children, adult or old) but at the same time, it is highly perishable because of soft skinned fruit, get easily bruised and readily infected with various fungi during handling and transit. In recent years, the production of guava has increased in the Punjab State due to the introduction of high-yielding varieties because of the great demand from local and distant markets. Some of the progressive traders and growers have taken up its long-distance marketing but with great difficulty.

The guava growers of Punjab have been concerned with the disposal of the fruit in the market, as a fresh fruit for the consumer. The marketing of guava in the Punjab state is not organised and transportation to distant markets and other such places, is beset with hazards of spoilage and pilferage. Moreover, during the glut season, which usually commences in January for winter crop, the guava prices crash and remuneration to the farmers is hardly tempting.

Hence, attempts have been made in the Department of Horticulture, PAU, Ludhiana to produce fresh, firm and mature guava fruit cv. Sardar which may be a possible alternative for efficient utilisation of fresh fruits because guava fruit is rich in essential vitamins, minerals and other nutritive factors and command universal appeal (Table 13.1). The dietetic value of real fresh fruit is far greater than that of processed. In a study conducted by the Director of Central Institute of Horticulture for Northern Plains on the Economics of Horticultural crop cultivation and assessment of post-harvest losses of fruits and vegetables, the loss of fresh guava in marketing channel was 14.32 per cent and only 65.37 per cent of total guava harvested reached to the consumer in healthy condition, and at processing level, 80 per cent product was destined to defence as a result of processing, the firm could recover 45 per cent guava slices and 30 per cent guava pulp while 15 per cent peel and 10 per cent seeds were dumped as waste. Substantial losses during the different stages of marketing is a common occurrence in case of guava and are result of a number of contributing factors such as:

- The overripe condition of the packed fruits.
- The unsuitability of the container.
- The use of imperfect packaging methods followed by mode of conveyance used for transport.
- The number of trans-shipments and handling during transit.
- The delay in the movement of the fruit from the despatching to consuming centre.
- The lack of adequate and proper storage facilities at the centre of consumption.

So, this perishable nature of guava fruit requires that their entire movement from their area of production to the ultimate consumer is carried out as expeditiously as possible so that the fruit may remain reasonably fresh by the time it reaches to the consumer. Therefore, in order to increase the marketing period:

- avoid economic losses, by providing an effective and cheap storage systems to slow down biological activity of the product by maintaining the lowest temperature that will not cause freezing or chilling injury.
- slow down the growth of microorganism by maintaining low temperature and minimise surface moisture.

Table 13.1: Food composition of guava fruit/100 g

	Guava, Country	*Guava, Hill*
Moisture (g)	81.7	85.3
Protein (g)	0.9	0.1
Fat (g)	0.3	0.2
Minerals (g)	—	0.7
Fibre (g)	—	5.2
Carbohydrate (g)	11.2	9.0
Energy (K cal)	51	38.0
Calcium (mg)	10	50.0
Phosphorus (mg)	20	28.0
Iron (mg)	0.2	0.27
Vitamins		
Thiamine (mg)	0.03	0.02
Riboflavin (mg)	0.03	0.02
Niacin (mg)	0.4	0.3
Vitamin C (mg)	212	15
Minerals & Trace Elements		
Magnesium	24	
Sodium	5.5	
Potassium	91	
Copper	0.14	
Manganese	0.14	
Zinc	0.16	
Chromium	0.009	
Sulphur	14	
Chlorine	4	

Source: Indian Council of Medical Research (ICMR, 1995).

- reduce product drying by reducing the differences between product and air temperature and maintaining high humidity in the storage room.

Therefore, the increased consumer demand for fresh guava fruit, there is a great need to extend the storage/shelf life of food commodities, under controlled/modified atmosphere technology which will avoid use of pesticides, additives and preservation but at the same time to provide a premium quality product to the consumer (Bhupinder *et al*, 2003).

Keeping this in view, the present study was undertaken to check absolute growth and growth rate of winter crop Guava cv. Sardar of different maturities during the ontogeny.

MATERIAL AND METHODS

Procurement of Raw Material

The freshly harvested guava cv. Sardar (L-49) is a high sugar variety of early maturity. The guava production in Punjab occupyies 60 per cent of the area. The winter season crop of guava was procured from the Experimental Horticultural Orchard, Punjab Agricultural University, Ludhiana. The guava fruit harvested on 90, 120 and 150 days of maturity after anthesis represented early, mid and late corresponding to unripe, half-ripe and ripe stages of the fruit (Table 13.2). The fruit was graded according to uniformity of size, colour, ripening, freedom from damage and blemishes and used for different studies. Twenty-five trees in the orchard were selected for studying the quality and variation in the crop.

Physical Characteristics of Guava cv. Sardar of Different Maturities

Mean weight and Shape Index

Length (cm), width (cm), girth (cm) and weight (g) of fully grown unripened and ripened healthy guavas of different maturities were determined by using ten randomly selected fruits from the bulk lot. The data was interpreted by calculating the range, mean, standard deviation (S.D.) and coefficient of variation (C.V.) of each parameter. The fruit shape, surface, external colour, flesh colour, the property of adherence to stalk, mouthfeel, flavour and taste were evaluated. The yielding pressure on guava fruit was measured by thumb pressure and pressure tester/fruit tester (McCormick FT-327) whereas the shape index was determined by dividing polar diameter (lengthwise) with transverse diameter (bredthwise).

Absolute growth and Growth rate

Absolute growth was measured on the basis of dry matter production. The fruit was dried in oven at a temperature of 100°C. The growth rate was calculated by the formula:

$$R/100 = \log_e W_2 - \log_e W_1$$

where W_1 and W_2 are the weights of the two successive maturities.
R is growth rate
e is the base of the natural logarithm

Colour

The colour of the pulp was measured by Universal Loviband Tintometer whereas external colour of the fruit was compared with the R.H.S. (The Royal Horticulture Society, London) colour chart (Wilson, 1938) (Table 13.3).

Texture

Texture studies were conducted on the fresh and stored guava of different maturities with the help of Pressure Tester (McCormick Fruit Tester FT-327). The range of fruit tester was from 1.5 kg/cm^2 (3 1b) to 12 kg/cm^2 (27 1b) and the unit of measurement was kg/cm^2 or 1b.

Chemical Composition of Guava cv. Sardar of Different Maturities

Freshly harvested guava fruit of three different maturities (unripe, half-ripe and ripe) were washed, peeled and cut into small pieces with a stainless steel knife and macerated in a pestle and mortar. Weighed aliquots were drawn for the chemical analysis according to AOAC method, 1985.

Table 13.2: Harvesting stages of winter season crop of guava cv. Sardar

Maturity	*Stage of ripening*	*Harvesting* days after anthesis*
Early (I)	Unripe (Dark green)	90
Mid (II)	Half-ripe (Light green)	120
Late (III)	Ripe (Greenish yellow)	150

*Harvested in November, 1996-97.

Table 13.3: Colour groups* according to scores and degree of variation

	Colour	
Group	*Score*	*Degree of variation***
Green	125-143	A-B-C-D
Yellow green	144-154	A-B-C-D
Green yellow	1-13	A-B-C-D

*R.H.S. (Royal Horticultural Society, London)

**Proceeding from A-D, depicts the lighter tone of the colour.

RESULTS AND DISCUSSION

The guava fruit of unripened, half-ripened and ripened stages of harvesting corresponding to early, mid and late maturity during ontogeny were analysed for mean weight, shape index, absolute growth and growth rate, colour, textural behaviour and sensory quality of fresh guava.

The results of this study are presented and discussed as under:

Physico-chemical Composition of Guava cv. Sardar of Different Maturities

Physical Characteristics

The physical characteristics of fresh guava fruits of three different maturities graded as unripe (I), half-ripe (II) and ripe (III) stages were measured and the results are summarised in Table 13.4-13.6.

The mean weight of guava fruit cv. Sardar at early, mid and late representing 1st, 2nd and 3rd maturities showed significant variation. The mean weight of 10 randomly selected guavas at 1st, 2nd and 3rd maturity was 92.38, 128.61 and 146.82 g, respectively. The coefficient of variation was higher i.e. 10.04 per cent at 1st maturity stage than 6.52 per cent at 2nd maturity and 3.35 per cent at 3rd maturity stage. The present studies are in accordance with the results reported by Singh *et al* (1995) regarding fruit weight i.e., 208 g in winter season crop, but contradicts the results of Kundu *et al* (1995) which showed 81.5 g weight of winter guava fruit.

Shape index as measured by polar/transverse diameter of fruits showed significant variation. The value of 1st, 2nd and 3rd maturity of winter season crop ranged from 0.97-1.06, 0.97-1.07 and 0.95-1.02 cm as compared to mean value 0.99, 1.01 and 0.99 cm, respectively (Table 13.4). These observations truly represent the spherical shape of fruit cv. Sardar (Lucknow-49) as described by Teotia and Pandey (1962).

The external colour of guava fruit changed from dark green to greenish yellow (Table 13.5) with the increase in the maturity days i.e., unripe guava fruit has dark green colour whereas half-ripe has light green and ripe fruit has greenish yellow colour which are in accordance with the Table 13.2 and 13.3.

The property of adherence to the stalk was also decreased with increase in the maturity day. Unripe (90 days) and half-ripe (120 days) guava had very strong adherence to stalk but ripe fruit (150 days) had moderate strong adherence to the stalk. Unripe and half-ripe guava yielded no pressure to thumb whereas the ripe guava yielded pressure to thumb which indicates the firmness of the fresh fruit of guava of different maturities (Table 13.5).

The unripe fruit (90 days) was hard and crisp on chewing as compared to half-ripe fruit (120 days) which was less hard but was crisp during mouthfeel. At ripe stage of maturity, the guava fruit became smooth and slight soft on chewing.

During increase in the maturity stage, the flesh colour changed from greenish white colour to greenish cream at unripe stage to half-ripe stage and then from greenish cream to creamish white from half-ripe stage to ripe stage.

Very little flavour was developed in the unripe stage as compared to half-ripe and ripe stage which had little and good flavour, respectively. The changes in taste were observed organoleptically from astringent to sweet, with increase in maturity stage from 90 days to 150 days. Similarly, increase in weight was also observed with increase in the maturity period as it increased 39.2 per cent from unripe to half-ripe stage, as the mean values were 92.38 to 128.61 g. Thereafter, it further increased from 128.61 g on half-ripe stage to 146.82 g on ripe stage, with per cent increase of 14.15.

The rate of increase of fresh weight of guava during each month on unripe, half-ripe and ripe stages was 1.02, 1.20 and 0.60 g/day, respectively. The per cent rate of increase of fresh weight of guava was observed during increase in maturity stage from unripe to ripe stage was 41.2 per cent.

The firmness decreased with increase in maturity days. The firmness decreased from unripe to half-ripe stage was 27.0 1b to 20.0 lb with per cent decrease of 25.9. Thereafter, further decrease in firmness was observed from half-ripe stage to ripe stage, with per cent decrease of

Table 13.4: Mean weight and shape index of fresh fruit of guava cv. Sardar of different maturities

Maturity/ Stage	*Weight, g*			*Shape Index*			*Visual*	
	Range	*Mean ± S.D.*	*CV%*	*Range*	*Mean ± S.D.*	*CV%*	*Shape*	*Colour*
Unripe (I)	74.62-103.50	92.38 ± 9.27	10.04	0.98-1.06	0.99 ± 0.03	3.06	Round	Dark green
Half ripe (II)	103.50-152.48	128.61 ± 4.52	3.52	0.97-1.07	1.01 ± 0.03	3.13	Round	Light green
Ripe (III)	140.84-155.58	146.82 ± 12.75	6.69	0.95-1.02	0.99 ± 0.01	1.91	Round	Greenish yellow

Table 13.5: Physical characteristics of fresh fruit of guava cv. Sardar of different maturities

Age Days	*Maturity*	*External colour*	*Adherence to stalk*	*Whether yielding to pressure*$^{+}$	*Mouthfeel*	*Flesh colour*	*Flavour*	*Taste*	*Weight of fruit**	*Rate of increase of fresh weight during each month (g/day)*	*Pressure test (lb)*
90	Unripe	Dark green	Very strong	Not yielding	Hard and crisp	Greenish white	Very little	Astrin-gent	92.38	1.02	27.0
120	Half-ripe	Light green	Very strong	Not yielding	Less hard but crisp	Greenish cream	Little	Slightly astringent	128.61	1.20	20.0
150	Ripe	Greenish yellow	Moderate strong	Yielding	Slight soft and smooth	Creamish white	Good	Sweet	146.82	0.60	12.2

*Mean of 10 fruits

$^{+}$Thumb pressure leaves mark on the fruit surface

39.0 as the mean values decreased from 20.0 lb to 12.2 lb. The results are in accordance with the result of Rodriquez *et al* (1971).

The relative growth rate increased rapidly during the early periods of the life-cycle i.e., during unripe and half-ripe stage whereas 50 per cent decrease in growth rate was observed in ripe stage within 30 days (Table 13.6). The increased growth rate during the early stages may be attributed with an increased activity of the rapidly multiplying cells in the formation of new tissues necessitating a rapid supply of nourishment from the tree. The studies are in accordance with the study conducted by Singh *et al.* (1937).

Chemical Composition

The chemical composition of absolute growth and growth rate of winter crop guava cv. Sardar of different maturities during ontogeny are presented in Tables 13.7–13.8.

Total soluble solids

The total soluble solids showed gradual increase from 90 days to 150 days of maturity. The TSS values of guava fruit at 1st, 2nd and 3rd maturity were 10.0, 11.0 and 11.8°B, respectively. Similar increase in TSS has been reported by Gulhane and Gupta (1974). The increase in amount of TSS is seen to be directly related with the increase in weight of the fruits as non-accumulation of carbohydrates during the earlier period of fruit growth on account of their simultaneous utilisation for the vegetative growth. The TSS values of guava fruit within maturities differed significantly.

Table 13.6: Absolute growth and growth rate of winter crop guava cv. Sardar of different maturities during the ontogeny*

Stage	*Age days*	*Dry weight %*	*Relative growth rate R*
Unripe	90	14.81	—
Half-ripe	120	15.88	6.975
Ripe	150	16.43	3.404

*The lifecycle of a single organism, biological development of the individual : distinguished from phylogeny.

Table 13.7: Proximate chemical composition of fresh fruit of guava cv. Sardar of different maturities

Parameters	*Maturities*			
	Unripe (I)	*Half-ripe (II)*	*Ripe (III)*	*CD (5%)*
TSS, ªBrix	10.0	11.0	11.8	0.73
pH	3.6	3.9	4.1	NS
Acidity, as citric acid, %	0.60	0.45	0.30	0.07
Sugar: Acid ratio	11.30	16.88	29.30	0.72
Ascorbic acid, mg/100 g	198.86	225.60	230.81	5.22
Total Sugars, %	6.8	7.6	8.8	0.73
Reducing Sugars, %	3.6	5.1	6.8	0.73
Non-reducing Sugars, %	3.2	2.5	2.0	0.52
Pectin, as ca-pectate, %	0.88	0.98	0.68	0.07

Table 13.8: Colour and textural studies of fresh fruit of Guava cv. Sardar of different maturities

Maturity	Tintometer Colour, TU*				Firmness
	Visual Colour	Red	Yellow	Blue	1b (Kg/ Cm^2)
Unripe (I)	144 – A	3.0	86.6	4.0	27.0 (12.8)
Half-ripe (II)	144 – B	2.0	30.0	3.0	20.0 (9.4)
Ripe (III)	150 – A	1.0	30.0	1.5	12.2 (5.5)
CD (5%)	-	0.73	3.22	0.90	5.24

*Tintometer Units

Acidity and pH

The acids showed a decreasing trend at three maturity stages. The mean value of titrable acidity was 0.60, 0.45 and 0.30 at 1st, 2nd and 3rd maturity, respectively. Results showed significant difference within the three maturities. In the same way pH increased with decrease in acidity within maturities. Total acids showed a decreasing trend with increase in ripening stage. An inverse relationship between the sugar and acid indicated that organic acids were converted to sugars during fruit ripening. Statistically significant results were obtained in acidity and pH within the maturities.

Sugar : Acid Ratio

The sugar acid ratio increased with the increase in the maturity stage as at unripe (1st maturity), half-ripe (2nd maturity) and ripe (3rd maturity) stage, the values of sugar : acid ratio were 11.3, 16.8 and 29.3, respectively. The results showed significant difference within the maturities.

Total Sugars

The trend in the change of the concentration of total sugars as well as reducing sugars was found to be same i.e., they showed continuous increase till the fruit was matured i.e., from 90 days to 150 days of harvesting stage. The mean value of total sugars at unripe (1st maturity), half-ripe (2nd maturity) and ripe (3rd maturity) stage was 6.8, 7.6 and 8.8, respectively. This gradual increase in the level of sugar in the mesocarp of fruits was mainly due to the augmented translocation of photosynthesis from leaves. With increase in ripening stage, starch content started declining till the fruit attained full maturity. This decrease in starch content may be attributed to its conversion into sugars and was reflected in higher corresponding total soluble solids. Similar trend was observed in reducing sugars i.e., 3.6, 5.1 and 6.8 at 1st, 2nd and 3rd maturity, respectively (Dhillon *et al*, 1987). Statistically significant results were obtained in sugars within the three maturities.

Pectin

Pectin content increase was maximum at firm ripe stage of maturity and decrease with advancement of maturity followed by softening. The mean value of pectin at unripe (1st maturity), half-ripe (2nd maturity) and ripe (3rd maturity) stage was 0.88, 0.98 and 0.68, respectively. The results showed significant difference within the three maturities.

Ascorbic Acid

The ascorbic acid content of guava fruit increased steadily in three maturity stages. It appears that the active synthesis of ascorbic acid commended with the beginning of maturity and it increased rapidly till ripening. In the unripe (1st maturity), half-ripe (2nd maturity) and ripe (3rd maturity) guava fruit, the mean value of ascorbic acid was 198.86, 225.60 and 230.81 mg/100g, respectively. The ascorbic acid content of guava fruit showed significant effect within the three maturity stages. Similar results have been observed by Sachan *et al* (1969).

Colour

According to Horticulture Colour Chart, the colour of the guava fruit at 1st, 2nd and 3rd maturity was 144-A, 144-B and 150-A, respectively. In Tintometer Units, the Red, Yellow and Blue colour Units of 1st, 2nd and 3rd maturity was 3.0, 86.6 and 4.0; 2.0, 30.0 and 3.0 and 1.0, 30.0 and 1.5, respectively. The change in colour from green to light yellow of fruits at maturity gave an indication of decomposition of chlorophyll with the start of the maturity (Gangwar, 1972). The Red, Yellow and Blue colour Units of Tintometer differed significantly with different maturity stages of guava fruit.

Firmness

The firmness of the guava fruit decreased with increase in the maturity stage as at unripe (1st maturity), half-ripe (2nd maturity) and ripe (3rd maturity) stage, the values of firmness were 27.0 lb (12.8 kg/cm^2), 20.0 lb (9.4 kg/cm^2) and 12.2 lb (5.5 kg/cm^2), respectively. The softening of fruit at maturity gave an indication that dissolution of cell wall cementing materials begin with the start of maturity. The firmness of guava fruit differed significantly within the maturity stages.

REFERENCES

A.O.A.C. (1985). Official Methods of Analysis, Association of official Analytical Chemists 13th Ed., USA.

Bhupinder K. Nirmaljit K. and Gursharan K. (2003). Temperature and Pressure effects on post-harvest storage of ripe guava (*Psidium guajava* L.). Indian Journal of Ecology. 30 (2): 228-236.

Dhillon B.S., Singh S.N. and Gill S.S. (1987). Development physiology of guava. Punjab Horticulture Journal. 27 (1-2): 28-33.

Gangwar B.M. (1972). Biochemical studies on growth and ripening of guava. Indian Food Packer 26 (5): 13-15.

Gulhane A.R. and Gupta P.K. (1974). Physico-Chemical changes in the developing Lucknow–49 Guava (*Psidium guajava, L.*) Haryana J. Hort. Sci. 3 (3 & 4): 134-139.

Kundu S., Ghosh S.N. and Mitra S.K. (1995). Physico-Chemical characters of Twelve Guava cultivars in the Laterite Tract of West Bengal. Indian Food Packer 49 (2): 11-13.

Rodriguez R., Agarwal P.C. and Saha N.K. (1971). Effect of maturity of Safeda guavas on flavour preference. Indian Food Packer 25 (1): 13-20.

Sachan B.P., Pandey D and Shankar G. (1969). Influence of weather on the chemical composition of guava fruit var., Allahabad Safeda. Punjab Horticulture Journal 9: 119-123.

Singh B.N., Seshagiri P.V. and Gupta S.S. (1937). Ontogenetic drifts in the Physiology and Chemistry of tropical fruits under orchard conditions. Indian Journal of Agricultural Sciences. 176-192.

Singh G., Sinha G.C., Pandey D. and Rajan S. (1995). Studies on the physico-chemical composition of fruits of twenty four Guava varieties. Indian Food Packer. 49 (3): 15-20.

Teaotia S.S., Pandey I.C., Agnihotri B.N. and Kapur K.L. (1962). Study of some guava varieties of Uttar Pradesh Indian Agriculture 6 (1982): 12-18.

Wilson R.F. (1938). Horticultural colour chart. Wilson Colour Ltd. in collaboration with Royal Horticultural Society and British Colour Council.

CHAPTER **14**

A Comparative Analysis of Vegetation and Soil Characteristics of Montane Broad-Leaved, Mixed Pine and Pine Forests of Northeast India

B.P. Mishra* and R.C. Laloo

INTRODUCTION

The northeast India is a treasure house of plant resources. Varied physiography, climatic conditions mainly temperature and rainfall have resulted in a wide range of vegetation from tropical to alpine. This part of the country is an extension of the eastern Himalayan complex and a hot spot of biodiversity. About 50% species of Indian flora is confined to this region only (Rao and Hajra 1986). Forest cover of this region (including Sikkim) is about 63% of the total geographic area of northeast India. It contributed about 5% to the total forest cover and about 8% to the total geographical area of India (FSI 1999).

The state of Meghalya (25° 2′-26° 7′ N and 89° 49′-92° 50′ E) occupies about 13% of the total geographic area of northeast India. With its great topographic and climatic diversity, the state has perhaps a wide range of natural vegetation communities than any other part of the country. The range of vegetation communities has resulted in very rich flora from tropical mixed deciduous type to subtropical broad-leaved and/or pine forests. Based on digital satellite data, the area under broad-leaved forest (tropical and subtropical evergreen and semi-evergreen forests and moist deciduous forests) is 48.3%. Mixed pine and pine forests are about 4.5% and 7.1% respectively, of the total geographical area of the state (Porwal *et al,* 2001). The climate of the state is monsoonic and the soil is lateritic and oxisol type (Pascoe 1950).

*Address for correspondence: Department of Botany, School of Life Sciences, North-Eastern Hill University, Shillong-793 022, India. Tel: +91-364-2722227 (O), 2535307 (R); Fax: +91-364-2722000;
E-mail: mishrabp111@yahoo.com, bpmishra111@rediffmail.com

Anthropogenic activities such as extraction of timber, collection of fuelwood, and grazing of cattle have altered the vegetational landscape of the state and as a result of this pine has gradually replaced broad-leaved species on the disturbed sites. The disturbance is one of the major factors which determines plant communities in natural ecosystems (Armesto and Pickett 1985).

Although, the vegetation and soils of Meghalaya have been studied extensively. But, there is paucity of data on plant community organisation and its impact on soil properties. Similar studies have been carried out in temperate conditions, where oak (*Quercus* spp.) forest is invaded by pine species. Rao *et al* (1990) have studied community composition and tree population structure in a subtropical broad-leaved forest along a disturbance gradient and they have reported invasion of pine in oak forest of Meghalaya.

The present study was conducted with an aim to study how colonisation of pine leads to changes in community organisation, biodiversity and soil characteristics (through litter input) and how these changes are taking place with altitude.

Study Sites

In Meghalaya, pine forests are confined between 800 and 2000 m asl. One stand each of subtropical broad-leaved, mixed pine and pine forests was selected between 1000 and 1100 m asl (Umroi area in Ri-bhoi district), and between 1900 and 2000 m asl (upper Shillong area in East Khasi hills district). The Umroi (25°40′ N and 91°54′ E) is 27 km towards northeast direction and the upper Shillong (25°32′ N and 91°54′ E) is 12 km south of Shillong, the state capital. The prevailing microenvironment differed markedly not only between the stands at two altitudes but also between three types of forests at same altitude. The annual rainfall varied greatly from Umroi (1850 mm) to upper Shillong (2880 mm) area (Tripathi 2002). The microenvironment and light interception in selected forest stands are given in Table 14.1. The abbreviations used for subtropical broad-leaved, mixed pine and pine forests at Umroi are BL-I, MP-I and P-I, and at upper Shillong are BL-II, MP-II and P-II, respectively.

METHODOLOGY

Vegetation Analysis

The field study was carried out during 2002 and 2003. Vegetation analysis was done using the methods given by Misra (1968) and Mueller-Dombois and Ellenberg (1974). For vegetation analysis, 20 quadrats (10 m × 10 m size) were laid in each stand. The frequency, density, abundance, basal area and IVI of trees were determined. Whitford index (Whitford 1948) was used for distribution pattern of tree species. The dominance-distribution of trees was determined on the basis of their IVI. Shannon diversity index (Shannon & Weaver 1949) and Simpson dominance index (Simpson 1949) were calculated.

Soil Analysis

The soil samples were collected during pre-monsoon (April to May) and post-monsoon (September to October) seasons from top layer nutrient uptake zone (0-15 cm) and sub-layer

Table 14.1: Forest microclimate of broad-leaved, mixed pine and pine forests of Meghalaya

Forest stand	Parameter						
	Air temperature (°C)		Soil temperature (°C)		Relative humidity (%)		Light interception (%)
	Pre-monsoon season	Post-monsoon season	Pre-monsoon season	Post-monsoon season	Pre-monsoon season	Post-monsoon season	
Broad-leaved-I	28.3±1.3	24.8±2.5	20.3±0.9	21.7±1.2	58±2.6	62±4.5	94±10.3
Mixed pine-I	31.4±1.5	26.5±1.7	21.8±1.2	22.3±2.3	54±2.6	59±2.5	61±8.9
Pine-I	33.7±1.1	29.6±1.9	22.0±1.5	22.8±1.6	43±2.9	42±2.7	55±9.4
Broad-leaved-II	18.9±0.9	16.4±1.7	12.9±0.7	15.7±0.7	84±3.1	98±6.5	92±12.4
Mixed pine-II	20.8±1.3	18.1±1.5	13.5±1.0	17.6±0.9	79±1.8	86±1.9	73±8.6
Pine-II	21.9±1.6	18.9±0.8	14.6±1.4	17.9±1.4	76±2.0	82±2.7	48±9.3

±, Standard error.

(15-30 cm). The samples were analysed for total kjeldahl nitrogen (TKN), available phosphorus (Molybdenum blue method) and organic carbon (Walkley and Black method) as outlined in Allen *et al* (1974) and Anderson and Ingram (1993). The C:N ratio was calculated using the organic C and total N values of soil in different forest stands. Soil moisture content was determined gravimetrically by drying 10 g of field-moist sample at 105°C for 24 hours in a hot air oven (Allen *et al,* 1974). Soil pH was measured electrometrically by digital pH meter using 1:2.5 suspension of soil and water (Anderson and Ingram 1993).

Accumulation of litter was determined by collecting litter samples from the forest floor by quadrat (0.5m × 0.5m size) method. Litter thickness was measured in each forest stand. The microenvironmental conditions such as ambient and soil temperature, humidity and light intensity were also measured during the time of sampling.

RESULTS

Tree Community Characteristics and Plant Diversity

Vegetation analysis showed that BL-II had highest (24 species) number of woody species followed by BL-I (20 species). Species richness was decreased from broad-leaved to pine forest. Tree density ranged from 1180 individuals ha^{-1} in mixed pine and pine forests to 1285 individuals ha^{-1} in broad-leaved forest at Umroi. On the contrary, it ranged between 1065 individuals ha^{-1} in broad-leaved to 1355 individuals ha^{-1} in pine forest at upper Shillong. The basal area varied from 33.2 m^2ha^{-1} in mixed pine to 45.3 m^2ha^{-1} in pine forest at Umroi and 39.6 m^2ha^{-1} in mixed pine to 50.8 m^2ha^{-1} in broad-leaved forest at upper Shillong (Table 14.2).

Shannon diversity index declined from broad-leaved to mixed pine and finally to pine forest and there was no marked variation in the values between same type of forests at two altitudes. On the contrary, Simpson dominance index showed a reverse trend, and it was increased from broad-leaved to pine forests (Table 14.2).

Table 14.2: Tree (gbh > 20cm) density, species and family richness, basal area, and diversity and dominance indices in broad-leaved, mixed pine and pine forests of Meghalaya

Parameter	Forest stands					
	Broad-leaved-I	Mixed pine-I	Pine-I	Broad-leaved-II	Mixed pine-II	Pine-II
Species richness	20	15	11	24	15	13
Density (individuals ha^{-1})	1285	1180	1180	1065	1245	1355
Basal area (m^2 ha^{-1})	33.7	33.2	45.3	50.8	39.6	42.8
Shannon diversity index	2.52	2.07	1.44	2.67	2.11	1.32
Simpson dominance index	0.12	0.21	0.43	0.12	0.22	0.50

Dominance and Distribution of Woody Species

The dominant species were determined on the basis of their importance value index (IVI). *Schima wallichii* (IVI 83.4, tree density 235 individuals ha^{-1}) in BL-I and *Phoebe attenuata* (IVI 92.3, tree density 285) in BL-II were the dominant species. *P. kesiya* replaced *Schima wallichii* of BL-I in mixed pine and pine forests. However, *Persea kingii* and *Eurya acuminata* were the co-dominant species of MP-II and P-II, respectively. Importance value index of *P. kesiya* was increased from mixed pine to pine forest (Table 14.3). The dominance-distribution curves were log-normal type in two broad-leaved forests. However, broken stick and short-hooked curves were obtained in mixed pine and pine forests (Fig. 14.1).

Majority of species showed contagious distribution that leads the forest stands patchy in nature. A few species, namely, *Persea gamblei* in P-I, and *Myrica esculenta* and *Rhus javanica* in MP-II showed regular distribution. Random distribution was seen only in case of *Eurya acuminata, Phoebe attenuata* and *Myrica esculenta* in BL-II. Two broad-leaved forests were highly dissimilar in species composition (dissimilarity index 0.68). However, forests of different kinds at same altitude were more similar. The similarity index was maximum between MP-II and P-II (0.85), followed by BL-II and MP-II (0.67) and BL-I and MP-I (0.57). *Callicarpa arborea, Eurya japonica, Persea gamblei* and *Schima wallichii* were present in all three types of forests situated at Umroi. However, *Breynia retusa, Eurya acuminata, Grewia multiflora, Lindera caudata, Litsea salicifolia, Myrica esculeta, Rhus javanica, Rhus succedaenea* and *Symplocos javanica* were common in forests at upper Shillong.

Forest Floor Characteristics

The litter thickness in different types of forests was about twofold higher at upper Shillong, and the values declined from broad-leaved to pine forest at both the altitudes. The forests at Umroi showed higher values during post-monsoon season. However, a reverse trend was recorded at upper Shillong where thickness of litter was more in pre-monsoon season. Accumulation of litter on forest floor also followed a similar trend in results (Table 14.4).

Table 14.3: Tree (gbh > 20 cm) density (individuals ha^{-1}) and basal area ($m^2\ ha^{-1}$) of component species in broad-leaved, mixed pine and pine forests of Meghalaya. (values in parenthetheses are species rank on the basis of IVI)

Species (Family)	Forest stand											
	Broad-leaved-I		Mixed pine-I		Pine-I		Broad-leaved-II		Mixed pine-II		Pine-II	
	Density Area	Basal	Density Area	Basal	Density Area	Basal	Density Area	Basal	Density Area	Basal	Density Area	Basal
Ardisia floribunda Wall in Roxb. (Myrsinaceae)	–	–	–	–	–	–	35	0.34 (14)	–	–	–	–
Breynia retusa (Dennst.) Alst. (Euphorbiaceae)	–	–	–	–	–	–	15	0.22 (23)	10	0.11 (15)	10	0.04 (13)
Bridelia retusa (Linn.) Spreng. (Euphorbiaceae)	60	0.62 (8)	30	0.12 (12)	–	–	–	–	–	–	–	–
Callicarpa arborea Roxb. (Verbenaceae)	145	3.36 (4)	20	0.22 (11)	20	0.14 (11)	–	–	–	–	–	–
Castanopsis indica A. DC in Seem (Fagaceae)	–	–	–	–	–	–	30	0.65 (16)	–	–	–	–
Castanopsis kurzii (Hance.) Biswas (Fagaceae)	80	1.18 (6)	15	0.11 (14)	–	–	20	0.63 (15)	–	–	–	–
Castanopsis tribuloides (Sm.) DC. (Fagaceae)	140	4.0 (3)	25	0.23 (13)	–	–	45	1.33 (9)	–	–	–	–
Cinnamomum sp. (Lauraceae)	20	0.16 (17)	15	0.1 (15)	–	–	–	–	–	–	–	–
Daphniphyllum himalense (Daphiphyllaceae)	–	–	–	–	–	–	25	1.54 (13)	–	–	–	–
Erythrina stricta Roxb. (Fabaceae)	–	–	85	0.98 (4)	–	–	–	–	–	–	–	–
Eurya acuminata DC. (Theaceae)	30	0.43 (5)	30	0.72 (3)	20	0.08 (10)	40	0.56 (5)	95	0.51 (7)	95	0.33 (2)
Ficus hirta Vahl. (Moraceae)	–	–	–	–	30	0.19 (8)	–	–	–	–	–	–
Fucus gibbosa Bl. Bijdr. (Moraceae)	–	–	–	–	45	0.37 (4)	–	–	–	–	–	–
Garcinia cowa Roxb. Ex DC. (Clusiaceae)	170	3.8 (2)	–	–	–	–	–	–	–	–	–	-

(Table 14.3 Contd.)

(Table 14.3 Contd.)

Grewia multiflora Juss. (Tiliaceae)	–	–	–	–	–	–	15	0.29 (22)	25	0.30 (14)	20	0.12 (8)
Itea chinensis Hk.f. & Arn. (Iteaceae)	35	0.32 (11)	–	–	–	–	–	–	–	–	–	–
Itea macrophylla Wall. (Iteaceae)	30	0.2 (15)	–	–	–	–	–	–	–	–	–	–
Knema linifolia (Roxb.) Warb. Mon. Myrist. (Myristicaceae)	–	–	–	–	–	–	40	1.99 (6)	–	–	–	–
Kydia calycina Roxb. Pl. Cor. (Malvaceae)	–	–	60	0.84 (8)	–	–	–	–	–	–	–	–
Lindera caudata Benth. (Lauraceae)	65	0.96 (7)	–	–	–	–	25	0.69 (18)	35	0.56 (12)	15	0.12 (9)
Lindera latifolia Hk.f. (Lauraceae)	–	–	–	–	–	–	60	2.47 (3)	–	–	–	–
Litsea monopetala (Roxb.) Pers. (Lauraceae)	30	0.38 (13)	30	0.34 (10)	–	–	45	1.42 (8)	–	–	–	–
Litsea salicifolia (Roxb. ex Nees.) Hk.f. (Lauraceae)	–	–	–	–	–	–	–	–	95	1.33 (6)	25	0.11 (6)
Lyonia ovalifolia (Wall.) Drude (Ericaceae)	35	0.36 (10)	–	–	–	–	20	0.50 (19)	60	0.8 (10)	–	–
Morinda angustifolia Roxb. Pl. Corom. (Rubiaceae)	10	0.11 (19)	–	–	25	0.12 (9)	–	–	–	–	–	–
Morus indica Thunb. (Moraceae)	–	–	–	–	–	–	25	0.87 (17)	85	1.03 (4)	–	–
Myrica esculenta Buch.-Ham. ex D.Don. (Myricaceae)	–	–	–	–	–	–	105	5.75 (2)	75	0.95 (3)	35	0.23 (4)
Neillia thyrsiflora D.Don. (Rosaceae)	–	–	–	–	–	–	–	–	25	0.35 (13)	15	0.08 (10)
Persea gamblei (King ex Hk.f.) Kosterm. (Lauraceae)	25	0.5 (14)	25	0.67 (6)	40	0.37 (3)	–	–	90	2.49 (2)	–	–
Persea kingii (Hk.f.) Kosterm. (Lauraceae)	–	–	–	–	–	–	35	1.10 (10)	–	–	–	–

(Table 14.3 Contd.)

(Table 14.3 Contd.)

Phoebe attenuata (Nees) Syst. Laurin. (Lauraceae)	–	–	–	–	–	–	285	24.1 (1)	–	–	–	–
Pinus kesiya Royle (Pinaceae)	–	–	390	19.4 (1)	790	42.3 (1)	–	–	465	28.0 (1)	990	40.82 (1)
Pithecellobium angulatum Benth. (Mimosaceae)	10	0.15 (20)	–	–	–	–	–	–	–	–	–	–
Prunus cerasoides D.Don. Prodr. (Rosaceae)	–	–	–	–	–	–	35	0.94 (7)	85	1.25 (5)	–	–
Quercus dealbata Hk.f & Th. ex Miq. (Fagaceae)	25	0.16 (16)	–	–	–	–	–	–	–	–	–	–
Quercus griffithii (Fagaceae)	–	–	–	–	–	–	50	2.55 (4)	–	–	–	–
Rhododendron arboreum Smith. (Ericaceae)	–	–	–	–	–	–	–	–	–	–	35	0.22 (5)
Rhus javanica Linn. (Anacardiaceae)	–	–	75	0.61 (7)	40	0.17 (7)	15	0.37 (21)	40	0.31 (9)	15	0.06 (11)
Rhus succedaenea Linn. (Anacardiaceae)	30	0.44 (12)	–	–	45	0.26 (5)	20	0.41 (20)	60	1.01 (8)	10	0.04 (13)
Saurauia roxburghii Wall. (Saurauiaceae)	–	–	55	1.18 (5)	–	–	–	–	–	–	–	–
Schima wallichii (DC.) Korth. (Theaceae)	235	15.8 (1)	170	7.01 (2)	85	0.98 (2)	10	0.34 (24)	–	–	75	0.57 (3)
Symplocos glomerata King. Ex Cl. (Symplocaceae)	–	–	–	–	–	–	40	0.89 (11)	50	0.6 (11)	15	0.07 (7)
Symplocos javanica (Bl.) Kurz. (Symplocaceae)	–	–	–	–	–	–	30	0.86 (12)	–	–	–	–
Syzygium balsameum (Wt.) Wall ex Am & Sm. (Myrtaceae)	40	0.56 (9)	45	0.63 (9)	–	–	–	–	–	–	–	–
Trevesia palmata (Roxb.) Vis. (Araliaceae)	15	0.17 (18)	–	–	–	–	–	–	–	–	–	–
Wendlandia ligustrina (Wall ex Hk.f.) DC Prodr. (Rutaceae)	–	–	–	–	40	0.33 (6)						

Abbreviation:- = absent.

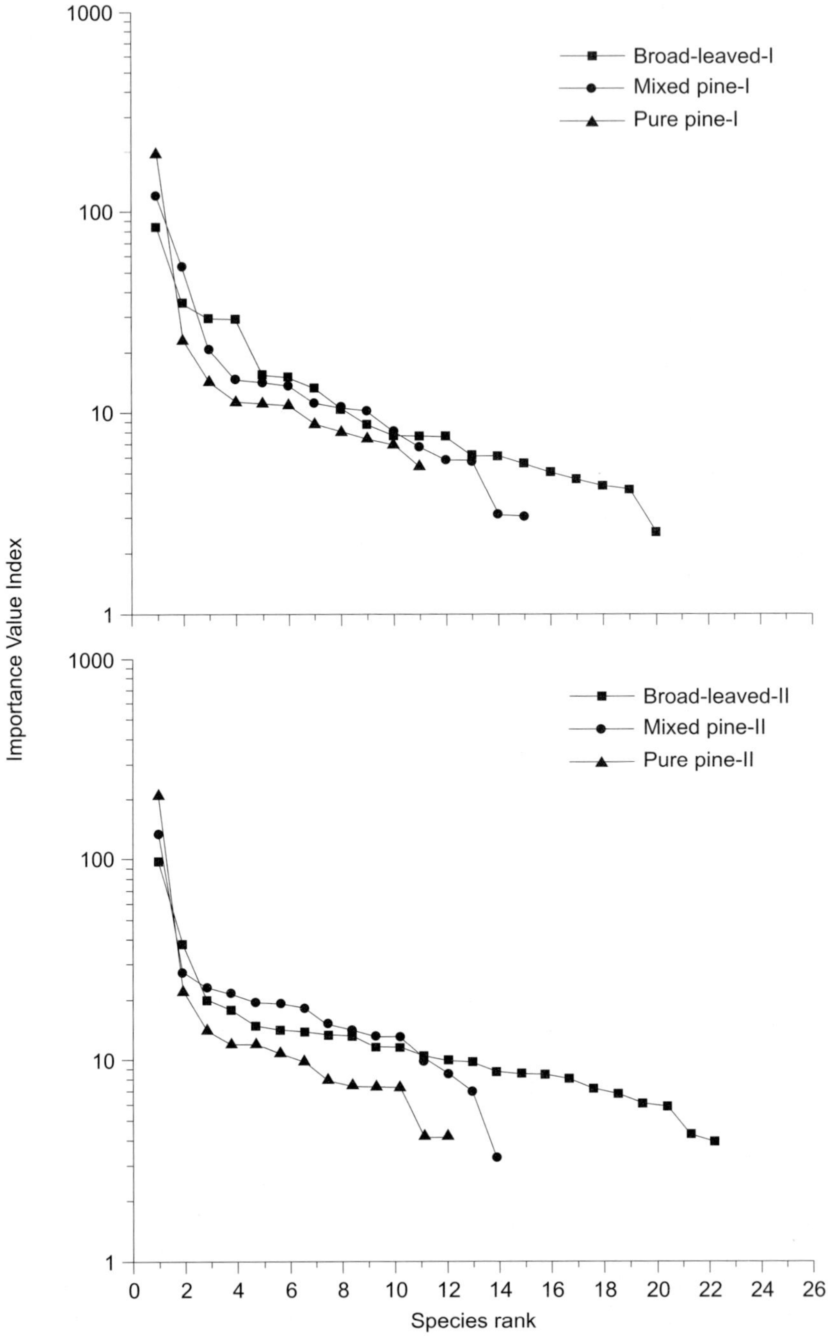

Fig. 14.1: Dominance-distribution pattern of trees in the broad-leaved, mixed pine and pine forests

Table 14.4: Forest floor characteristics in broad-leaved, mixed pine and pine forests of Meghalaya

Forest stand	*Parameter*			
	Litter thickness (cm)		*Litter accumulation (kg ha^{-1})*	
	Pre-monsoon season	*Post-monsoon season*	*Pre-monsoon season*	*Post-monsoon season*
Broad-leaved-I	1.9 ± 0.4	2.3 ± 0.6	7835 ± 54	5732 ± 89
Mixed pine-I	1.6 ± 0.7	2.0 ± 0.4	5764 ± 72	4117 ± 67
Pine-I	1.0 ± 0.5	1.2 ± 0.5	2612 ± 49	2942 ± 54
Broad-leaved-II	4.5 ± 0.9	3.9 ± 0.6	10450 ± 84	8232 ± 79
Mixed pine-II	3.9 ± 1.2	3.2 ± 0.4	8540 ± 96	7413 ± 86
Pine-II	2.7 ± 0.7	2.4 ± 0.9	5490 ± 72	3203 ± 94

Soil Characteristics

Soil moisture content was markedly higher in forests at upper Shillong. Top-soil had more moisture content than sub-soil, and greater values were recorded during post-monsoon season. There was a marked variation in values from top-soil to sub-soil in the forests at upper Shillong. The soil pH was in acidic range and acidity decreased from top-soil to sub-soil during pre-monsoon season, and increased from top-soil to sub-soil during post-monsoon season. Top-soil had higher Organic C in both the seasons, however, TKN and available phosphorus contents were high in top-soil during pre-monsoon season and in sub-soil during post-monsoon season. The C:N ratio varied from 8.8 to 16.6 (top-soil) and 7.1 to 12.6 (sub-soil) in the forests at Umroi, and from 12.1 to 13.9 (top-soil) to 9.2 to 15.4 (sub-soil) in the forests at upper Shillong. The soil properties are shown in Table 14.5.

DISCUSSION

The present study revealed that pine gradually replaces broad-leaved species from broad-leaved to mixed pine and finally to pine forests. As a result of this, there was a sharp decline in species richness from broad-leaved to pine forests. Species richness (0.2 ha area) of two broad-leaved forests is comparable with tropical semi-evergreen forests of Eastern Ghats, India (Kadavul and Parthasarathy 1999). Greater species richness in three types of forests at upper Shillong (high altitude) could be due to cumulative effect of topography and favourable edapho-climatic conditions. The lowest tree density and highest basal area in BL-II was mainly attributed by the presence of a large number (about 45%) of trees with gbh >100 cm. *Phoebe attenuata*, the dominant species of BL-II contributed about 27% and 47% of total tree density and basal area, respectively. This indicated presence of a large number of old trees. Decreased diversity index from broad-leaved to mixed pine and finally to pine forests was mainly due to increased dominance of *P. kesiya*. In pine forests vegetation was homogenous and poor in species content. Connel and Lowman (1989) who had also suggested that in species poor and low diversity forest about 50-80% of the canopy trees are represented by only one tree species. Increased dominance index from broad-leaved to pine forest indicated high dominance and low equitability in later case.

Table 14.5: Soil characteristics of broad-leaved, mixed pine and pine forests of Meghalaya

Parameter	Soil Depth (cm)	Forest stands											
		Pre-monsoon			Post-monsoon			Pre-monsoon			Post-monsoon		
		Broad-leaved-I	Mixed pine-I	Pine-I	Broad-leaved-I	Mixed pine-I	Pine-I	Broad-leaved-II	Mixed pine-II	Pine-II	Broad-leaved-II	Mixed pine-II	Pine-II
Soil moisture content (%)	0-15	27 ±2.3	22 ±1.9	18 ±2.1	39 ±3.2	36 ±4.6	39 ±2.7	54 ±2.9	45 ±1.7	42 ±2.3	72 ±4.9	69 ±5.4	58 ±2.7
	15-30	21 ±1.6	17 ±2.0	12 ±0.8	24 ±1.9	27 ±2.5	28 ±2.3	42 ±1.9	34 ±1.8	29 ±1.3	52 ±1.8	46 ±2.4	41 ±1.7
Soil pH	0-15	4.3 ±0.10	4.9 ±0.14	5.0 ±0.10	5.4 ±0.12	5.5 ±0.10	5.2 ±0.09	4.8 ±0.07	4.6 ±0.07	5.1 ±0.17	5.1 ±0.06	5.0 ±0.09	5.4 ±0.12
	15-30	4.6 ±0.09	5.2 ±0.13	5.4 ±0.10	4.9 ±0.09	4.9 ±0.14	5.0 ±0.08	5.3 ±0.17	5.2 ±0.14	5.9 ±0.18	4.9 ±0.04	4.8 ±0.09	5.1 ±0.09
Organic carbon (%)	0-15	3.40 ±0.14	2.84 ±0.08	2.50 ±0.10	2.74 ±0.13	2.78 ±0.10	2.57 ±0.22	4.81 ±0.10	4.30 ±0.22	3.57 ±0.29	4.96 ±0.14	4.60 ±0.08	4.03 ±0.18
	15-30	2.39 ±0.18	2.18 ±0.13	2.10 ±0.10	2.23 ±0.14	2.32 ±0.08	2.16 ±0.19	4.40 ±0.12	3.61 ±0.26	2.68 ±0.16	4.23 ±0.07	3.91 ±0.10	3.23 ±0.19
Total kjeldahl – N (%)	0-15	0.39 ±0.01	0.24 ±0.01	0.24 ±0.01	0.30 ±0.02	0.25 ±0.02	0.22 ±0.02	0.37 ±0.01	0.32 ±0.01	0.28 ±0.01	0.38 ±0.02	0.39 ±0.02	0.30 ±0.02
	15-30	0.33 ±0.01	0.19 ±0.01	0.19 ±0.01	0.33 ±0.02	0.29 ±0.03	0.25 ±0.02	0.30 ±0.01	0.23 ±0.01	0.20 ±0.01	0.47 ±0.02	0.44 ±0.02	0.33 ±0.08
C/N ratio	0-15 13.9	8.8 ±0.3	11.8 ±0.5	10.6 ±0.6	9.9 ±1.2	11.8 ±1.1	16.6 ±1.3	12.9 ±0.6	13.6 ±0.6	13.1 ±1.3	13.4 ±1.0	12.1 ±0.8	13.9 ±1.0
	15-30	7.1 ±0.4	11.5 ±0.9	11.7 ±1.1	6.1 ±0.3	8.4 ±0.6	12.6 ±0.8	14.9 ±0.6	15.4 ±0.9	13.8 ±0.8	9.2 ±0.6	9.1 ±0.5	10.3 ±1.2
Available – P – $\times 10^{-2}$ (%)	0-15	0.28 ±0.01	0.22 ±0.02	0.16 ±0.02	0.32 ±0.02	0.33 ±0.02	0.22 ±0.03	0.25 ±0.02	0.32 ±0.02	0.27 ±0.02	0.43 ±0.03	0.45 ±0.03	0.31 ±0.04
	15-30	0.26 ±0.02	0.21 ±0.02	0.12 ±0.02	0.35 ±0.01	0.38 ±0.02	0.26 ±0.04	0.19 ±0.01	0.28 ±0.03	0.23 ±0.01	0.52 ±0.03	0.48 ±0.02	0.37 ±0.03

The log-normal dominance-distribution curves in two broad-leaved forests suggested stability of the community. The broken stick and short hooked curve in mixed pine and pine forests indicated loss of species due to disturbance. The dominance-distribution curves depicted that distribution of species was more uniform in broad-leaved forests than mixed pine and pine forests. The IVI in the broad-leaved forests was distributed more or less evenly among the different species i.e. high equitability or low dominance, indicating that these forests are undisturbed. While disturbance resulted in low equitability or high dominance in rest two types of forests and it increased from mixed pine to pine forest (Rao *et al,* 1990). Mishra *et al,* (2004) have also reported log-normal dominance-distribution curve in the undisturbed and short-hooked curve in the highly disturbed forest stands of a sacred grove of Meghalaya. Predominance of contagious distribution showed interaction of biotic and micro-climatic conditions acting together on the population. The findings of the present study are in conformity with the work of Jamir (2000). Clumping is the characteristic of natural forests of tropical and temperate climate. Variation in community composition between same type of forests at two altitudes was mainly due to differed microenvironment (mainly temperature, humidity and rainfall) and altitude.

The nutrient release from litter through natural decomposition process is recognised as an important part of the nutrient cycle whereby essential mineral elements tied up in the plant biomass are made available for further plant growth (MacLean and Wein 1978). Low temperature adversely affected microbial activity in the forests at upper Shillong (high altitude). As a result of this, litter thickness and its accumulation on forest floor of these forests were markedly high. Litter layer plays an important role in forest growth on soils (Perala and Alban, 1982) and leaf litter accumulation enhances soil chemicals (Singh and Mudgal 2000). The finding of present study was supported by the above statement.

Soil pH determines microbial growth and activity (Alexander 1977). The pH (4.3-6.0) recorded during the present study indicated high microbial biomass and it was more favourable in the forests at upper Shillong. Arunachalam and Pandey (2003) have also argued that the pH range of 4.5-6.0 supports optimum bacterial and fungal growth. Availability of organic C is important in controlling nutrient cycling and soil biological activity. Greater organic C content in soils at upper Shillong indicated more microbial biomass, which may be due to increased moisture content (Arunachalam *et al,* 1996, Arunachalam and Pandey 2003). Fine particles (clay) help in retention of more organic C and TKN in soil of upper Shillong than sandy soil of Umroi. This result was in conformity with the work of Arunachalam and Pandey (2003). Increased TKN in broad-leaved forest depicted high rate of litter decomposition, the decomposition rate is more rapid on nitrogen-rich site (Gosz 1981, Prescott 1995). Nutrient enrichment in broad-leaved forests indicated enhanced rate of decomposition of mixed litter, which could be attributed to its diverse chemical composition as compared to single-species litter. Chemical composition of the leaf litter determines decomposition rate (Vitousek *et al,* 1994). Conn and Dighton (2000) and Zimmer (2002) have also opined that nutrient supply to the soil due to decomposition of litter is enhanced from monoculture to polyculture. High C:N ratio indicated low nitrogen content and/or high organic C content in the soil. Generally nutrient poor soils showed high C:N ratio (Arunachalan and Pandey 2003).

Low TKN and available phosphorus in top-soils during post-monsoon season may due to (i) high nutrient uptake from surface soils by herbaceous species and tree seedlings abundantly grown, and (ii) reduced mineralisation and/or increased nutrient loss from top-soil through leaching and runoff (Vitousek *et al*, 1982). During post-monsoon season, the most abundant ground flora were *Daphne cannabina* spp., *Eupatorium* spp., *Osbeckia* spp., *Pteris* spp., *Rubia* spp., *Rubus* spp. and *Smilax* spp. in forests at Umroi, and *Asperagus* sp., *Asplenium* sp., *Glychenia* sp., *Eriosema chinense*, *Hedera helix*, *Impatiens chinense*, *Impatiens* sp., *Jasminum* spp., *Lyonia ovalifolia*, *Melastroma malabathricum*, *Oxalis corniculata*, *Paspallum* sp., *Pogostemon strigosus*, *Rubia* spp., *Rubus* spp., *Smilax* spp. and *Viola* sp. at upper Shillong. *Cyathea* sp., *Hedychium* sp. *Lantana camara*, *Poulzolzia sanguinea* were present in the forests at both the altitudes.

REFERENCES

Alexander M. (1977). Introduction to soil microbiology. 2nd Ed. John Wiley & Sons, New York.

Allen A.E., Grimshaw H.M., Parkinson J.A. & Quarmby C. (1974). Chemical Analysis of Ecological Materials. *Blackwell Scientific Publications, Oxford.*

Anderson J.M. & Ingram J.S.I. (1993). Tropical Soil Biology and Fertility - A Handbook of Methods (2nd Ed), *CAB International*, Wallingford, U.K.

Armesto J.J. & Pickett S.T.A. (1985). Experiments on disturbance in old-field plant communities: impact on species richness and abundance. *Ecology*, 66: 230-240.

Arunachalam A., Maithani K., Pandey H.N. & Tripathi R.S. (1996). The impact of disturbance on detrital dynamics and soil microbial biomass of a *Pinus kesiya* forest in northeast India. *Forest Ecology and Management*, 88: 273-282.

Arunachalam A. & Pandey H.N. (2003). Ecosystem restoration of jhum fallows in northeast India: microbial C & N along altitudinal and successional gradient. *Restoration Ecology*, 11: 168-173.

Conn C. & Dighton J. (2000). Litter quality influences on decomposition, ectomycorrhizal community structure and mycorrhizal root surface acid phosphatage activity. *Soil Biology & Biochemistry*, 32: 489-496.

F.S.I. (1999). The State of Forest Report. FSI, Ministry of Environment & Forests, New Delhi.

Gosz J.R. (1981). Nitrogen cycling in coniferous ecosystems. In: Terrestrial nitrogen cycles, F.E. Clark & T. Rosswall (ed.), *Ecol. Bull.* (Stockholm), 33: 405-426.

Jamir S.A. (2000). Studies on plant biodiversity, community structure and population behaviour of dominant tree species of some sacred groves of Jaintia Hills, Meghalaya, *Ph.D Thesis*. North-Eastern Hill University, Shillong, India.

Kadavul K., & Parthasarathy N. (1999). Structure and composition of woody species in tropical semi-evergreen forest of Kalarayan hills, eastern ghats, India. *Tropical Ecology*, 40: 247-260.

MacLean D.A. & Wein R.W. (1978). Weight loss and nutrient changes indecomposing litter and forest floor material in New Brunswick forest stands. *Canadian Journal of Botany*, 56: 2730-2749.

Mishra B.P., Tripathi R.S., Tripathi O.P. & Pandey H.N. (2003). Effect of disturbance on the regeneration of four dominant and economically important woody species in a broad-leaved subtropical humid forest of Meghalaya, northeast India. *Current Science*, 84: 1449-1453.

Mishra B.P., Tripathi O.P., Tripathi R.S. & Pandey H.N. (2004). Effects of disturbance on plant diversity and community structure of a sacred grove in Meghalaya, northeast India, *Biodiversity & Conservation*, 13: 421-436.

Misra R. (1968). Ecology Work Book, *Oxford Publishing Company*, Calcutta.

Mueller- Dombois, D. & Ellenberg H. (1974). Aims and Methods of Vegetation Ecology, *John Wiley & Sons*, USA.

Pascoe E.H. (1950). A manual of the geology of India & Burma. Geological Survey of India, *Govt. of India Publication.*

Perala D.A. & Alban D.H. (1982). Biomass, nutrient distribution and litterfall in Populus, *Pinus* and *Picea* stands on different soils in Minnesola. *Plant Soil*, 64: 177-192.

Porwal M.C., Talukdar G., Singh, H., Tripathi O.P, Tripathi R.S. & Roy P.S. (2001). Biodiversity characterization at landscape level using Remote Sensing & Geo-spatial Modeling in Meghalaya. *In:* Biodiversity & Environment: Remote Sensing and Geographic Information System Perspectives, P.S. Roy, S. Singh & A.G. Toxoleus (eds.), Indian Institute of Remote Sensing, Dehradun, 206-215.

Prescott C.E. (1995). Does nitrogen availability control rates of litter decomposition in forests? *Plant & Soil*, 168-69: 83-88.

Rao P., Barik S.K., Pandey H.N. & Tripathi R.S. (1990). Community composition and tree population structure in subtropical broad-leaved forests along a disturbance gradient. *Vegetatio*, 88: 151-162.

Rao R.R. & Hajra P.K. (1986). Floristic diversity of eastern Himalaya in a conservation perspective. Proceedings of Indian Academy of Sciences (Supplementary – Nov.), PP. 103-125.

Shannon C.E. & Weaver W. (1949). The Mathematical Theory of Communication. Urbana University, *Illinois Press.*

Simpson E.H. (1949). Measurement of Diversity. *Nature*, 163-688.

Singh J.N. & Mudgal V. (2000). Assessment of mineral content of tree leaf litter on Nokrek Biosphere Reserve and its impact on soil properties. *Tropical Ecology*, 41: 225-232.

Tripathi O.P. (2002). Study of distribution pattern and ecological analysis of major forest types of Meghalaya. *Ph.D. Thesis.* North-Eastern Hill University, Shillong, India.

Vitousek P.M., Gosz, J.R., Grier, C.C. Mellilo, J.M. & Reiners, W.A.A. 1982. Comparative analysis of potential nitrification and nitrate mobility in forest ecosystems. *Ecological Monograph*, 52: 155-177.

Vitousek P.M., Turner D.R., Parton W.J. & Sanford R.L. (1994). Litter decomposition on the Mauna Loa environmental matrix, Hawin patterns, mechanisms, and models. *Ecology*, 75: 418-429.

Whitford P.B. (1948). Distribution of woodland plants in relation to succession and clonal growth. *Ecology*, 30: 199-208.

Zimmer M. (2002). Is decomposition of woodland leaf litter influenced by its species richness ? *Soil Biology & Biochemistry*, 34: 277-284.

CHAPTER 15

Macronutrient Stresses and Their Management in Crop Plants

A.L. Singh

INTRODUCTION

In the coming 25-30 years of this century, the world population is predicted to become nearly 10 billions. Of this population, more than 60% people will be living in Asia and hence food shortage may become an important problem in future in Asia due to rapid increase in population. Though the food scarcity has been repeatedly claimed, it has been evaded by the tremendous progress of agricultural technology in several countries. Nowadays, environment-conservative or environment-harmonic agriculture is required and the research on plant nutrition has a very important role to achieve this type of agriculture with no reduction of yield. The 21st century, as wealth and prosperity of human beings all over the world, should be guaranteed by the sustainable food production through plant nutrition and agricultural development.

The economic value of any crop is determined by its yield and quality, which are the resultant in part of the grower's ability to exploit the plant genetic make-up and part of less tractable components of the environment in which it is growing. If the growing conditions provide all that the plant needs for full expression of genetic potential, yield and quality will be maximised, but in practice due to environmental constraints and cultural shortcomings this objective is rarely achieved. However, within the constraints posed by site and climate the genetic potential of a crop can be exploited to a greater extent by maintaining adequate water and mineral supply which often become the limiting factors for most of the crops in the world. There are records for exceptionally high yield such as 'Melhite Kenye', a rare rice species, gave a record yield of 21 t/ha in Nagaland, India (Agricomplex Newsletter, 2001). The achievable yield potential for many crops has been worked out which exceeds 7 t/ha for oil seeds, 10 t/ha

National Research Centre for Groundnut, P.B.5, Junagadh 362001, India
E-mail: alsingh@nrcg.res.in and alsingh_ad1@sancharnet.in

for cereals, 20 t/ha for maize and 80 t/ha for potatoes, but the average yield of these crops is only around one-third of these achievable yields or even less than that in most of the countries except a few developed one. The inadequate water and mineral nutrient supply is the principal factor for such a low yield and it is not economical to tolerate the nutrient deficiencies in standing crop because these can be readily and cheaply eliminated without ecological imbalance by fertiliser application.

Thus, it is high time to re-look the mineral stress aspects of crops for achieving high-yield and advocate the suitable package of practices for its improvement. The intervention of man is important and he must use the processes taking place in the soil and physiological process going on in the plant. The present chapter highlights some of the current views on the macronutrient deficiencies and toxicities and their interactions in crop plants where the emphasis has been given on the mineral deficiency symptoms, yield losses and remedial measures to enhance the yield and quality.

NUTRIENT REQUIREMENT AND THEIR STRESSES

The science of plant nutrition has made a great progress during the 20th century. It became an established scientific discipline, defining the role of chemical elements supporting plant growth, and has contributed to improvement in plant production and environmental quality. By 1900, Plant physiologists such as Knop and Sachs had established that plants require just seven chemical elements—6 required in relatively large amounts (N, P, K, Ca, S and Mg), and 7th (Fe) in small amounts. The period of 1900-1930 was the period of curiosity and by the end of 1940s established the essentiality of most of the micronutrients and the criteria for essentiality of an element was fixed (Arnon and Stout, 1939). After 1945, sophisticated water culture research by plant physiologists confirmed their old findings and added a few more beneficial elements to the list and the criteria for essentiality was recasted by Marschner (1995). Using the present criteria, the total number of chemical elements in addition to C, H, and O are 17 (three essential for some plants), although most plants require only 14.

Plants require all these elements for their growth and development. Carbon, hydrogen and oxygen, assumed to be plentiful and mainly required for photosynthesis, are taken up by the plant from air, air space of the soil and water and hence called non-mineral nutrients. The essential elements are assimilated into plant through absorption by root or other plant parts as ions from the soil and hence described as mineral nutrients. Among these nutrients N, P, K, Ca, Mg, and S are considered major or macronutrients, because they are required in large quantities that range between 1 to 150 g per kg of plant dry matter. The Fe, Zn, Mn, Cu, B, Mo and Cl are minor or micronutrients that are required at rates of 0.1 to 100 mg per kg of plant dry matter (Marschner, 1995). Chloride is essential in micro quantities but can accumulate in the plant in large quantities when present in high concentrations in the soil solution, (Xu *et al*, 2000). In addition nickel (Ni) and cobalt (Co) are also beneficial for legume crop. Also there are several evidences to include Ni to the list of essential elements. In English usage the term 'plant nutrition' generally refers only to 'mineral nutrition'—the title chosen by Mengel and Kirkby (1987).

The nutrition of crop plants and trees varies depending upon their annual or perennial nature. The dry matter accumulation in crop plants follows the growth pattern characterised by a lag phase in early growth, exponential increases in weight from vegetative to flowering stage, a linear and maximum growth rate during late vegetative to early seed filling, and leveling of weight during late seed filling stage. Similar is the trend of nutrient absorption and uptake in crop plants and its maximum amount is needed at peak stage of growth and development. However, in trees once the full growth is achieved the maximum amount is needed every year. The biological limitations for low yields in oilseeds is due to conversion of photosynthates to lipids as one gram of glucose manufactured through photosynthesis produces 0.83 g starch, 0.40 g protein and only 0.32 g of lipids. Unless these crops are provided with extra input, it is not expected to produce the same high-yield as in case of cereals. As the oilseeds and pulses are much remunerative crops, the fertiliser response in these need not be compared with those of cereals because of the large differences in the energy or market values. These responses varied with states to states and soil types.

Several scientists spread around the world have worked out the nutrient requirement of crop plants and the critical stages and tissues have been identified for assessing the nutrient status of plants (Jones *et al*, 1991; Rengel, 1993; Bergmann, 1992). The concentration of a nutrient in tissues just below the level, which gives maximum growth, is the critical concentration and above the critical concentration additional increment in the nutrient content have no particular effects on growth and the nutrient content is said to be adequate (Epstein, 1972). The growth falls off sharply below the critical concentration as the nutrient content becomes deficient. When the nutrient level exceeds critical concentration, it is no longer limiting, however, there is a limit for the adequacy and at sufficiently high level virtually all the nutrients become toxic. In most of the cases the first fully developed leaves are the main indicator for assessing the nutrient status of the plants, however, for the plants storing their reserve food underground, the analysis of underground parts becomes essential especially for the phloem immobile elements. However, concentrations of macronutrients in leaves at 60 DAE and seeds at harvest were the main indicators to assess the mineral nutrient status of groundnut crop (Singh, 1999). In macronutrients, the toxic levels are seldom achieved but are common in case of micronutrients. Hence, there is no problem of macronutrient toxicity and as these are needed in high quantity deficiency is the major problem.

Inadequate and imbalanced use of nutrient is one of the major factors responsible for low yields in crop plants worldwide. India is the world's largest producer of many crops and vegetables and depending upon the soil types, nutritional status of soil and crop varieties, the mineral disorders cause negligible to severe yield and quality losses, however, there is inadequate information on the quantification of the same in India as well as in the world. As most of these are location specific effort needs to be made to measure the yield losses due to mineral stresses in problem soils. Providing the adequate quantity of nutrients can increase 50% production and thus meeting the challenges of the future. The sulphur and iron deficiencies cause 15-29 and 14-40% yield losses in groundnut in a medium black calcareous soil (Singh *et al*, 1991a; 1995). The Fe deficiency sometimes causes complete crop failure with no pod formation in calcareous soil. The average yield increase due to application of calcium in the pegging zone was 24% (Singh, 1999). In an effort, the yield losses due to deficiencies of N, P, K, Ca, S and Mg, in groundnut grown in the calcareous soils, were 17.6, 30.7, 29, 27.1, 24.7 and 29.6%, respectively indicating that all these macronutrients are essential for harvesting high-yield in this soil (Singh, 1999). However, when all these macronutrients were combined could increase 27.4-47.5% pod yield (NRCG, 1995). The nitrogen deficiency is a major problem and non-supply of the same during sowing showed severe yellowing during the early growth stages in all crops, but in legumes as soon as the nodules were developed this yellowing was reduced.

As the fertiliser management practices and their response vary considerably from situation to situation, the fertiliser management in sole crop, sequential crop, vary and need to be studied cautiously. Majority of the crop in the world are grown as sole crop, and most of the research work on fertiliser management have been dealt considering sole crop.

MAJOR CAUSES OF MACRONUTRIENT STRESSES

Plants require all the essential nutrients in balanced proportion and deviation from this may result in mineral disorders, which may be due to deficiency or the toxicity of a particular nutrient or multi-nutrients. When two or more elements are deficient or toxic simultaneously, the composite picture of symptoms may resemble no single known symptoms. Generally, nutrient deficiency in the plants occurs when a nutrient is insufficient in the growth medium and or cannot be absorbed and assimilated by the plants due to unfavourable environmental conditions. Nutrient disorders limit crop production in all types of soil around the world. Plant grows in an environment facing different climatic conditions and soil types and, hence, there are several factors causing these nutritional deficiencies and toxicities in plants. Some of these are:

- continuous withdrawal and inadequate supply of nutrients in the soil;
- leaching run-off and nutrient fixation in the soil;
- edaphic factors (soil, water, temperature and environmental) preventing absorption of nutrients by plants;
- changes in soil physico-chemical conditions such as pH and EC;
- imbalance use of fertilisers;
- induced deficiency;
- interaction between minerals during uptake;
- antagonism;
- use of intensive nutrient (response) requiring crops and nutrient inefficient crops;
- biotic (disease and pests) factors.

These factors are interrelated and interaction between them, for the absorption and utilisation of nutrient by plants, is very complex. However, early detection of nutritional deficiency stress is important as these stresses might extend to the entire plant if relief to stress is not employed and continuous shortage of a nutrient or nutrients might cause plant death. Some of these factors are discussed here in details.

Problem Soils and Major Macronutrient Stresses

Acid and alkali are the two major problem soils causing nutritional stresses in crops. World-wide these soils occupy about two-thirds (each one-third) of the land surface area. In general, a soil with less than pH 7.0 is an acid soil, but the optimum pH for most of the plant species is below pH 7.0 and, hence, for plant growth and development the acid soils are the one whose pH is much below and may vary with plant species. Looking to the difficulties of defining acid soil on the basis of relationship between pH, exchangeable hydrogen and aluminium, soils low in pH and in responding to application of lime are considered acidic soils, but for characterisation and classification, acid soils may be defined as those having pH less than 5.5 ((Murthy *et al*, 1997). The saline and sodic soils, acid sulphate soils, degraded saline soils, coastal soils, and marshy soils are also problematic soils causing nutritional stresses in plants. The alluvial soils are neutral to alkaline, rich in P and K but poor in organic carbon and N, but the coastal alluvials are alkaline and poor in N, P and organic carbon. The black soils are generally alkaline, rich in lime Ca and Mg high CEC (40-60 meq/100 g soil) and K but poor in N, P and Organic matter.

Acid soils

Acid soils, with pH less than 5.5 in their surface zones, occupy approximately 30% (3950 m ha) of the world's land area and occur mainly in two global belts where they have developed under udic or ustic moisture regimes (Von Uexkull and Mutert, 1995). Sixty-seven per cent of the acid soils support forest and woodlands, 18% covered by grasslands and only 4.5% (179 m ha) of the acid soil area is used for arable crops (FAO, 1991). America has the world's largest acid soils accounting for 41% (1616 m ha) of world's acid soils and Asia (26.4% and 1044 m ha) has the second largest area of acid soils. Humid tropics account for 60% of the acid soils of the world. The net cultivated area of India is nearly 145 million ha and about 25 m ha soils show pH below 5.5 which are found in hill terrains such as hill tops, hill slopes, terraced or unterraced uplands, medium lands, low lands or valley floors in the Himalayan region, eastern and north-eastern plains, peninsular India and coastal plains under varying environmental conditions of landscape, geology, climate and vegetation (Murthy *et al*, 1997; Singh, 2000a).

The acid soils are deficient in Ca, P and Mg, degree of saturation of CEC is low (below 25%) and have excess Al, Mn and Fe. Poor crop growth in acid soils can be correlated directly with Al saturation. Poor fertility of acid soils is due usually to a combination of toxicities of Al, Fe and Mn and deficiencies of P, Ca, Mg and K. Besides these the acid soils have less water-holding capacity, susceptible to crusting erosion and compaction making them less productive. Liming is a viable solution and is cost effective. As these are not easily rectifiable the concepts of fitting plant to soil may be more economical than the soil rectification, which can be achieved by screening the crop genotypes for their tolerance of these toxicities. Soil acidity, which is most detrimental to crop production, once considered a problem confined to tropical agricultural region or high rainfall areas with highly weathered soil, has now attracted global attention. Excellent attempts have been made to review the mechanism of tolerance and crop production in acid soils and put forth the views of plant physiologists, agronomists, breeders and biotechnologists on a problem basically of soil science (Singh, 2000a).

Calcareous and alkaline soils

Generally the soils having pH value above 7 are associated with carbonates and are calcareous. The calcareous soils are mainly due to the presence of high amount of lime and limestone parental rocks below ground. Calcium, as Calcite is the most common form of carbonate present in the soil, however, argonite ($CaCO_3$), dolomite ($CaCO_3MgCO_3$), and magnesite ($MgCO_3$) also occur in calcareous soils. Calcium occupies most of the exchange sites on clays and organic matter. In calcareous soil high HCO_3^- is associated with increased pH or CO_2 concentration. This increased pH reduces the solubility of Fe and Mn. The management practices such as deep ploughing, which brings up carbonate of higher solubility to increase the amount in the surface layers and increases the soil pH should be avoided. Phosphorus availability is often low in calcareous soils and adsorption of added P depends on the number of weakly adsorbing sites. The high bicarbonate content directly affects the uptake of P. Thus, utilisation of P depends on plant tolerance to alkalinity.

About one-third of the soils in the world, are calcareous where iron deficiency is the most common problem (Vose, 1982). India has about 33 m ha of calcareous land, most of which is spread in Gujarat, Maharashtra, Karnataka, Rajasthan, Bihar, Tamil Nadu and Punjab. The plants that grow well on calcareous soil are called 'calcicoles', and those which do not are 'calcifuge'. Among various soils, the medium to deep black calcareous soils of Gujarat, Maharashtra, Karnataka and Rajasthan, and calcareous alluvial soils of Punjab, Haryana, U.P. and Bihar are productive mainly due to high Ca and cation contents and their easy accessibility (Singh and Joshi, 2000). However, because of Ca and Mg in calcareous soils, the plants show multi-nutrient deficiencies, which are very difficult to diagnose and rectify. The chlorosis due to iron and sulphur deficiencies is a major problem of calcareous and alkaline soils of Saurashtra (Gujarat), Marathwada (Maharashtra) and part of Tamil Nadu, Karnataka and Rajasthan (Singh *et al*, 1990a b; Singh, 1999). The high free $CaCO_3$, HCO_3^-, moisture, pH and poor aeration, heavy manuring and low organic matter content in calcareous soils (Chen and Barak, 1982; Mengel and Geurtzen, 1986; Singh, 1999; Wallace *et al*, 1976).

Light and Temperature

The concentrations of elements are also influenced by light intensities and duration and extreme temperature. Light affects the photosynthates produced and alter the ratio of element to dry matter due to dilution effects. Though the light, through photosynthesis, provides energy for active uptake of elements and enhancing concentration, this positive effect is overridden by the dilution effect. Thus, increasing light exposure reduces the N, P, K, concentration, but Ca concentration may increase (Jones *et al*, 1991), however, genotypic differences are obvious as changes are genetically controlled. While shading increases the P, K, Al, Ca, Fe, and Mn concentrations and decreases the Cu, Mg, and Zn concentrations in leaves. The adverse effects of excess Na and K associated with high salinity and low light have been partially corrected by high light by helping to maintain plant cation concentrations.

The temperature influences the movements, translocation and utilisation of elements by plants, however, it is difficult to distinguish the effect of temperature under normal field conditions. There is optimal temperature for plant growth and development which varies from species to species and deviation from it definitely causes abnormality in nutrient uptake. Some of the controlled condition study reports that the elemental concentration at high temperature is less than that of at low temperature, because high temperature, like light, increases the dry matter production and here also causing dilution effect. But in addition high temperature also enhances respiration, depletes photosynthates and increases transpiration which allow the more absorption of elements from soil. However, the effects of temperature are masked by concomitant effects of light and moisture prevailing during the cropping season (Jones *et al*, 19991).

Water

The rainfall increases soil moisture, by bringing the levels within the beneficial range between wilting and field capacity, stimulates the plant gowth and hence tends to lower the elemental concentration again due to dilution effects. Though the favourable moisture condition also influences the soil nutrients availability and their movement in plant, all these are overcome by fast growth and dry matter accumulation. The humidity influences the rate of transpiration and thus indirectly affects the nutrient content in plants.

The flood irrigation aggravates problem of root aeration in many crops resulting in abnormal respiration, inhibiting root growth and altering metabolic functions. Because of this plant becomes chlorotic due to deficiency of N caused by inability of roots to take up N and ineffectiveness of nitrogen-fixing bacteria, deficiency of S due to leaching in course texture soil and deficiency of Fe in calcareous soil due to conversion of ferrous to ferric form. Broad-bed furrow method is beneficial especially where water stagnation during rainy season is a problem (NRCG, 1994, 1995). The furrow made in system helps to drain of excess water thus improves the micro-environment especially aeration in the field. Cultivation of groundnut in set-furrows is a common practice in Saurashtra region of Gujarat where groundnut is being grown in same rows for the past 40-50 years. In this system, the FYM, fertilisers and ties (Calcium-rich Murrhum) are applied in the furrows making it more fertile and airy compared with remaining soils (Singh *et al*, 1997).

Mineral deficiency symptoms are sometimes confused with other complex field events such as damage caused by insect-pest, disease, salt stress, water stress, pollution, light and temperature injury (Bennett, 1993) and herbicide damage. Toxicity of Mo or Se is similar to P deficiency (Bennett, 1993), Fe deficiency in mango is similar to chloride toxicity (Xu *et al*, 2000). Therefore, it is necessary to critically observe and define these deficiency symptoms. The deficiency symptoms might be distinguished based on the plant part that shows deficiency symptoms, presence or absence of dead spots and entire leaf or interveinal chlorosis. A description of initial appearance of deficiency symptoms on leaves is given in and the associated text below.

Interaction of Macronutrients

Increasing the levels of any macronutrient, upto certain level in the nutrient solution, increases the concentration of that particular element in the leaves, stems and seeds and their uptake by plant, but low or excess of any element influences the uptake of other nutrients also that may have synergistic (beneficial) or antagonistic (detrimental) effects. Though the interaction of individual macronutrients is discussed under the next heading, in general N along with N also increased P, K, Ca and Mg concentration, K decreased Ca and Mg, and Ca decreased K and Mg concentrations in plant tissues. The P and S, however, did not show any of such interaction with other macronutrients. Singh (1996) in a nutrient solution culture, reported that at sufficiency levels of macronutrient in nutrient solution, the elemental concentrations in leaves of groundnut at 60 DAE and seeds at harvest were 3.14-3.54 and 4.75-4.82% N, 0.32-0.35 and

0.4-0.41% P, 1.84-2.25 and 0.78-1.08% K, 2.5-2.6% and 440-470 ppm (mg/kg) Ca and 0.2-0.32 and 0.22-0.25% S, respectively. The concentrations of Ca and Mg were high in all the plant tissues except seeds at very low levels of K in the nutrient solution which decreased with increasing the levels of K. Similarly, the concentrations of K and Mg in plant tissues were high at low levels of Ca and increase with Ca levels.

FUNCTIONS, DISORDER SYMPTOMS AND INTERACTIONS

This section deals with the macronutrients, functions, deficiency and toxicity symptoms and their rectification measures and also the interactions of these macronutrients. Each element has one or more specific structural or functional roles in the plant and in absence of that the plant exhibits certain morphological or biochemical symptoms of that deficiency. Though the deficiency symptoms for some elements are not always common in all plants, there are certain generalised deficiency symptoms for each element. The deficiency symptoms also depend on the mobility of the element in the plant. If the elements are mobilised within the plants and exported to the young developing tissues, deficiency symptoms tend to occur first in the older tissues. On the other hand, where the elements are immobile, once located in a tissues they are not readily mobilised for use elsewhere, the deficiency occurs first on the young tissues. These all have been discussed individually for each element in the following text.

Nitrogen

Nitrogen is a constituent of proteins, chlorophyll, amino and nucleic acids, nucleotides, IAA, cytokinin and several secondary substances. It accounts for 1-6% of the dry matter in various plant parts with sufficiency values from 2.5-3.5% in leaf tissues. Only prokaryotes can utilise atmospheric nitrogen which represent 80% of gases. It is an important constituent of protoplast, cell constituents involved in storage and transfer of genetic informations and enzymes and play active role in the enzyme reactions and energy metabolism. It is required for the vegetative and reproductive growth, nutrient absorption, photosynthesis and production of assimilates for sink development.

Most plants absorb nitrogen from the soil primarily as nitrate ion (NO_3^-) and in a few cases as ammonium (NH_4^+) ion. But once entered into plant, the NO_3^- is reduced to NH_4^+ before incorporation into amino acids, protein and other nitrogenous substances. However, the plants that contain urease, such as phaseolus bean are able to use urea directly as nitrogen source (Polacco, 1977). There is a positive correlation between nitrogen status and nitrate reductase activity (NRA), however, several other factors such as OH^- production, pH value and HCO_3^-, are also responsible for high NRA (Mengel et al, 1983).

Nitrogen is most often limiting nutrient in agriculture and its deficiency causes slow, stunted growth and chlorosis of older leaves. Nitrogen is very mobile and as the older leaves die, it is mobilised from older to younger leaves in the form of amines and amides. The characteristic deficiency symptom of nitrogen is the appearance of uniform yellowing of leaves including the veins due to decomposition of chloroplast, this being more pronounced on older leaves. The leaves become stiff and erect. In dicotyledonous crops the leaves detach easily under extreme deficiency condition. Cereal crops show characteristic 'V'-shaped yellowing at the tip of lower leaves. O'Sullivan *et al*, (1993) observed relatively small and pale green leaves with dull appearance in sweet potato. If such condition of nitrogen stress do persist, the result is a decreased foliage and shoot growth. Nitrogen stress also leads to anthocyanin pigments causing purplish colour of stems, petioles and lower surface of leaves in many species.

Most of the soils in Asian and African countries and particularly in India are deficient in nitrogen. Nitrogen deficiency accelerates the reproductive development causing low yield and protein levels. The nitrogen requirement of legumes is much higher than cereals because of its high protein content and they are capable of meeting their nitrogen requirements both from symbiotic nitrogen fixation (60-80%) by root nodules and from soil nitrogen (20-40%). Though it fixes atmospheric nitrogen to meet the N requirement of plant, the nitrogen supply in legume and more particularly of groundnut is very crucial and deficiency is observed in initial stages and at pod formation stages (Singh *et al*, 1991; Singh and Joshi 1993). During reproductive stage, N is mobilised continuously from leaves to the developing pods (Kvien *et al*, 1986) and hence deficiency may occur at this stage too. The Rhizobia fix atmospheric nitrogen and supplies to the host plant in the form of amide to be incorporated into proteins through glutamyl phosphate pathway. Williams (1979) reported that groundnut accumulated 2.39 kg N/ha/day during vegetative growth and 3.77 during first half of the reproductive growth after which N accumulation ceased with the cessation of vegetative growth. He further reported that the crop accumulated a total of 30 g N/m^2 and 66% of this was in the kernel.

The recommended doses of nitrogen vary from crop to crop and place to place and hence range from 20-200 kg/ha. Application of 10 kg N/ha as basal dose followed by two foliar sprays of 1% aqueous solution applied at 30 DAE (500 l/ha) and 70 DAE (1000 l/ha) produced highest pod yield in bunch type groundnut (Singh *et al*, 1991a) and increased the efficiency of N fertiliser for per kg pod production. Walker *et al* (1984) found that yield of non-nodulating Florunner and Tifrun cultivars were linearly increased with number of foliar application of N at 13.5 kg/ha beginning from 28 DAE. However, the studies conducted at ICRISAT (Nambiar, 1990) and NRCG, India by authur did not find the response of nitrogen application on the non-nodulating groundnut even up to 800 kg N/ha.

Fertilisation with NO_3^- also enhances uptake of Mg^{2+} and Ca^{2+}, but inhibts phosphate uptake. However, excess of NO_3^- in cucumber causes Ca deficiency. The relationship between N and P and N and K is well known and the ratio of these is used as DRIS norms. In general uptake of nitrate stimulates the uptake of cations, but the uptake of excess ammonium restricts the cations causing Ca and K deficiencies. Singh (1996b) reported that increasing the levels of N, in the nutrient solution, increased the concentrations of P, K, Ca and Mg in groundnut leaves, but did not affect S and Mg concentration.

Though there is not a clear-cut demarcation between the optimal and excess nitrogen supply, excess of nitrogen causes excessive vegetative growth resulting in high shoot/root ratio, delayed onset of flowering and low harvest index in agricultural crops. Plants with excess N are dark green in colour with succulent leaves and susceptible to water stress and diseases. Due to excess N sometimes crop fails to flower and yield. Bergmann, (1988) has reported that in an arable farming the N fertilisation is economic, if the 'net efficiency coefficient' (quotient of the economically utilisable yield or the bound energy by photosynthesis and the expenses for N fertilisers) is less than one.

Nitrogenase and nitrate reductase are the main enzymes influenced by Mo. The Mo deficiency disrupts the nitrogen metabolism and the plant shows nitrogen deficiency because of role of Mo in nitrogen fixation. The requirement of Mo is reduced greatly by the availability and utilisation of ammonium.

Phosphorus

Phosphorus accounts for 0.1-1.0% of the total dry matter of plant with a sufficiency range from 0.20-0.40 % in leaves tissues. It is present largely in inorganic form as a component of phosphate esters (sugar-phosphates, nucleotides and phospholipids), certain enzymes and proteins, ATP, are involved in various energy transfer reactions and genetic informations. The sugar-phosphates have an important role in photosynthesis, nucleotides make DNA and RNA and the phospholipids are present in the membrane. Other P compounds are AMP, ADP, NAD, NADP and NATP. The alternate synthesis (photosynthesis and phosphorylation) and breakdown (transfer of P ion and energy liberation) process of energy storage and release, respectively in plants are governed by the P. Phytin, which accounts for up to 80% of the total P in grain and seed and only less than 2% in green plants, is the one form in which P is stored in plant. It decomposes rapidly releasing phosphoric acid for enzyme and substrate reactions in metabolism of germinating seeds.

On global level, P is the most deficient element, and hence P deficiency is predominant in the unfertilised areas. From the 500 soil samples collected from 42 countries, the World Phosphate Institute classified 65% samples as acutely deficient in P, and only 8% were classified as not deficient (Koala, 1988). Most of the Indian soils are either deficient in P or having medium P due to its fixation and low availability in the soil due to several factors. The extensive survey concludes that about 45 and 50% of the Indian soil have low and medium available P, respectively, and only 5% soils were high in available P (Tandon, 1987). In soil, the P is present primarily as polyprotic phosphoric acid (H_3PO_4) and soil pH plays a major role in its availability. However, this P is present either as Ca-, Fe- and Al-phosphate, and organic P. The ratio of inorganically and organically bound P in the soil is about 1:1, however, the ratio varies in organic soil. The form of P present in the soil is in a state of dynamic equilibrium which differs from soil to soil and depends upon various factors such as pH, carbonate content, sequoxide content, clay and humus content, heavy metals, moisture contents, water : air ratio, temperature and microbial activity and the speed with which equlibrium is attained with in 'labile pool' of P compound. The mobility of P is highest at pH between 6.0 and 6.5. The monovalent $H_2PO_4^-$ is the predominant form of P in a soil having pH below 6.8, which is readily absorbed by plants, however, between pH 6.8-7.2 the HPO_4^{2-} is the predominant anion, which is less readily absorbed. In alkaline soil with pH more than 7.2, the predominant form of P is trivalent HPO_4^{3-} which is virtually unavailable for uptake by plants. Also, P tends to form an insoluble complex with Al and Fe in neutral pH while in basic soil Ca and Mg complexes precipitate the P. Thus P is always a limiting factor in calcareous soils.

The P deficiency causes purpling of leaf margins and stunted growth but more dark green in colour. The leaves remain small, erect, unusually dark green with greenish red in sweet potato (O'Sullivan *et al*, 1993), bluish green in chili (Balakrishnan, 1999), brown in sugarcane (Nautiyal *et al*, 2000). As P is readily mobilised and redistributed, its deficiency first occurs on older leaves causing rapid senescence and later spread to other leaves from the bottom, but it takes minimum of 4 weeks to appear the deficiency symptoms on the plants. The underside of leaves develops bronzy appearance. The root growth is also restricted in black pepper (Nybe and Nair, 1986) and anthocyanin pigment increases in leaves of barley (Hamy, 1983) and under phosphorus stress. Some older leaves also show yellow symptoms of P deficiency. The older leaves become orange yellow, then brittle and finally shed. In legumes P limits N_2-fixation either directly by affecting nodule initiation, nodule development and N_2-fixation or indirectly by affecting plant growth. In general the P content less than 0.15% indicates lower concentration of P in most of the crop plants, however, this value differs from crop to crop and species to species.

In the present agriculture system P is a key nutrient. It is a valuable but scarce resource and, hence, needs efficient management. Due to the transformation of P into unavailable form after coming into contact with soil colloids, the response of P to crop is not consistent which give an indication to tailor the management practices to P application cautiously. Also the P deficiency does not necessarily mean that levels of available P in the soil are low as several other factors such as soil moisture, physical condition, temperature and ability of the crop genotypes to extract P from the soil plays a major role. There are number of reports about the non-responsiveness of P application even in apparently low available P mainly due to low yield levels, P-efficient crops which can absorb P from its low concentration (Dwivedi, 1987; Patel and Kanzaria, 1985, Singh, 1999) and the beneficial effect of native mycorrhizal fungi and phosphorus solubilising microbes. Using the root "contact reactions" of groundnut, soybean and sorghum for solubilising Fe- and Al-bound P, Ae and Otani (1997) reported that cell wall activity and P solubilising activity of groundnut were superior indicating that groundnut has superior ability to take up P from soils with low P. Using "a double pot system", it was demonstrated that Ca-P and Al-P fractions of accumulated P in the acid soil around mycorrhizal root are mobile and available to groundnut plant. In acid soils, P deficiency is induced by Al-toxicity which vanishes with application of lime (Singh *et al*, 2001).

Many of the crops and tree form symbiotic association with certain zygomycetous fungi known as vesicular arbuscular mycorrhiza (VAM) in their root. The VAM fungi are known to augment plant phosphorus (P) uptake ability from soils deficient in this element. The positive response of VAM has been reported in groundnut with inoculation of *Gigaspora calospora, Glomus mosseae* in red soil (ICRISAT, 1986), and with *Glomus fasciculatus* a marginal response in calcareous soil (Singh and Chaudhari, 1996b). As P deficiency is very common in Indian soil, Dwivedi *et al* (1987) proposed a model, $R = (A_y \times 1.32)(A_n \times 1.09)$ for selection and prediction for phosphorus deficiency resistant plant types. Where R is resistance value of P deficiency or low supply, A_y is yield and A_n is P uptake of adopted cultivars on P deficient soil (The constant value of y and n are 1.32 and 1.09, respectively).

There is a considerable loss of yield if the P deficiency persists for a longer period during the early stages of crop growth and to avoid yield losses the crop must be fertilised with adequate amount of P fertiliser. Dryness and low temperature inhibit P uptake and the plant suffers from P deficiency. The P uptake by the plant depends greatly on the form of P in the soil, cation absorption capacity of the plant, root extension (volume), presence of VAM and PSM and more particularly of the plant species and genotypes.

Excess P does not have any direct toxic effects on crop plant. It stimulates root growth and reduce shoot/root ratio. However, in perennial plant application of excess P during early stages cause strong root development. The relationship between N and P and P and micronutrients exists and the ratios of 3:1 between N and P and 200:1 between P and Zn are considered critical (Jones et al, 1991). Heavy dressing of P fertiliser however induces deficiencies of Fe and Zn and prevents toxicities of Al and heavy metals (Loneragan, 1978).

Potassium

Though potassium is not a constituent of any compound or structurally bound in plants, its concentration in plants varies from 0.5-6% of the dry matter, which is generally higher than any other cations, with a sufficiency value from 1.5-3.0% in leaves. Potassium ion activates a number of enzymes involved in photosynthesis, and respiration. It is required for translocation of assimilates, involved in maintenance of water status of plant especially the turgor pressure of cells, osmoregulation and opening and closing of stomata, increase the solar energy harvesting efficiency and the availability of metabolic energy for the synthesis of starch and proteins (Mengel and Kirkby, 1987; Dwivedi, 1989). It also plays an important role in the daily changes in the orientation of leaves or sleep movement, increases fruit sets, synthesis of sugar and starch and helps in growth and filling of grain and pods. The nitrate reductase, EDTA-osmoticum, and productivity of groundnut under water deficit conditions were increased due to K application (Dwivedi *et al*, 1997). Since high concentration of monovalent cations is needed to keep the active centres of enzymes free so that reaction can take place, the K is believed to be the only candidate for the same (Bergmann, 1992). The K is available as monovalent cation K^+ in the soil. The K deficiency occurs mostly on light, usually acid soils with low CEC. The K is frequently deficient in sandy soils due to its high solubility and frequent leaching. Though Indian soils were reported to be rich in K, yet the deficiency occurs in almost all states and more frequently in Orissa, entire NEH region and part of A.P., U.P., Gujarat and Maharashtra. This is mainly due to very little or no potassium application in the soil, in spite of its very high removal by intensive cropping. This requires greater attention of soil scientists and mineral nutritionist to revise their recommendations looking to the yield and quality aspects of various crops.

Potassium is required in large amounts by most of the plants and as long as other growth factors favour more optimal the K uptake, better the yield. The saturation level of the CEC of the soil for K is 3-5% and its deficiency occur when the K levels of the soil and plant goes down. The K^+ is highly mobile in both xylem and phloem which enables the plant to regulate their K budget easily. The K deficiency symptoms first appear in the older leaves characteristically developing mottling or chlorosis. Under potassium stress condition, yellowing of leaves starts from the tips or margins of leaves extending towards the centre of leaf base. Drying up of leaf margin with hallow yellowed margin and necrotic symptoms and reddish colouration of tip of branches. The yellowing is interveinal and irregular in the leaves of tomato (Besford, 1978) and blueberry (Tamada, 1989). These yellow parts become necrotic (dead spots) with leaf curling in tobacco (Arnold *et al*, 1986) and sugarcane (Nautiyal *et al*, 2000). There is a sharp difference between green, yellow and necrotic parts. The stem becomes red accompanied by excess storage of starch and leaves become light green. In monocots especially maize and other cereals, the necrotic lesions begin at the older tips of the leaves and gradually progress along the margins near the leaf base. Stems are shortened and weakened and susceptibility to disease increases.

Even if there are no visible symptoms plants lacking in K can be identified by their tendency to wilt more rapidly on dry, warm and sunny days than plants receiving sufficient K. Owing to dehydration and accumulation of toxin substances such as putrescin or peroxides, the leaf tissues suddenly break down and die causing dot-like or small patchy lesion first appearing on the tips and along the margines of the leaves. In monocots the K deficiency symptoms always start at the tip and spread downwards making inverted "V" shape due to faster progress at the margin. This is in contrast to N deficiency where V-shape symptom occurs. Sudden collapse of the palisade tissue form necrotic lesions at the leaf tips and along the margins occur which is preceded by edge scorch a typical K deficiency. The K deficiency reduced flowers, peg and kernels in groundnut (Singh, 1999).

The plants are highly efficient in obtaining K from the soil. Thus the indication of optimal soil K levels for high yield and quality vary from crop species and soil types. In order to get high yield, there is need to maintain optimal N:K ratio when increasing the rates of N fertilisers. It should be about 1:1 but never more than 2:1. The adequate K contents in leaves have been reported to be in between 1.0-3.0% for most of the crops (Jones *et al*, 1991). Haeder and Beringer (1982) demonstrated that K supply improves the grain filling by increasing the length of milk and dough development stages in cereals. The abscisic acid (ABA) concentration in grain, which is important for ripening of the grain and its increase at the time of ripening, could be reduced by K supply thus prolonging ripening which helps in increasing grain size. For groundnut, the soil having less than 150 kg/ha exchangeable K was deficient and has to be fertilised with 50 kg K_2O/ha (Dwivedi, 1986).

The recommended doses of K vary from 20-100 kg K_2O/ha depending upon the crop variety soil types and many other edaphic factors. It should be applied as basal in the furrows. The K^+, due to their mobility, are important wherever rapid changes in osmotic pressure are necessary. With influx of K^+, into guard cells opens the stomata by increasing the osmotic pressure and efflux of the K^+ closes them and this mechanism does not function properly if K is lacking. Thus K deficiency reduces the plant's ability to prevent excessive transpiration during hot weather as a result plants lacking this elements are prone to wilting, cause low net assimilation and yield. Application of potassium decreased transpiration and increased the stomatal resistance, solar energy harvesting efficiency and energy partitioning in kernel in rainfed conditions in groundnut (Dwivedi, 1989). The shelling percentage, 100 kernel weight and oil content was increased with K application in groundnut (Singh, 2000b). Potassium is also known to increase frost resistance during winter. Thus, high yield of desired quality can be obtained by optimal supply of potassium.

Plants luxuriantly absorb K and its uptake gradually increases with growth and development of plant. The excess of K fertilisation rarely causes any toxic symptoms, because it is not absorbed in toxic concentration by plants, however 'salt damage' occurs and it induces Ca deficiency in various plants. Singh (1996b) has shown that increasing the levels of K, in the nutrient solutions, increased the concentrations of K, and N in groundnut leaves and only K in seeds, stems and shells, but decreased Ca and Mg concentrations of leaves, stems and shells. At low levels of K, the cations like Ca and Mg were more absorbed by plant to balance the anions, but these were proportionally taken up at 50 ppm K. High K levels in the soil depress the uptake of Mg, Ca and NH_4 first and then B, Zn, Mn, inducing deficiencies of these elements. Among the Ca and Mg, high K concentration will first result in Mg deficiency and when K is greater it causes Ca deficiency (Jones *et al*, 1991). Excess K causes delayed growth, scorching and scabbing in lemon, however in orange it delayed ripening.

At their higher levels of K in groundnut, it accumulates in leaves, stems and shells, but not in seeds, as seeds do not require much K unlike N (Singh, 1996). However, at high concentration of K (above 1.6%) in leaves of apple tree negatively affect the quality and storage of fruits. Liming, depending upon K status of the soil, increases and reduces the K uptake of plants (Kamprath and Foy, 1971). Application of NH_4 also has similar effects. Thus in the soil with poor K, it is necessary to take into account the antagonism between Ca and K and between NH_4 and K ions. The K:Mg ratio in the soil should range between 2:1 and 5:1 because excessive concentrations of either element can negatively affect plant growth (Doll and Lucas, 1973). The K to Mg and K to Ca ratios are used as DRIS norms. Ammonium plays an important role in the balance between three cations K, Ca and Mg.

Calcium

Calcium is required for physical maintenance of the cell integrity and membrane permeability, enhances pollen germination, activates the number of enzymes for cell division and takes part in protein synthesis and carbohydrate transfer. It is most important for dividing cells as it plays a role in the mitotic spindle during cell division and forms calcium pectates in the middle lamella of the cell plate forming between daughter cells. Recently it has been implicated as a second messenger in certain hormonal and environmental responses and regulating enzyme activities. In its physiological effects, Ca is usually regarded as counterpart to K. Calcium influences the stability of structure of protoplast and owing to its dehydration properties, it opposes the plasma-expanding action of K. One of the important roles of Ca is that of stabilising the pectin of the middle lamella in the primary cell wall by forming calcium pectinate. Plant species though differ greatly in their Ca need, uptake and translocation ability, it accounts for 0.2-3.0% of the dry matter with sufficiency range from 0.3-1.0%, however in most of the plants 0.5% Ca content in the tissues is generally considered adequate (Yamada, 1975). The herbaceous plants, dicots and legumes need 5-6 times more Ca than the cereals and monocots. It is taken up by plants as a divalent cation (Ca^{2+}), which is abundant in soil and rarely deficient under natural conditions. However, its availability from soil to plant is affected by the concentration of available Ca^{2+}, the pH of the soil, the cation exchange capacity (CEC) of the soil and plants and Ca^{2+} saturation of the soil colloids. The acid soils, because of their low base saturation due to leaching of bases owing to high rainfall, are deficient in Ca and require liming. About one-third of Indian soils are acidic, still calcium is least researched in India (Singh, 2000a).

The calcium ion (Ca^{2+}) is transported exclusively in xylem tissue upward with transpiration stream but its downward movement from the leaves through phloem is practically nil (Mengel and Kirkby, 1978). Since calcium plays a major role in cell division its deficiency appears in meristematic and growing region of plants. Moreover once it is absorbed, Ca is not mobilised from the older leaves, hence the deficiency occurs on the fresh and emerging leaves. However, under field conditions, the visual deficiency of Ca in plant is rarely observed except in extreme conditions. Calcium deficiency in plants results in chlorosis of young leaves along the veins of blueberry (Tamada, 1989) and bird's foot trefoil (Russelle and McGraw, 1986). The calcium deficiency in leaf is characterised by development of localised pitted area on lower surface of leaves which later on converts into large necrotic spots. Cracking of basal stem and dieback of shoot at later stage of growth are also reported. If deficiency persists longer, bleaching of upper half leaf followed by leaf tip curling do occur in black pepper (Nybe and Nair, 1987) and sugarcane (Nautiyal *et al*, 2000). In groundnut, the Ca is taken up directly from the soil by pods, and inadequate supply results in pods without seeds called "Pops" or blackened plumule inside the seed known as "Black heart" (Cox and Reid, 1964). As soon as peg penetrates soil it ceases to transfer root absorbed water and hence loses access to root absorbed Ca, thus developing fruit absorbed Ca from the soil. So the developing groundnut fruit absorbs phloem-immobile Ca^{2+} directly from the soil as the fruit develops underground and it will not transpire root-absorbed water.

The growing bud leaf becomes chlorotic white with base remaining green, the distortion of the tips of shoots i.e. dieback in peach seedlings (Edwards and Hortan, 1997). Similarly, Spehar and Galway, (1997) found brown spots on leaves, reduced expansion and premature leaf senescence under Ca deficiency in soybean. In tomato Ca deficiency increases susceptibility to blossom end rot (Adams and El-Gizawy, 1988; Sonneveld and Voogt, 1991 and Ho *et al*, 1999). Calcium deficiency also show disorders such as bitter pit in apple (Ford, 1979; Monge *et al*, 1995 and Silva and Rodriguez, 1996); leaf tip burn in cabbage (Miao *et al*, 1997) and lettuce; black heart of celery; cavity spot of carrots (Scaife and Clarkson 1978); vitrescence in melons (Jean-Baptist *et al*, 1999). Ca deficiency also causes poor root gowth and slippery structure due to deterioration of middle lamella. Tomato requires 10-15 time more Ca than wheat. Since little or no Ca is redistributed in the plant the Ca content of plant cannot be regarded as Ca reserve and to maintain metabolism and growth, plant must continuously take up Ca and transport it to wherever needed.

Calcium is more important for groundnut, often lack of Ca reduces the yield and quality more than any other element. The calcium requirement is very high especially for gyanophore development and pod filling. In the field, calcium and potassium levels in the fruiting zone affected seed quality (Cox *et al,* 1982; Zharare *et al,* 1997) and early ovule abortion was prevented by adequate calcium supply. The reported leaflet Ca concentration for maximum fruit yield is 1.8% in 12-week old groundnut plant and the minimum Ca content in seed needed for maximum germination and seedling survival is 400 ppm Ca (Singh (1996b). In general, the calcium requirement, of groundnut, is greater for pod filling than flowering, and it is greater for flowering than vegetative growth. "The high calcium is required in the 5-10 cm of soil. In 1940's the North Carolina Agricultural Experiment Station, USA conducted a comprehensive study and found that the primary response to lime was due to its effect on available calcium in the pegging zone. The lime has to be added in a manner to ensure that fine lime in the pegging zone (in the top 7 or 9 cm of soil) during fruit development. Kvien *et al,* (1988) reported that the accumulation of Ca in pod is positively correlated with the pod surface area, days required to maturity, and negatively correlated with the pod thickness."

The plant receiving no Ca are completely unviable and do not survive after germination. Lack of Ca in legume prevents the development of nodulation and nitrogen fixation, thus inducing N deficiency. Lack of Ca also inhibits the transport of carbohydrates. Both lime stones and dolomites are useful as a Ca source in field crops and shallow incorporation of 1000 kg/ha lime stone (90% $CaCO_3$), before planting produced maximum yield. The critical level of soil Ca is about 250 ppm (1.25 meq/100 g), (Cox et al., 1979; Walker *et al*, 1979). Ca has antagonistic effect with excess K in the soil. About 1 meq/100 g soil in the root zone and 3 meq/100 g of soil in the pod zone. 5 cm surface soil are considered to be the threshold concentration of calcium sufficiency in Indian condition.

To provide optimal nutrition for plants, it is important to have a harmonious balance between Ca and K and between Ca and Mg. A close functional relationship also exist between Ca and B and their deficiency symptoms are also sometimes similar. Besides these high concentrations of H^+, K^+ Na^+, Mn^{2+}, Al^{3+}, Mg^{2+} and NH_4^+ ions inhibit the uptake of Ca in plant and thus induce Ca deficiency (Bergmann, 1992). For proper plant growth, the CEC of the soil should be saturated with 65-85% Ca^{2+}, 6-12% Mg^{2+}, and 3-5% K^+ (Doll and Lucas, 1973).

Calcium has antagonistic effect with excess K and Mg in the soil. The ratio of Ca and K and Ca and Mg serves as DRIS norms. Excess liming or Ca cause imbalance between Ca and Mg and induce Mg deficiency in acid soil, and hence need to be replaced with dolomite. The ratio of exchangeable Ca^{2+} and Mg^{2+} therefore should not be higher than 6 (Jokinen, 1981). Singh (1996b) in a nutrient solution culture experiment reported that increasing the levels of Ca increased the concentration of Ca and N in leaves and seeds and only Ca in stems, but decreased K and Mg concentrations of leaves, stems and seeds. The concentrations of K and Mg in leaf tissue were normal and sufficient at 50 ppm of Ca, but it dropped down and below the sufficiency levels with increasing Ca levels further. At 50 ppm of Ca, in the nutrient solution, the Ca concentration was 2.5-2.6% in leaves at 60 DAE and 440-470 ppm in seeds at harvest, respectively. Ca deficiency leads to nitrite accumulation (Yamada, 1975) due to disruption of intracellular nitrite transport.

The excess calcium in the form of calcium carbonate increases the alkalinity and decreases the availability of Mn, Fe, Zn, Cu, B and P. Lime-induced chlorosis is the major problem commonly observed in the calcareous soil mainly due to deficiencies of Fe and Zn (Singh and Dayal, 1992; Singh and Joshi, 1997).

Sulphur

Sulphur is present in both organic and inorganic compounds and constitutes methionine, cysteine and cystine amino acids. Sulphur is also a constituent of vitamin thiamine and biotin and of coenzyme A and is also a required for respiration and oil and fatty acid synthesis. As Fe-S protein it imparts electron transfer reactions in both photosynthesis and nitrogen fixation. In plants, S accounts for 0.1-0.5% of the dry matter, however, cabbage and crucifers sometimes contain higher than 1%. The S, in soil, is present in both organic and inorganic form, but the organic-S is dominant in most of the Indian soil. There are various forms of S present in the soil (Tandon 1991) and iron sulphide and elemental S is common. However, plant takes up sulphur mainly as sulphate ions (SO_4^{2-}). The amount of total S and in different forms depends on the parental material, organic matter, temperature, moisture regime, texture, type and level of management. The available S consists of easily extractable SO_4-S which is the immediate supplier of SO_4^{2-} to root. The coarse-texture and low organic matter soil contains less available S. In well aerated soil, S is present as SO_4^{2-} and as component of organic matter, but in arid soil where sulphate concentration is high it occurs as gypsum ($CaSO_4.2H_2O$). The mineralisation of S is fast in neutral to alkaline soils than acid soils. As plants take up S as SO_4^{2-}, all the S sources like pyrite and elemental sulphur added to soil are subjected to microbial transformation by *Thiobacillus* sp. into SO_4^{2-} before the plants take it up.

Though it is difficult to demonstrate S-deficiency in experiments in industrial areas, the increasing role of S application in raising crop production has been now well established (Tandon, 1991; Singh, 1999). Except for coastal region and those close to industrial places which receive sufficient S from the atmosphere, in india, the S-deficiency is reported to be very severe in Bihar, Gujarat, Punjab, M.P., U.P., Karnataka and A.P. mainly due to use of S-free NPK fertilisers, adoption of high-yielding varieties and loss of S through leaching and erosion (Tandon, 1991, Singh 1999). The crop grown on coarse-textured sandy and calcareous soils generally suffers from S-deficiency due to leaching of SO_4^-S.

The S-containing thiocyanates and isothiocyanates are responsible of pungent smell of mustard and many other plants of Brassicaceae (cruciferae) family. Also due to the presence of these compounds in crucifers some species prove fatal to livestocks grazing on them and it serves defence against insects etc. It improves nodulation, and pod yield besides reducing the incidence of diseases and is as important as phosphorus for oilseed crop. Sulphur increases chlorophyll and decreases chlorosis in calcareous soil (Singh *et al*, 1990a).

The S deficiency is like nitrogen but occurs on young leaves. This is due to reduced protein synthesis rather than impairment of chlorophyll synthesis and plants become smaller. It causes leaves to become yellowish in black pepper (Nybe and Nair, 1987); potato (Gupta and Sanderson, 1993) and *Brassica oleracea* (Stuiver *et al*, 1997) and it appears similar to nitrogen deficiency (Russelle and McGraw, 1986). The Fe and S deficiency symptoms in groundnut appear together in most parts and distinction is not possible. The affected leaves are narrow and the veins are paler and chlorotic more than interveinal portion, especially towards the base with marginal necrosis in sugarcane (Nautiyal *et al*, 2000). Though S is mobile, unlike nitrogen it is not readily mobilised in most of the crop species and the symptoms tend to occurr initially on young leaves and extend to middle showing pale yellow colour. The calcareous soil is a S deficient soil as their available sulphur (heat soluble S) level is in between 10-12 ppm. The various vegetable plants should have at least 0.1% S to avoid deficiency (Maynard, 1979), however, the critical levels vary from crop to crop and species to species. The N:S ratio is the reliable indicator of S deficiency and critical N:S ratio was found to be in between 10-15 for various crop (Saalbach, 1973).

Sulphur requirement of crop can be met through a number of S-containing materials such as gypsum, elemental S, pyrite and phosphogypsum (Biswas *et al*, 1988; Singh and Chaudhari, 1995). Generally 20-40 kg of S/ha is sufficient to meet the nutrient requirement of most of the crop (Singh and Chaudhari, 1995, Tandon 1991). Application of 1 kg of nutrient sulphur produced 12.9 kg more groundnut pod and 4.3 kg more oil/ha in the experimental plot (Singh *et al*, 1991b). In the demonstration plots 250 kg/ha of gypsum applied in the pegging zone at flowering increased 20.5% pod yield (Singh *et al*, 1991b). The natural deposits of gypsum and pyrites are found in abundance in India and are effective sources of S for almost all crops

(Singh, 1999). Pyrite is reported to be a good and cheap sulphur source for calcareous soil. The combined application of S with micronutrient is useful (Singh *et al*, 1990).

Sulphur is synergistic with N, while antagonistic with B, Mo, Fe etc. (Jones *et al*, 1991). However, Singh (1996) found S having synergistic effect with N, P and Fe in groundnut. Though there is a rare occurrence of excess sulphate in the soil, excess S causes discolouration of leaves with intercostals chlorosis until the whole leaves become yellow with scorching symptoms and fruit ripen before maturity. The yellow and scorch leaves fall prematurely.

Though the plants fulfil their requirements from atmospheric SO_2 gas, but when the atmospheric concentration of SO_2 gas achieves 0.2-1.0 mg/m^3, in the industrial areas and nearby it causes damage to plant and crops due to SO_2 toxicity with acute and chronic symptoms. Excess S in the food and fodder affects the copper metabolism in man and animals. Besides damage by direct exposure, to excessive SO_2 gas, damage results from "acid rains" (wet deposition) and long term exposure to acid dew, fog and snow at high altitudes. However, acid rain is not caused by SO_2 alone, but also by atmospheric pollutants such as HCl and HNO_2. Accumulation of H_2S in soil rich in sulphate causes chlorosis similar to that induced by lack of iron and application of iron or heavy metals to bind the toxic H_2S in form of metal sulphides is the remedy, however, it is better to improve the structure of the soil to reduce the accumulation of H_2S.

Magnesium

Magnesium is a component of chlorophyll and porphyrin in plants. Besides, it is also required for stabilising ribosome structure and cell membranes, serves as a cofactor in most of the enzymes that activate phosphorylation processes, and acts as a bridge between pyrophosphate structures of ATP or ADP and the enzymes molecules. According to Gunther (1981), about 300 enzyme reactions are influenced by Mg^{2+}. It is taken up by plant as the divalent cation (Mg^{2+}) from the soil and accounts for 0.1-0.5% of DW in plants. The deficiency of magnesium is also a problem of sandy and strong acid soils as under high rainfall, the Mg is leached out more easily in these soils. In India, the Mg is deficient in acid soils particularly of Karnataka, Tamil Nadu, part of A.P. and Eastern and North Eastern Hill Regions. Magnesium is present as carbonate in the form of dolomite ($(CaMg)CO_3$) and magnesite in calcareous soils. A magnesium ion resembles Ca in its behaviour on ion exchange and as an exchangeable ion, Mg is second in abundance to calcium. But with increase of pH in alkaline soils the Mg becomes non-exchangeable. Magnesium also influences the phytohormone balance and nitrate reduction.

In Mg deficient plant, the most pronounced symptom is interveinal chlorosis due to breakdown of chlorophyll in the region that lies between the veins. The chloroplasts in the vein are less susceptible to Mg deficiency and retain their chlorophyll much longer. As Mg is mobile the basal leaves are first affected and the deficiency starts from leaf margin and advances towards midrib and under extreme stress the young leaves are also affected. In dicotyledons the intercostals chlorotic lesion is the typical Mg deficiency, which are usually bright yellow and in some species violet tint is also observed. However, in monocotyledons, strings of beads along the veins of the older leaves occur, subsequently pale to yellowish-green chlorotic blotch appear between the veins. The concentration of Mg in leaves depends upon the crop genotype age and position of leaf.

According to Jokinen (1981) the CEC, in the fertile soils, should be saturated as 25% H^+, 60% Ca^{2+}, 10% Mg^{2+} , and 5% K^+ and Na^+ and on this basis the ideal ratio between exchangeable cations are Ca^{2+}:Mg^{2+} = 6 and K^+:Mg^{2+} = 0.5. Thus Mg deficiency is induced not only by low concentration of Mg, but also by high concentrations of H^+, $NH4^+$, K^+, Ca^{2+}, Al^{3+}, and Mn^{2+}. The Mg deficiency is conducive to the occurrence of tikka disease in groundnut. There is wide variation in the critical levels of Mg in leaf tissues and varies from crops and soils. The Mg^{2+} can be replaced by Mn^{2+} in phosphorylation reactions, but the later are less efficient. The Mg content of legume is usually twice as high as of cereals and much higher in oil plants. The Mg concentration in grains and seeds are usually high (0.16-0.49%).

Though Mg is required in high amount by most of the crops, there are hardly reports available on the response of Mg fertilisation mainly because of unawareness of the use of Mg for increasing yield. In Mg deficient soils, applications of 20-50 kg $MgSO_4$/ha are helpful in alleviating the deficiency and offsetting yield losses. However, acid soils with high available Al^{3+} concentrations always need liming with dolomite to recover the Mg deficiency and Al-toxicity. The groundnuts are very efficient in extracting Mg from the soil, and rarely show its deficiency, however, Mg deficiency has been found to cause up to 25% yield losses of groundnut depending upon the varieties in calcareous soil and suggest its fertilisation (Singh and Joshi, 2001).

Under moderate soil Mg concentration, heavy application of K fertilisers and liming reduce Mg uptake and induce deficiency due to their antagonistic effects. Therefore, the ratios of K:Mg and Ca:Mg ions are important factors and it should not exceed 5 (Bussler, 1979) and 3 (Bergmann, 1992). Excess ammonium ion also induces Mg deficiency (Kafkafi *et al*, 1971). Thus Mg deficiency is induced by the high concentration of either of NH_4, K or Ca as Mg is the poorest competitor among these. High Mg concentration disrupts the Ca:Mg and K:Mg ratios, thus reducing the uptake of Ca and K and plant growth. It also reduces the Mn uptake owing to competition between these two ions.

MANAGEMENT OF MACRONUTRIENTS STRESSES

To maintain a healthy crop proper nutrition is a must. These nutrients help the crop to grow well and provide the resistance to sustain against the disease, pests and abiotic stresses. Applications of gypsum help the drought management by delaying the wilting and senescence of leaves (Rajendrudu and williams, 1983; Singh and Chaudhari, 1996) thus establishing the role of Ca in drought management. Though, during drought the uptake of most of the mineral nutrients is restricted, application of Ca and K and upto some extent S is beneficial. During winter season when the nutrient uptake and growth are reduced, application of N helps the crop to grow well. However, there are evidences that plants given abundant nutrients become susceptible to diseases, insects, and abiotic stresses such as drought than the plants given adequate nutrition (Marschner, 1995).

The disease incidence during rainy season causes severe yield losses and the field experiments conducted by authors have shown that application of Fe reduces the incidence of tikka and rust diseases in groundnut (Singh, 1999). It is of general observation that in malnourished crop the disease incidence is comparatively higher than the properly nourished crop. The chlorotic and dying leaves during degradation invite other microflora to develop and hence bring new invasion to the plant. If such deficiencies are prevented or recovered timely such invasion could be avoided.

The role of K, Ca and S in disease management is well known and application of these reduce their occurrence. The elemental S or copper sulphate when mixed with the seed act as a seed dresser and hence seeds do not require additional treatment with fungicides. In general greater Ca and B is required for large seeded groundnut than the small seeded one and improper nutrition of these elements causes cracking of pods and seed inviting the soil fungi especially *Aspergillus flavus* to attach and develop aflatoxin. Fernandez *et al,* (1997) in a recent study reported that application of Ca increased the thickness of seed coat leading to decrease in seed infection by *Aspergillus* spp., *Penicillium* spp. and *Rhizopus* spp. Thus providing proper nutrition to the crop, most of the diseases could be prevented to occur in the field as prevention is better than cure.

Soil and Plant Analysis and Crop Response

The soil analysis provides a base but many a times mislead as the deficiency of the element, showing their adequate concentration in soil, occurs in plant. The cultivars are developed for ideal condition of soil fertility and pH. But when grown on soil even slightly different from those of originating centres developed mineral deficiency or toxicity problems which are corrected with soil amendment or foliar application of nutrient. Though the soil samples from various parts of the world have been analysed and sufficiency, deficiency and toxicity areas of certain elements have been worked out, but these studies are based on the limited soil samples and crops and hence require location specific delineation of the nutrient deficiency and toxicity areas for various crops and their visual symptoms. Till date the soil and plant analysis are the main basis for the fertiliser recommendation for a particular crop on that particular soil. But, these recommendations vary from variety to variety as some of the cultivars response very high to fertiliser application whereas a few do not. Plant analyses data can be used to guide future soil test calibration work and agronomic research (Peters and Schulte, 1996). The responses of nutrients are very high on deficient and marginal soils, but require systematic work. The soil and plant analysis can adequately meet the need of the crop growers by providing efficient and profitable site specific fertiliser recommendations for increasing crop production. The desired yields only could be achieved with suitable fertiliser applied on soil test-crop response basis. This will enhance the fertiliser efficiency and overall productivity.

Singh (1999) reported that, in groundnut, the leaves are the major sink for nutrients accumulation during vegetative stage as a result high nutrient concentration was noticed at 60 DAE. But soon as pod development start, the developing pods become the major sink and the macronutrients accumulated in leaves translocate to seeds. As a result the nutrient concentration in leaves was lowest, except for K and Ca which were more in leaves and stems than seed, at harvest. The high concentration of Ca in leaves at harvest is due to its phloem-immobility and whatever the pod absorbed directly from the solution was available to the seed. The K, however, is required in low quantity by seeds.

Though the nutrient deficiency and the responses of nutrient are reported, only a few systematic approaches has been made to find out the yield losses due to deficiency or toxicity of a particular macroelements and target the yield. In groundnut, the yield targeting field experiments conducted by Kadam *et al* (1985) in clay soil at Rahuri with 157, 10 and 410, kg/ha. Available N, P, K, respectively indicated highest return of Rs 47 per rupees of fertiliser applied in 1.5 t/ha yield target. But, the highest monetary return was observed under 2.0 t/ha yield target and these all depend upon the soil available N, P, K. However, such results are location specific and need to be conducted on other crops at several locations.

Application of soluble or liquid fertiliser through micro-irrigation system as ferti-rrigation substantially increases fertiliser use efficiency and yields (Ghetia, 1995; NRCG, 1995). The field experiment conducted by Ghetia (1995) in calcareous soil of Saurashtra indicated that application of N and P as DAP and urea through drip irrigation increased the fertiliser use efficiency, pod and kernel yields, shelling per cent and kernel weight of groundnut when compared with its soil application. The higher fertiliser use efficiency in fertilisation was mainly attributed to reduced leaching and denitrification loss of nitrogen and higher concentration of available fraction soil P.

Selection of Nutrient Efficient Crop Genotypes

The first option, in the soil low in available nutrients, is to correct the mineral deficiencies by supplying fertilisers. As the plant-available nutrients are generally a small fraction of the total nutrient, the second option is the use of nutrient efficient genotypes which is very efficient and allows the resource poor farmers to harvest reasonable yield without any extra inputs. Ae *et al* (1993) reported that 'efficient' pigeonpea could acquire P and produced reasonable yield on a soil where the inefficient sorghum failed without P fertilisers. Though, nutrient efficiency has been defined by Clark (1990) and Blair (1993) in many ways in diverse contexts, there is no general agreement as to what constitutes nutrient efficiency in agriculture, because any definition is likely to be specific to particular agricultural systems, soils and the plant products. The mechanism of nutrient efficiency in plant appears to be very complex and governed by several factors and work is initiated to understand the mechanism of nutrient efficiency in plants and screening methodology for the search of nutrient efficient genotypes (Blair, 1993; Dunlop *et al,* 1992, Goldstein, 1991, Randall, 1995).

The concept of the nutrient efficiency is the ability of a system to convert inputs into desired outputs, or to minimise the conversion of inputs into waste. The supply or availability of the mineral nutrient is the input and plant growth, physiological activity, and yield are the outputs. Thus the efficiency is the relationship of output to inputs and expressed as simple ratio, such as kg yield per kg fertiliser or kg dry weight per g of nutrient supply. However, this efficiency depends upon the uptake efficiency (uptake of nutrient per unit supply of nutrients) and utilisation efficiency (dry matter production per unit of nutrient taken up). The nutrient efficiency should be distinguished from the responsiveness, which is the capacity of a plant to increase uptake and yield as nutrient supply increases. The responsive plants are most desirable in fertilised high-input systems, while the nutrient efficient plants, which produce high yields at low levels of nutrients, are most valuable in low-fertility situations. However, if the responsiveness and efficiency are combined in one genotype through breeding programme or achieved through natural selection, it is the best and efforts are needed in this direction in the present-day modern agriculture. Randall (1995), while giving stress on the improvement of agronomy and selection and breeding for nutrient efficient genotypes, express the need to quantify the effects of rhizosphere modification.

This nutrient efficiency has its greatest validity when the system in question has organic integrity and when the input has a direct relationship with output of interest. For example, the ability of a leaf to utilise N for photosynthesis is a useful index of nutrient efficiency since leaf is an obvious functional unit and N content is a primary determinant of photosynthesis (Lynch and Rodriguez, 1994), but the ability of leaf to utilise Ca for photosynthesis is less meaningful as leaf Ca content is less directly related to photosynthesis. The crop grown in P-fixing soils may be inefficient at converting applied P fertiliser into yield, simply because most of the applied fertiliser is unavailable to the crop, regardless of the inherent ability of the crop to acquire and utilise available P in producing yield.

It is difficult to devise the selection criteria for nutrient efficiency in plants as the definition, even for single crop, vary depending upon the situation in which plant is being grown. Blair (1993) while working on the P efficiency in clover, listed 18 operational definitions of the various aspects of the P efficiency including efficiency of uptake and utilisation and the rank position of any genotype varied considerably depending on the definition. Gourley (1994) suggested comparisons of lines at nonlimiting nutrient supply levels as well as under low-nutrient conditions. Though it is true that different species require different amounts of nutrients to achieve maximum yields, marked differences exist between the genotypes of the same species in external nutrient requirements.

As the nutrient efficiency refers to the ability of a plant to convert nutrient inputs into desired outputs, every plant has a nutrient efficiency. Accordingly a 'nutrient efficient' genotype is the one that better converts nutrient inputs into desired outputs than other genotypes, which by comparison are 'nutrient inefficient' (Lynch, 1998). In their excellent review Lynch (1998) made an effort to critically analyse the role of nutrient efficient crops in the modern agriculture, and through response curve, demonstrated three strategies for the improved nutrient efficiency. These were, improving yield response to high inputs, improving yield response to low nutrient availability and improving yield response to both low and high nutrient inputs. When nutrient efficiency is defined as the function of desired output over nutrient input, an efficient genotype is the one that has greater output over some significant range of nutrient input. This was the basis for the 'green revolution' where short-stature high-yielding varieties of rice and wheat responsed to high rates of N fertilisation without lodging (Borlaug, 1972). In contrast, traditional cultivars were adopted to low input conditions, and would lodge when excessively fertilised, thereby reducing yield. The role and value of the nutrient-efficient crops in the modern agriculture are clearly shown by the fact that the green revolution was basically an improvement in the nutrient efficiency of wheat and rice. However, the demerits and often-criticised aspects of the green revolution varieties were that these new varieties were less efficient in low levels of nutrients. Thus the green revolution technology disproportionally benefited larger or rich farmers that could afford fertilisers and hence accentuated resource inequities among agricultural producers (Shiva, 1991). An alternative-approach to enhancing nutrient efficiency is the one where yield response to nutrients over the low range of nutrient availability is increased without affecting the response to high rates of nutrient uptakes. This type of nutrient efficiency is more useful in situations of low nutrient availability and also in low-input agriculture. The legume crops are the typical examples that are preferably grown on marginal soils. The ability to grow and yield at low nutrient availability has been called 'efficiency', while the capacity to respond to increasing levels of nutrients is called 'responsiveness' (Whiteaker *et al*, 1976). The third types of nutrient efficiency are superior yield response at all levels of nutrient availability and genotypes showing this response are highly valued in selection and breeding programme. However, such a broad adaptation across diverse environments may be due to several other factors, that are not directly related with nutrient acquisition or utilisation, such as better temperature and photoperiod adaptation, better root and shoot architecture, more appropriate phenology, indeterminancy, broad disease resistance and Al-tolerance etc.

Off late, in modern agriculture, both the high and low input production systems are being given due importance. With the introduction of 'organic agriculture' there is preferably less use of nutrients as crops produced with high fertilisation are less healthy and less nutritious than plants produced more naturally. In the poor and developing countries the most important role of nutrient efficient crops is the improvement of crop yield quality, or profitability, however, in the wealthy and developed countries such as USA, Canada, Western Europe, Australia, New Zealand, Japan and Israel, the most important role of nutrient efficient crops is the reduction of nutrient wastes. The introduction of nutrient-efficient crops in low-input system for a longer time will 'mine the soil' thereby degrading the productivity in the long run.

As discussed earlier, the acid and alkali are the two major problem soils causing nutritional disorders in most of the crops. These are not easily rectifiable through soil amendments and hence the concepts of fitting plant to soil may be more economical than the soil rectification. This can be easily achieved by screening crop genotypes tolerant of these toxicities. In India, the eastern and north eastern parts of the country have acid soils where the prospects for crop improvement and yield are very high due to high rainfall. However, the crop faces Al-toxicity and Al-induced P, Ca, K and Mg deficiencies. In contrast the western part of India have calcareous soil which are rich in cations due to low rainfall but low organic matter, N, P and S and lime-induced Fe, P, S and deficiencies are the main problems. Thus, as plant scientists, our strategy should be to screen crop genotypes for their tolerance to low P, K, Ca and Mg and Al and Fe toxicities for acid soil and for low P, S, Fe in calcareous soil. In response to plant adaptation in low pH soil, it is reported that native plant growing in acid soil requires Al for growth to a certain extent; however, the characteristics of Al accumulation in organs are different among different species. The ameliorative effects of calcium, phosphate and silicate on Al-toxicity may be effective in groundnut as silicate prevents Al ions from entering into root tissues by decreasing the concentration of Al. Ownby and Dees (1985) reported that soil amended with $AlCl_3$ showed complete cessation of growth which was attributed primarily to inhibition of uptake and translocation of Ca rather than accumulation of Al in plant tissues. Secretion of citric acid is reported to be involved in the Al-resistant mechanism in many tropical leguminous trees.

The field experiment conducted by this centre in collaboration with ICAR Res Complex in NEH Region report that, under low pH condition, the groundnut varieties ICGS 76, ICGV 86590 and TKG 19A performed well, however, other high yielders failed. Further these were most suitable for NEH Region and, hence, recommended (NRCG, 2000). One hundred groundnut genotypes were screened and the Al-toxicity and acid soils tolerant and sensitive genotypes were as follows:

i. Tolerant : ICG 813, 1001, 1021, 1048, 1056, 1064, 1355, 3606, 10964,11183.

ii. Sensitive: ICG 2120, 4407, 6727, 6855, 7288, 7600, 7787, 7821, 10580, 11748.

An experiment at Manipur has shown that groundnut genotypes ICGV 88401, 86708 and 87281 were high-yielder in acid soils (ICAR NEH, 1995). Recently a rice named 'Melhite Kenye', a rare paddy species, growing near Dimapur in, Nagaland, India has been found to have 220 cm height, 30 cm panicle length with 510-570 grains/panicle and record yield of 21 t/ha (Agricomplex Newsletter, 2001).

In calcareous soil, due to high Ca and Mg, there are multi-nutrient deficiencies, which are very difficult to rectify through soil amendments. Selection of nutrient efficient genotypes and soil amendments with low costing materials such as pyrite or gypsum are the only solution. Efforts were made at NRCG, Junagadh to identify the groundnut genotypes having tolerance of lime-induced iron-chlorosis and the tolerant (Fe-efficient) and sensitive (Fe-inefficient) groundnut genotypes/cultivar (Singh and Chaudhari, 19991, 1993). Some of the Fe-efficient genotypes are high-yielding too and that may be grown in the areas where the problem of Fe-deficiency is very serious. The Fe-inefficient (sensitive) genotypes can be grown in Fe-toxic areas of acid soils. The yield evaluation for consecutive five years have revealed that the genotypes having Acc. No 7085-1, 7085-3, 2588 and 6919 of Spanish group and 7599 of Valencia group have good yield potential. Similar efforts are needed for all the crops and the calcareous soils of Saurashtra are a hot spot for such screening.

Alkaline and sodic soils also limit crop cultivation specially in coastal and highly irrigated areas where the salinity is building. Unfortunately, less work has been done on these aspects as most of the crops are susceptible to salinity. Germination and seedling vigour is suppressed by different salt concentrations, and sodium carbonate is most toxic followed by sodium sulphate and magnesium sulphate (Nautiyal *et al*, 1989). Based on the seedling mortality, the groundnut genotype NRCG 7537 was found tolerant while NRCGs 5507, 7548, 7356 and 7202 were moderately tolerant to salinity (NRCG 1990-91). Ammonium application in the hydroponics resulted in less absorption of nitrogen under saline (NaCl) condition than NO_3 for groundnut (Leidi *et al*, 1992).

In an effort to screen groundnut genotypes for macronutrient using minus nutrient, the genotypic variations were noticed and the minus treatment of all the macronutrient produced less pod and fodder yields. The groundnut genotype 6919 did not show yield loss due to minus N, and 7085-1 due to minus Ca, P and S and hence were categorised as efficient to those nutrients. The minus P treatment showed maximum yield reductions in all the genotypes except 7085-1 which was identified as a P-efficient genotype (Singh, 1999). A few of the nutrient-efficient, high-yielding, groundnut genotypes identified by authors are given below (NRCG, 2000).

- Ca-efficient: NRCG 7085-1, 6820, 6919, MOR 161, ICGHNG 88448,
- Ca-inefficient: TG 26, BAU 13, NRCG 7472 and 162
- P-efficient: NRCG 7085-1, 1308, 6155, PBS 13, PBS 18057, 20016, MOR 204, PBS,11037, 20016, 20057 and MOR 139,
- P-inefficient: VRI3, SG 84-1, B 95

In view of the increasing problems of the nutrition, time bound efforts needed to test the hypothesis that genetic manipulation can contribute significantly to mineral nutrition in plants and bring another green revolution. Further to achieve this a number of laboratories having expertise in various components of research should work with a consortium approach.

Selection for Optimum Nutrient and Water Use Efficiency

The nutrients are absorbed by root in soluble form, in water, and distributed to various plant parts through transpiration stream. Thus all through the water is the main medium right from absorption to its transport and distribution and hence water and mineral nutrients go hand in hand and one system cannot exist without the other. The nutrients and cultivars can affect WUE by altering yield, photosynthesis, evaporation and transpiration (Wong *et al*, 1991). The water use efficiency (WUE), which is normally defined as: WUE= Y/ET (where Y= yield of grain or biomass, and ET=evapotranspiration) and also can be defined as photosynthesis/respiration (which is generally transpiration efficiency) is affected much due to fertilisation as the photosynthesis and transpiration are controlled by nutrients. The nutritional effects on flower and seed development are either direct (deficiencies) or through influence on plant hormone. And its impacts on plant reproduction influences yield and subsequently influence WUE. Many cereals crop N application prior to flowering increases number of grain/seeds per plant. The nutrient can influence ET of the crop by altering the available water supply through root growth, and the demand of water as the deficiency or toxicity symptoms of nutrients appearing in the aboveground parts influence the demand of water. Further improving rooting volume and surface area and soil water balance can increase the water and nutrient supply. Rooting depth is an estimate of rooting volume at one point of time.

The cultivars can be selected for traits that improve water use efficiency, and fertilisers and other nutrient sources also can be managed to optimise WUE of a crop. The plant nutrients affect the yield and, subsequently, WUE through its influence on flower initiation, flower fertilisation, and seed development. Nitrogen fertilisation increases WUE in wheat (Campbell *et al*, 1992), Corn and sorghum (Varvel, 1995), however, increasing WUE may also be limited by N supply. Flower initiation in apple increases with ammonium application due to its effect on cytokinin transport from roots to shoots (Buban *et al*, 1978). Supply of P and K also influences cytokinin levels and the flower number in many plant species (Marschner, 1995). K deficiency elevate Abscisic acid levels in the leaves, resulting in premature ripening and reduced size (Haeder and Beringer, 1981). As Ca plays a critical role in cell extension cell wall structure its deficiency limits root extension (Marschner, 1995). On the other hand, the nutrient starvation for a longer period, such as of P in maize and many other crops, increases the root surface areas (Anghinoni and Barber, 1980) and number and length of root hairs (Fohse and Jungk, 1983) and lack of N increased both root length and biomass in Maize (Eghball and Maranville, 1993). However, high nitrate concentration is reported to reduce the number and length of root hairs (Bhat *et al*, 1979), but ammonium causes abundant and long root hairs (Bhat, 1983). Fertiliser placement, also have marked effects on rooting pattern, and sub-surface fertiliser banding increases root mass and subsequently WUE (Murphy and Zaurov, 1994).

Increased rooting depth due to proper fertilisation reduces the deep drainage of water out of the rooting zone thus increasing WUE. Improvement in the nutrient status of plant increases the leaf area index and influences transpiration rate. Leaf area duration influences transpiration differently throughout the growing season. Leaf growth and senescence rates influence the time required to reach the maximum leaf area and the time for the leaf area to decline. Under limited nutrient supply the growth rates of leaves are limited. Nitrogen and P deficiencies caused smaller leaves thus reducing transpiration, however, application of these increase leaf area index and delay senescence, resulting in increased transpiration. Also the nutrient deficiencies lead to more rapid senescence. Nutrient disorders also influence leaf colour which may influence light absorption, leaf temperature and photosynthesis rates. Chlorosis is a common symptom of deficiencies of N, Mo and Fe, while P deficiency results in dark green leaves and Mg deficiency causes reddening of leaves.

The stomatal functions are governed by K^+ balance in guard cells and nutritional effects on abscisic acid production. The K^+ is also a major solute responsible for water potential gradients between guard cells and surrounding epidermal cells (Zeiger,1983). Thus K deficient plants are low tolerant for water stress. During water stress, the Abscisic acid level in leaves increases, and stomatal closure occurs, thus reducing transpiration. Nitrogen deficiency causes more rapid closure of stomata and increases resistance to water vapour diffusion diminishing transpiration rates (Radin and Ackerson, 1981). The P-deficient plant also accumulates more abscisic acid in leaves than P-sufficient plants and the stomata closes at higher leaf water potentials reducing transpiration (Radin, 1984). Improved plant nutrition increases stomatal conductance as well as assimilation rate, and the ratio between these two determine the impact on WUE (Radogolou *et al*, 1992). There is report that increasing soil P levels increased photosynthetic rate/ transpiration rate (transpirational WUE) in pearl millet (Payne *et al*, 1992). However, improvement in N nutrition resulted in increased photosynthetic rates but does not affect stomatal control of water loss (Michell and Hinckley (1993).

Davis and Quick, (1998) in their excellent review emphasised that in order to optimise WUE, the cultivar and nutrient decision should be made jointly as this integrated approach will lead to improvement in WUE, increase productivity and profitability per unit water and nutrients. The cultivars tolerant to rooting limitations such as compaction, Al-toxicity, and salinity indirectly increase the available water supply and optimise the potential of a crop to respond to improved nutrition (Richards, 1993). Both cultivar selection and crop fertilisation practices can be chosen that will increase rooting volume or mass, and this increased rooting enhances the potential for a yield response due to additional water supply (Eghball and Maranville, 1993). As leaf senescence plays a major role in water use and is affected by fertilisation, as cultivars with delayed senescence under drought condition may also delay senescence under nutrient-deficient conditions which prolong photosynthetic productivity and increase WUE. Specific leaf area (SLA) is positively correlated with WUE and negatively correlated with leaf N concentration (Nageswara-Rao and Wright, 1994). Thus greater leaf area may increase productivity per unit land or unit water, but the nutrient in plant is distributed throughout a larger plant resulting in their lower concentration.

Light absorbance and leaf colour varies with the cultivar and nutrient status. The chlorophyll meter has been used to determine the N status of the plant and recommendation of fertiliser. This need to be standardised across the cultivars and other crop species. Number of gains/kernel is related with pre-anthesis environmental stress and wheat cultivars having differences in WUE were related to their ability to produce a high kernel number inspite of a pre-anthesis environmental stress (Entz and Flower, 1990). The B, K and water stress interfere with grain filling, thus although a cultivar may produce larger number of grain the same may not mature due to nutrient stress. Under saline conditions the Na, present in high concentration, competes with K for plant uptake and stomatal conductance is reduced as there is high correlation between K^+ levels in the shoots and stomatal conductance (Leidi *et al*, 1992).

The low-input agriculture system is prevalent in most part of the world where the nutrient efficient crops have an important role to improve the productivity (Lynch, 1998). Different genotypes show maximum yield at different fertiliser rates, the water regime also alter the optimum fertiliser rate. Eghball and Maranville (1991) found that genotypic selection of corn for nutrient use efficiency was not influenced by water regime, but it closely paralleled WUE. This clearly shows the potential for genetic improvements in WUE through selection for nutrient use efficiency.

Plant Modifications and Molecular Aspects

Plants can increase their access to soil nutrient either by increasing the absorption areas or by modifying the rhizosphere to increase nutrient availability and absorption mechanism to increase the uptake from low nutrient concentration in the soil. The absorption area can be increased either by increasing the total root surface area through root proliferation and root hair developments or by mycorrhizal associations which allows to capture nutrients beyond the rhizosphere region. The rhizosphere in the plant is modified through root exudates varying from protons to complex organic molecules, which influence the nutrient availabilities and uptake either directly, or through microbial inhabitants of the rhizosphere. The modification in the absorption mechanism uptake requires modification in the genotypes as variation in the abilities to extract nutrients vary among the genotypes. Further with the advancement of research, work on understanding of the genetic control of the nutrient acquisition mechanism started (Barrow, 1993; Bassam *et al,* 1990; Randall *et al,* 1993) opening new avenues for favourable genetic manipulation either through standard breeding procedures or through identifying and transferring genes controlling nutrient ion uptake mechanisms and root exudation processes.

The opportunities exist for applying molecular biology in tackling problems of nutrition in plants. From the genetic improvement point of view, root exudates that could mobilise sparingly soluble soil nutrients especially P from Al-P, Fe-P, and Ca-P permit better use of less processed fertilisers and amendments in developing countries of tropical and subtropical region. Exudation of piscidic acid by pigeonpea roots which release P from normally plant-unavailable Fe-P (Ae *et al,* 1991) is one such example. The root hair length in *Ttrifolium repens* was positively correlated with phosphate uptake and plant growth in P-deficient soil and was heritable (Caradus, 1979). It is possible to quantify genotypic variation in root system, but this genotypic differences in root morphology seems under multi or polygenic control and hence difficult to transfer the desirable root characteristic into agronomic background through traditional breeding approaches (Zobel, 1991).

The molecular markers such as isozymes, restriction fragment length polymorphism (RFLP), and random amplified polymorphic DNA (RAPD) have an excellent potential as a tool for gene mapping of root characteristics or other traits associated with nutrient uptake (Tanksley *et al,* 1989). The each quantitative trait loci (QTL) can be studied as a discrete entity and its properties measured. Torres *et al* (1993), in an attempt to develop a complete gene map for *vicia faba*, found linkage relationship of 9 isozymes, 1 RFLP marker, and 43 RAPD markers. There is also report that RFLP have been also used to map QTL in tomato for WUE (Martin *et al,* 1989) and tolerance to low-P stress (Reiter *et al,* 1991).

Root colonisation by indigenous VAM fungi differed among genotypes, but the same has not been related with gene (Smith *et al,* 1993). Besides these biochemical activities such as acid phosphatase may be a potential criteria for high root colonisation by VAM as these are correlated (Kesava Rao *et al,* 1990). Recently, the molecular structure of ion transport system in cell membranes is studied and it is possible to identify gene sequences associated with proteins of the ion channels and H^+-ATPase (Smith *et al,* 1993, Tyerman and Schchtman, 1993). However, very little is known about the genetic control of exudates.

CONCLUSIONS AND FUTURE STRATEGIES

In most of the developing countries, soil infertility limits its productivity, where the plant nutrition research can raise the productivity by diagnosis of nutrient deficiencies and toxicities of crops on previously unfertilised soils, their correction with minimal fertiliser and treatment costs, and development of cultivars with high nutrient efficiency in deficient soils. Application of fertilisers has contributed substantially to the increase in world food production in past century and the plant nutritionists have played a key role.

In USA and other developed countries the move towards sustainability has resulted in reduced fertiliser application rates, but in most of the developing countries the fertiliser application rates are much below the requirement and increasing the same will definitely improve the productivity. However, long-term sustainability must be kept in mind and it should not be disturbed for short-term benefit. A balance approach with long-term sustainability should be the main objective.

The productivity of most of the crops, in developing countries and also in India, is still much low mainly due to low consumption of fertiliser, inspite of prominent nutrient deficiencies. The coarse-textured soils, on which most of the crops are grown in most parts of India, are prone to leaching losses of S, K, Ca and Mg thus their application becomes essential. The nutrient leaching in shifting cultivation and acid soil due to high rainfall is quiet high and requiring high fertiliser doses.

The role of macronutrients in future crop production particularly in developing countries, breeding for nutrient efficiency, process of nutrient absorption, interaction of root with soil and ion absorption are the current and future research areas. It is necessary to maintain expertise in the diagnosis and treatment of macronutrient deficiencies and toxicities.

The P deficiency enhances root exudation of organic acids to mobilise sparingly soluble P on acid and calcareous soils. In acid soil, tartaric acid is the main component of exudates while malic and citric acids are in calcareous soil. The root "contact reactions" have shown that groundnut has superior ability to solubilising Fe- and Al-bound P, than many other crops and take up P from soils with low P.

Once the breeders have essentially exhausted the possibilities for raising the genetic yield potential and farmers are using the most advanced agronomic practices, the yield potential of a crop, in a given country, is determined largely by the physical environment, soil moisture and nutrients, temperature, day length and solar intensity. The most promising areas in which manipulation of nutrient absorption could improve the efficiency of crop production in the near future are the absorption of macronutrients under stress. Some of the areas requiring immediate attention of mineral nutritionist are mentioned here.

- The information on the available nutrient status and their critical levels for growing soils are limited. As the critical levels vary from soil to soil and varieties an intensive delineation is needed on countrywise and also at global level for problem soils and plant grown under stress conditions. There is an urgent need to work out the physiologically-based critical value for diagnosis of macronutrients in various plant parts and its correlation with agronomic yield and for making fertiliser recommendation.
- The information on the rate, frequency, time, mode, source and method of nutrient application are differing which need to be standardised in concise form and recommended for particular soil, crop variety and area. Evaluation of locally available, cost-effective sources of minerals and their economics should be studied.
- Screening of crop genotypes resistant in calcareous soil and lime-induced deficiencies and tolerant of acid soil on Al-induced Ca and P deficiencies should be extensively done. The ameliorative effects of Ca, P, on Al toxicity and molecular mechanism of Al toxicity and tolerance should be studied. Effort is needed for the selection of mineral stress resistant lines using tissue culture.
- Screening and identification of nutrient efficient genotypes with special reference to P, K, Ca, and S. Emphasis should be placed on efficient use of nutrient and yield targeting and seed quality under stress. Relationship between nutrient use efficiency and tolerance to water stress need to be studied.
- Diagnosis of deficiency in plants and soil through visual symptoms, development of standard norms for plant analyses and soil test. The biochemical diagnostic technique offers a new avenue for detecting nutrient status, these needs to be standardised and used. The new soil extractants should be used for predicting the nutritional status of acid and alkaline soils. The low-altitude aerial photography through computer-based image analysis can help in the assessment of plant growth and N status non-destructively. The remote sensing technique can be a new nutritional diagnostic tool, if the satellite data are used to estimate and improve the productivity by appropriate fertilisation in groundnut.
- For immobile element particularly of Ca, and S the leaf concentration does not provide the correct nutrient status and demand of economic parts growing underground such as sweet potato root and groundnut pods. Moreover, as these plant parts also absorb nutrient from the soil, the nutrient concentration of seed or other plant parts of these elements should be taken as the criteria for the critical levels.
- Finally, on global level, most of the scientists are working on the mineral nutrition of only a few crops and only a few scientists are working on the other important crops. Moreover, most of the study is under refined conditions and does not corroborate with the problem soils. As a result most of the crops in most part of the world is grown under low fertility. A group of scientists should work in close collaboration to improve the nutrition of crops particularly in acid and calcareous soils.
- In view of the increasing problems of the nutrition, time bound efforts are needed to test the hypothesis that genetic manipulation can contribute significantly to mineral nutrition in plants and bring another green revolution. Further to achieve this a number

of laboratories having expertise in various components of research should work with a consortium approach.

REFERENCES

Ae, N. and T. Otani, 1997. The role of cell wall components from groundnut roots in solublizing sparingly soluble phosphorus in low fertility soils. *In* Plant Nutrition-For Sustainable Food Production and Environment (T. Ando et al Eds.) pp. 309-314. Kluwer Academic Publishers, Printed in Japan.

Ae, N., J. Arihara, and K. Okada, 1991. Phosphorus uptake mechanisms of pigeonpea grown in Alfisols and Vertisols. In Phosphorus nutrition of grain legumes in the semi-arid tropics (C. Johansen et al. eds.) pp. 91-98. ICRISAT, Patancheru, India.

Ae, N., J. Arihara, and K. Okada, T. Oshihara, T. Otani and C. Johansen, 1993. The role of piscidic scid secreted by pigeonpea roots grown on a Alfisols with low-P fertility. *In* Genetic aspects of plant mineral nutrition (P. J. Randall et al. eds.) pp. 279-288. Kluwer Academic publisher, Dordretch, The Netherlands.

Adams, J.F., D.L. Hartzog and D.B. Nelson. 1993. Supplemental calcium application on yield, grade and seed quality of runner peanut. Agron.J. 85: 86-93.

Adams, P. and El-Gizawy, A.M. 1988. Effect of calcium stress on the calcium status of tomatoes grown in NFT. *Acta Horticulturae*. 22: 15-22.

Anghinoni, I. And S. A Barber, 1980. Phosphorus influx and growth characteristics of corn roots as influenced by phosphorus supply. Agronomy Journal, 72: 685-688.

Agricompex news Letter, 2001. News letter of ICAR Res. Complex For NEH Region, Umiam, Barapani, Jan-March 2001.

Arnold N., Chong, C. and Binns, M. 1986. Seasonal nutrient contents in grey and non-grey flue-cured tobacco. *Beitrage zur Tabakforschung Int*. 13(5): 265-269.

Arnon, D. I. and P. R. Stout , 1939. The essentiality of certain elements in minute quantity for plants with special reference to copper. Plant Physiology 14:371-375.

Balakrishnan, K. 1999. Studies on nutrients deficiency symptoms in chilli (*Capsicum annum* L.). *Indian J. Plant Physiol*. 4(3): 229-231.

Barrow, N.J. 1993. ed. Plant Nutrition –from genetic engineering to field practice. Kluwer academic press, Dordrecht, The Netherlands, 806 pp.

Bassam, N. El., M. Dambroth amnd B.C. Loughman, (eds). 1990. Genetic aspects of plant mineral nutrition, Kluwer academic press, Dordrecht, The Netherlands, 560 pp.

Bennett, W. F. 1993. Plant Nutrient Utilization and Diagnostic Plant Symptoms, In: *Nutrient Deficiencies and Toxicities in Crop Plants*. The American Phytopathological Society, St. Paul Minnesota, pp. 1-7.

Besford, R.T. 1978. Effect of sodium in the nutrient medium on the incidence of potassium deficiency symptoms in tomato plants. Plant and Soil. 50(2): 427-432.

Bergmann, W. 1992. Nutritional Disorders of Plants: Development, Visual and analytical diagnosis. Gustav Fischer Verlag Jena, 741 p.

Beringer, H. (1978): Functioning of potassium in plant metabolism with particular reference to yield. In: G. S. Sekhon (Editor), Potash in Soil and Plants. Potash Research Institute of India, India, pp. 185-202.

Bhat, K.K.S, P.H. Nye, and A.J. Brereton, 1979. The possibility of predicting solute uptake and plant growth response from independently measured soil and plant characteristics. IV. The growth and uptake of rape in solution of constant nitrate concentration. Plant and Soil 53: 137-167.

Biswas, B.C., Maheswari, S. & Yadav, D.S. (1986). Use of indigenous sources of sulphur in Indian Agriculture. Fertilizer News 31 (9): 59-65.

Blair, G. 1993. Nutrient efficiency–What do we really mean? *In* Genetic aspects of plant mineral nutrition (P.J. Randall et al. eds.) pp. 205-214. Kluwer Academic publisher, Dordretch, The Netherlands.

Borlaug, N.E. 1972. The green revoluton, peaceand humanity. Speech delivered upon receipt of the 1970 Noble Peace Prize. CIMMYT reprint and translation series No 3. CIMMYT, Mexico.

Buban, T.A., J. Verga, J. Tromp, E. Kenegt and J. Bruinsma, 1978. Effects of ammonium and nitrate nutrition on the level of zeatin and amino nitrogen in xylemsap of apple rootstocks Zeitschrift fur Pflanzenphysiologie 89:289-295.

Bussler, W. 1979. Anatomical-histological investigations on plants suffering from micronutrient deficiencies. Proc 2nd workshop 'Micronutrients and Plant Nutrition', Mariut, Egypt 25-46.

Campbell, C.A., R.P. Zentner, B.G. Mcconkey and F. Selles, 1992. Effect of nitrogen and snow management on efficiency of water use by spring wheat grown annually on zero-tillage. Canadian J. of Soil Science. 72: 271-279.

Chen, P. and P. Barak. 1982. Iron nutrition of plants in calcareous soils. Adv. Agron. 35: 217-240.

C.F.A., 1985. Methods of applying fertilizer. In: *Western Fertilizer Handbook* (8th Eds). Interstate Publishers, Inc. Danville, USA. 161-182.

Clark, R.B. 1990. Physiology of cereals for mineral nutrient uptake, use and efficiency. In Crops as Enhancers of Nutrient Use (V.C. Valigar and R.R. Duncan eds.) pp. 131-209. Academic Press, California USA.

Cox. F.R., J.F. Adams and B.B. Tucker. 1982. Liming, fertilization and Mineral nutrition. pp. 139-163. In (H.E. Pettee and C.T. Young eds.). Peanut Science & Technology. American Peanut Research and Education Society Inc. Yoakum, Texas, USA.

Cox, F.R. and P.H. Reid. 1964. Calcium-boron nutrition as related to concealed damage in peanuts. Agron. J. 56: 173-176.

Cox, F.R., G.A. Sullivan and C.K. Martin. 1976. Effect of calcium and irrigation treatments on peanut yield, grade and seed quality. Peanut Sci. 3: 81-85.

Caradus, J.R. 1979. Selection for root hair length in white clover (*Trifolium repens* L.) Euphytica 28: 489-494.

Davis, J.G. and J.S. Quick, 1998. Nutrient management, cultivar development and selection strategies to optimize water use efficiency. Journal of Crop Production. Vol 1 No. 2, pp. 221-240.

Doll, E.C. and R.E. Lucas 1973. Testing soil for potassium, calcium, and magnesium. In Soil testing and Plant analysis. Pp 133-151. Publ. Soil Sci. Soc. America, Madison Wisc, USA

Dunlop, J., M.G. Lambert, J. VanDEn Bosch, J. r. Caradus, A.L. Hart, G.S. Wewela, G.S. Mackay and M. J.M. Hay, 1992. A programme to creed a cultivar of trifolium repens L. for more efficient use of phosphate. *In* Genetic aspects of plant mineral nutrition (El Blassam et al. eds.) pp. 547-552. Kluwer Academic publisher, Dordretch, The Netherlands.

Dwivedi, R.S. 1988. Mineral nutrition of groundnut. Metropolitan Book Co. New Delhi, India 135 pp.

Dwivedi, R.S. 1989. Potash nutrition in relation to energy partitioning and harvesting efficiency of groundnut (*Arachis hypogaea* L.) under rainfed conditions. Oleagineux 44: 413-417.

Dwivedi, R.S., Y.C. Joshi, S.N. Saha, A.N. Thakkar and V.G. Koradia. 1987. Modeling of peanut (*Arachis hypogaea* L.) for higher yield on phosphorus deficient soil. Oleagineux 42: 165-168.

Dwivedi, R.S., Y. Misra and K.K. Srivastava, 1997. Effect of potassium on EDTA-osmoticum, nitrate reductase activity and productivity of groundnut-sugarcane intercropping under water deficit conditions. In Plant Nutrition-For Sustainable Food Production and Environment (T. Ando et al Eds.) pp. 93-94 Kluwer Academic Publishers, Printed in Japan.

Edwards, J.H. and Horton, B.D. 1979. Response of peach seedlings to calcium concentration in nutrients solution. *J. American Soc. Hort. Sci.* 104(1): 97-99.

Eghball, B. and J.W. Maranville, 1991. Interactive effects of water and nitrogen stresses on nitrogen utilization efficiency, leaf water status and yield of corn genotypes. Communications in soil science and Plant Analysis 22: 1367-1382.

Eghball, B. and J.W. Maranville, 1993. Root development and nitrogen influx of corn genotypes grown under combined drought and nitrogen stresses. Agronomy Journal 85: 147-152.

Entz, M.H. and D.B. Fowler, 1990. Differential agronomic response of winter wheat cultivars to preanthesis environmental stress. Crop Science 30: 1119-1123.

Epstein, 1972. Mineral Nutrition of Plants: Principles and perspectives. New York, Wiley.

FAO 1991. FAO production year Book 1990. Rome, Italy.

Fernandez, E.M., C.A. Roseleni, A.C. Manngoni and D.M.T. Oliveira, 1997. Fungus incidence on peanut grains as affected by drying method and Ca Nutrition. Field Crop Res. 52:9-15.

Fohse, D. and A. Jungk, 1983. Influence of phosphate and nitrate supply on root hair formation of rape, spinach and tomato plants. Plant and Soil 74: 359-368.

Fohse, D. Classen, N. and Jungk, A. 1988. Phosphorus efficiency of plants. I. External and internal P requirement and P uptake efficiency of different plant species. *Plant and Soil*. 11: 101-109.

Fords, E.M. 1979. The distribution of calcium in mature apple fruits having bitter pit disorder. *J. Horticultural Sci.* 54(1): 91-92.

Gascho, G.J. and J.G. Davis 1994. Mineral nutrition. In The Groungnut Crop: a scientific basis for improvement (Edt J. Smart) Chapman and Hall, London. pp. 214-254.

Ghetia, N.R. 1995. Response of Summer groundnut (*Arachis hypogaea*) to fertiirrigation and methods of irrigation under varying fertility levels. M.Sc. Thesis Department of Agronomy Gujarat Agriculture University. Junagadh.

Goldstein, A.H. 1991. Plant cells selected for resistance to phosphate starvation show enhanced P use efficiency. Theoritical and Applied genetics 82: 191-194.

Gourley, C. J. P. 1994. Plant nutrient efficiency: a comparison of definitions and suggested improvement. Plant and Soil. 158:287-299.

Gunther, T.H. 1981. Biochemistry and pathobiochemistry of magnesium. Magnesium Bulletin 3 H. 1a: 20-21.

Gupta, U.C. and Sanderson, J.B. 1993. Effect of S, Ca and B on tissue nutrient concentration and potato yield. *J. Plant Nut*. 16(6): 1013-1023.

Haeder, H.E. and H. Beringer 1981. Influence of potassium nutrition and water stress on the abscisic acid content ingrains and flag leaves during grain development. Journal of the Science of Food and Agriculture 32:552-556.

Ho, L.C., Hand, D.J., Fussell, M. and Papadopoulos, A.P. 1999. Improvement of tomato fruit quality by calcium nutrition. *Acta Horticulturae*. 481(2): 417-423.

ICAR, NEH 1995. Annual Report ICAR Res Complex for North Eastern Hill Regions, Barapani Shillong, India.

ICRISAT. 1990. Annual Report, 1989. ICRISAT, Patancheru, A.P. 502 324, India.

Jean-Baptist, I.; Morard, P.; Bernadac, A. and Papadopoulos, A.P. 1999. Effects of temporary calcium deficiency on the incidence of a nutritional disorder in melon. *Acta Horticulturae*. 481(1): 417-423.

Jones, Jr J.B., B. Wolf and H.A. Mills (1991). Plant Analysis Handbook. Micro-Macro Publishing Inc. Georgia, U.S.A., 213 pp.

Jokinen, R. 1981. The magnesium status of Finnish mineral soils and the requirement of the magnesium supply. Magnesium Bulletin 3 H. 1A: 1-5.

Kadam, J.R., K.R. Sonar and N.D. Patil. 1985. Studies on fertilizer use based on soil test and yield target in kharif groundnut. Fert. New (Sept.): 25-27.

Kafkafi, U., Walerstein, I. and Feigenbaum, S. 1971. Effect of potassium nitrate and ammonium nitrate on the growth, cation uptake and water requirement of tomato grown in sand soil culture. *Israel J. Agric. Res.* 21:13-30.

Kamprath, E.J. and C.D. Foy. 1985. Lime-fertilizer-Plant interaction in Acid soil. pp. 91-151. In: O.P. Engelstad (ed.), Fertilizer Technology and Use. 3rd ed. Soil Sci. Soc. Amer., Madison WI USA.

Kesava Rao, P.S., K.V.B.R. Tilak, and V. Arunachalam, 1990. Genetic variation for VA mycorrhiza-dependent phosphate mobilization in groundnut (*Arachis hypogaea*). Plant and Soil 122:137-142.

Kvien, C. S., W.D. Branch, M.E. Sumner and A.S. Csinos (1988). Pod characteristics influencing Calcium concentration in the seed and hull of peanut. Crop Sci., 28, 666-671.

Koala, S., J.R. Sims, H. El-Altar and M. El-Halfawi, 1988. Phosphorus deficiency in the semi-arid tropics and implications for grain legume production. In World CropsCool season legumes (R.J. Summerfield ed.). pp. 205-216. Kluwer Academic London, UK.

Leidi, F.O., M. Silberbush, M.I.M. Soares and S.H. Lips. 1992. Salinity and nitrogen nutrition studies on peanut and cotton plants. J. Plant Nutr. 15: 591-604.

Loneragan, J.F, 1978. Anamolies in the relationship of nutrient concentrations to plant yield. Proc. 9th International Colloquium Plant Anal. and Fertil. Problems, pp. 283-289. Auckland, New Zealand.

Lynch, J. 1998. The role of nutrient efficient crops in modern agriculture. Journal of Crop Production. Vol 1 No. 2, pp. 241-264.

Lynch, J. and N. Rodriguez 1994. Photosynthetic nitrogen efficiency in relation to leaf longeivity in common bean crop science 34:1284-1290.

Marschner, H. 1995. Mineral Nutrition of Higher Plants. 2nd ed. 889 p. Academic Press, New York.

Martin, B., J. Neinhuis, G. King and A. Schaefer, 1989. Restriction fragment length polymorphism associated with water use efficiency in tomato. Science. 243:1725-1728.

Maynard, D. N. 1979. Nutritional disorders of vegetable crops: a review. J. Plant Nutrition. 1: 1-23.

Mengel, K. and E.A. Kirkby 1980. Potassium in crop production. Adv. Agron. 33: 59-110.

Mengel, K. and E.A. Kirkby (1987). Principal of plant nutrition, Fourth edition. International Potash Institute, Berne, Switzerland.

Mengel, K. and G. Gurtzan. 1986. Iron chlorosis on calcareous soils. Alkaline nutritional conditions as the cause for the chlorosis. J. Plant Nutr. 9: 161-173.

Mengel, K., P. Robin and L. Salsac, 1983. Nitrate reductase activity in shoots and roots of maize seedlings as affected by the form of nitrogen nutrition and the pH of the nutrition solution. Plant Physiology 71: 618-622.

Miao, Y., Jiang, Y.T.; Zeng, G.H. and Cao, J.S. 1997. Changes of Ca^{2+} compartmentation and ultrastructure in inner leaves of chinese cabbage (*Brassica compasteris* sp. Pekinensis olson) during the development of calcium related physiological disorder. *Acta Horticulturae Sinica.* 24(2): 145-149.

Mitchell, A.K. and T.M. Hinckley, 1993. Effects of foliar nitrogen concentration on photosynthesis and water use efficiency in Douglas fir. Tree Physiology 12:403-410.

Monge, E.; Val, J.; Sanz, M.; Blanco, A. and Montanes, L. 1995. Calcium as a plant nutrient. Bitter pit in apple. Annale de la Estaction Experimental de Aula Dei. 21(3): 189-201.

Murphy, J.A. and D.E. Zaurov, 1994. Shoot and root growth responses of perennial ryegrass to fertilizer placement depth. Agronomy Journal, 86:828-832.

Murthy, R.S., H.S. Shankaranarayana and L.R. Hirekerur. 1997. Distribution, genesis and classification of acid soils. pp. 25-36. In: I.C. Mahapatra, et al (eds.) Acid Soils of India. ICAR, New Delhi.

Nageswara Rao, R.C. and G.C. Wright, 1994. Stability of the relationship between specific leaf area and carbon isotope discrimination across environments in peanut. Crop Science, 34: 98-103.

Nambiar, P.T.C. 1990. Nitrogen nutrition of groundnut in Alfisols. Information bulletin No. 30, ICRISAT, Patancheru, A.P., 502 324, India.

Nautiyal, N., Dubey, B.K. and Chatterjee, C. 2000. Symptoms of nutrient deficiency (N, P, Ca, Mg and S) in sugarcane grown in refined sand. Sugarcane Intr. March.; 12-17.

Nautiyal, P.C., V. Ravindra and Y.C. Joshi. 1989. Germination and early seedling growth of some groundnut (*Arachis hypogaea* L.) cultivars under salt stress. Indian J. Plant Physiol. 32: 251-253.

NRCG, 1991. Annual Report 1990-91. National Res. Centre for groundnut (ICAR), Junagadh.

NRCG, 1994. Annual Report 1993-94. National Res. Centre for groundnut (ICAR), Junagadh.

NRCG, 1995. Annual Report 1994-95. National Res. Centre for groundnut (ICAR), Junagadh.

NRCG, 2000. Annual Report 1999-2000. Natl Res. Centre for groundnut (ICAR), Junagadh.

Nybe, E.V. and Nair, P.C.S. 1986. Nutrient deficiency in black pepper (*Piper nigrum* L.). 1. Nitrogen, phosphorus and potassium. Agri. Res. J. Kerala. 24(2): 132-150.

O'Sullivan, J.N., Sher, C.J.; Blamey, F.P.C. and Edwards, D.G. 1993. Mineral nutrient disorders of root crops of the Pacific: Preliminary observations on sweet potato (*Ipomoea batatas*). Plant and Soil. 155/156: 263-267.

Otani, T; Ae, N. and Tanaka, H. 1996. Phosphorus uptake mechanism of crops grown in soils with low P status II. Significance of organic acids in root exudates of pigeon pea. Soil Sci. Plant Nutr. 42: 553-560.

Ownby J. D. and L. Dees 1985. Growth and mineral nutrient status in peanut and sorghum in response to Acid precipitation and aluminium. New phytol 101, 325-332.

Patel, M.S and Kanzaria M.V. 1985. Factors affecting response of groundnut to P fertilizer in medium black calcareous soil of Saurashtra. Fer. News 30(3): 31-36.

Payne, W.A., M.C. Drew, L.R. Hossner, R.J. Lascano, A.B. Onken and C.W. Wendt, 1992. Soil phosphorus availability and pearl millet water use efficiency. Crop Science, 32: 1010-1015.

Peters, J.B. and E.E. Schulte, 1996. Using plant analysis to guide research and extension programmes in the Gambia. Comm. Soil Sci. Plant Anal. 27: 1311-1329.

Polacco, J.C. 1977. Is nickel a universan component of plant ureases? Plant Sci. Letters 10: 249-255.

Prasad, M.V.R. 1993. Genetic enhancement of groundnut for higher productivity based on canopy development. Oilseeds Research and Development in India: Status and Strategies. National Seminar 2-5 August 1993, DOR, Hyderabad, India p. 2-5.

Radin, J.W. 1984. Stomatal responses to water stress and abscisic acid in phosphorus–deficient cotton plants. Plant Physiology 76: 392-394.

Radin, J.W. and R.C. Ackerson, 1981. Water relations of cotton plants under nitrogen deficiency, III. Stomatal conductance. Plant Physiology 67:115-119.

Radoglou, K.M., P. Aphalo, and P.G. Jarvis, 1992. Response of photosynthesis, stomatal conductance and water use efficiency to elevated CO_2 and nutrient supply in acclimated seedlings of *Phaseolus vulgaris* L. Annals of Botany 70: 257-264.

Rajendrudu, G. and J.H. Williams. 1983. Effect of gypsum and drought on crop yield in early maturing groundnut (*Arachis hypogaea* L.). Experimental Agriculture.

Randall, P. J. 1995. Genotypic differences in phosphate uptake. In: Genetic manipulation of crop plants to enhance integrated nutrient management in cropping systems. I. Phosphorus: Proceedings of an FAO/ICRISAT Expert Consultancy Workshop. Johnson et al. (Eds.). pp. 31-47. International Crop Research Institute for the Semi-Arid and Tropics, Patancheru, Andhra Pradesh, India.

Randall, P.J., E. Delhaize, R.A. Richards and R. Munns, 1993. Genetic aspects of plant mineral nutrition. Kluwer Academic publisher, Dordretch, The Netherlands. 416 pp.

Rangel, Z. 1993. Mechanistic simulation models of nutrient uptake: A review. Plant and Soil, 152: 161-173.

Reiter, R.S. , J.G. Coors, M.R. Sussman and G.H. Gabelman 1991. Genetic analysis of tolerance to low phosphorus stress in maize using restriction fragment length polymorphoism. Theoretical and Applied Genetics 82: 561-568.

Richards, R.A. 1993. Increasing salinity tolerance of grain crops: is it worthwhile? Developments in Plant and Soil Science 50:117-126.

Russelle, M.P. and McGraw, R.L. 1986. Nutrient stress in birdsfoot trefoil. Can. J. Pl. Sci. 66: 933-944.

Saalbach, E. 1973. The effect of sulphur, magnesium, and sodium on yield and quality of agricultural crops. Pontificiae Acad. Sci. Scripta Varia Nr. 38: 541-589.

Shiva, V. 1991. The green revolution in the Punjab. The Ecologist 21: 57-60.

Silva, H. and Rodrigue, J. 1996. Physiological disorders of Pomaceae fruit induced by calcium deficiency. An integrated focus on the problem. *Revista Fruticola*. 17(1): 5-17.

Singh, A.L. 1996. Macronutrient requirement of groundnut and their interaction. In: National Symposium on Modern Trends in Plant Physiology, Department of Plant Physiology G.B. Pant Univ. Agrl. Technology March, 18-20, 1996 p. 19.

Singh A.L. 1999. Mineral Nutrition of Groundnut. In: Advances in Plant Physiology (Ed. A. Hemantranjan), Vol II pp. 161-200. Scientific Publishers (India), Jodhpur, India.

Singh, A.L. 2000a. Mechanism of Tolerance and Crop Production in Acid Soils. In: Plant Physiology, Biochemistry and Plant Molecular Biology in 2000 (Ed. A. Hemantranjan), Vol II. pp. 353-394. Scientific Publisher (India), Jodhpur, India.

Singh, A.L., 2000b. Potassium, calcium and boron fertilization of bold-seeded groundnut in calcareous soil. Proceedings of the GAU-PRII-IPI National Symposium on Balanced Nutrition of Groundnut and other Field Crops Grown in Calcareous Soils of India., 19-22 Sept. 2000. (B.A. Golakiya, J.D. Gundalia, S.K. Bansal. And Patricia Imas ed.) Vol. 2 pp. 199-204. Gujarat Agricultural University, Junagadh.

Singh, A.L. 2002. Potassium influences kernel filling of large-seeded groundnut in calcareous soil. J. Potassium Research 18: 47-52.

Singh A.L. 2002. Effect of macro- and micro-nutrients on yield and pod filling of bold-seeded groundnut. In: K.K. Vora, Karan Singh and Arvind Kumar (Eds.) Production and Developmental Plant Physiology, pp. 381-86 Pointer publishers Jaipur, India.

Singh A.L., M.S. Basu and N.B. Singh, 2004. Mineral Disorders of Groundnut. National Research center for groundnut (ICAR), Junagadh India. p. 85.

Singh A.L, M.S. Basu and N.B. Singh, 2003. Iron deficiency chlorosis and its management in groundnut. National Research Center for groundnut (ICAR), Junagadh India. p. 30.

Singh, A.L. and Vidya Chaudhari, 1991. Screening of groundnut cultivars tolerant to iron chorosis. Indian Journal of Agricultural Sciences. 61: 925-927.

Singh, A.L. and Vidya Chaudhari, 1993a. Screening of groundnut germplasm collection and selection of genotypes tolerant of lime-induced iron-chlorosis. J. Agric. Sci. Camb.121:205-111.

Singh, A.L. and Vidya Chaudhari, 1995. Source and mode of sulphur application on groundnut productivity. J. Plant Nutr. 18: 2739-2759.

Singh, A.L. and Vidya Chaudhari, 1996. Interaction of sulphur with phosphorus and potassium in groundnut nutrition in calcareous soil. Indian J. Plant Physiology New Ser. 1: 21-27.

Singh, A.L. and Vidya Chaudhari 1997. Sulphur and micronutrient nutrition of groundnut in a calcareous soil. Journal of Agronomy and Crop Science, 179(2): 107-114.

Singh, A.L. and D. Dayal, 1992. Foliar application of iron for recovering groundnut plants from lime-induced iron deficiency chlorosis and accompanying losses in yields. J. Plant Nutr., 15: 1421-1433.

Singh, A.L. and Y.C. Joshi, 1993. Comparative studies on the chlorophyll content, growth, N uptake and yield of groundnut varieties of different habit groups. Oleagineux 48:27-34.

Singh A.L. and Y.C. Joshi, 1997. Prevention and correction of iron-deficiency chlorosis of groundnut in India. In: Plant Nutrition for Sustainable Food Production and Environment (Eds. T. Ando et al.), pp. 271-272. Proceedings of the 13th International Plant Nutrition Colloqüium, Tokyo University of Agriculture, Tokyo, Japan 13-19 Sept. 1997. Kluwer Academic Publisher, printed in Japan.

Singh, A.L. and Y.C. Joshi, 2001. Dynamics of sulphur, iron and magnesium and their nutrition in groundnut in calcareous soils of India. Proceedings of the GAU-PRII-IPI National Symposium on Balanced Nutrition of Groundnut and other Field Crops Grown in Calcareous Soils of India, 19-22. Sept. 2000, pp. 103-122. Gujarat Agril. University, Junagadh.

Singh, A.L., Y.C. Joshi, Vidya Chaudhari, and P.V. Zala, 1990a. Effect of different sources of iron and sulphur on leaf chlorosis nutrient uptake and yield of groundnut. Fertilizer Research 24: 85-96.

Singh, A.L., Y.C. Joshi, Vidya Chaudhari and P.V. Zala (1990b) Effect of different sources of iron and sulphur on nutrient concentration of groundnut. Fert. Res., 24, 97-103.

Singh, A.L., Vidya Chaudhari and V.G. Koradia, (1991). Foliar nutrition of nitrogen and phosphorus in groundnut. In D.N. Tyagi et al. (editors) Physiological Strategies for Crop Improvement: Proceedings of the International Conference of Plant Physiology. pp. 129-133. B.H.U., Varanasi, India.

Singh, A.L., Vidya Chaudhari, V.G. Koradia and P.V. Zala, 1995. Effect of excess irrigation and iron and sulphur fertilizers on the chlorosis, dry matter production, yield and nutrient uptake by groundnut in calcareous soil. Agrochimica 39: 184-198.

Singh, A.L., P.K. Ghosh and Devi Dayal 1997. Nutrient management in groundnut and groundnut-based cropping systems. In: Nutrient Management Practices in Crops and Cropping Systems (Eds. C.P. Ghonsikar, and V.S. Shinde.), pp. 157-190. Scientific Publication (India), Jodhpur.

Smith, S.E., M.J. Hawkesford, M.I. Prosser and D.T. Clarkson 1993. Approaches to cloning genes encoding for nutrient tansportation in plants. In Plant Nutrition –from genetic engineering to field practice (N.J. Barrow, ed.) pp. 151-154. Kluwer academic press, Dordrecht, The Netherlands.

Sonneveld, C. and Voogt, W. 1991. Effects of Ca-stress on blossom-end rot and Mg-deficiency in rockwool grown tomato. *Acta Hort*. 294: 81-88.

Spehar, C.R. and Galwey, N.W. 1997. Screning soybean (*Glycine max* L. Merrill) for calcium efficiency by root growth in low Ca nutrient solution. *Euphytica*. 94(1): 113-117.

Stuiver, C.E.E., Kok, L.J., Westerman, S. and Dekok, L.J. 1997. Sulphur deficiency in *Brassica oleracea* L: Development, biochemical characterisation and sulphur/nitrogen interactions. *Russian J. Plant Physiol*. 44(4): 505-513.

Tamada, T. 1989. Nutrient deficiency of rabbiteye and highbush blue barries. *Acta Hort*. 241: 132-138.

Tanksley, S.D., N.O. Young, A.H. Paterson and M.W. Bonierbale. 1989. RFLP mapping in plant breeding: new tools for an old science. Biotechnology 7: 257-264.

Tandon, H.L.S. 1987. Phosphrus Research and Agriculture Production in India. FDCO, New Delhi, India. 160 pp.

Tandon, H.L.S. 1991. Sulphur Research and Agriculture Production in India 3rd edition FDCO, New Delhi, India.

Torres, A.M., N.F. Weeden and A. Martin 1993. Linkage among isozyme, RFLP and RAPD markers in *Vicia faba*. Theoretical and Applied Genetics 85:937-945.

Tyerman, S.D. and D.P. Schachtman, 1992. The role of ion channels inplant nutrition and prospects for their genetic manipulation. Plant and Soil 146: 137-144.

Varvel, G.E. 1995. Precipitation use efficiency of soybean and grain sorghum in monoculture and rotation. Soil Science Society of America Journal 59: 527-531.

Von Uexkull, H.R. and E. Mutert 1995. Global extent, development and economic impact of acid soils. Plant and Soil 171: 1-15.

Vose P.B. 1982. Iron nutrition in plants: A world overview. J. Plant Nutr. 5: 233-249.

Walkar, M.E., R.A. Flowers, R.J. Henning, T.C. Keisling and B.G. Mullinex. 1979. Response of early bunch peanuts to Calcium and Potassium fertilization. Peanut Sci. 6: 119-123.

Walkar, M.E., W.D. Branch, T.P. Gaines and B.G. Mullinex. 1984. Response of nodulating and nonnodulating peanut to foliar applied nitrogen. Peanut Science. 11: 60-63.

Wallace, A., E.M. Romney and C.V. Alexander. 1976. Lime-induced chlorosis caused by irrigation water. Commun. Soil Sci. Plant Anal. 7: 47-49.

Whiteaker, G., G.C. Gerloff, W.H. Gabelman and D. Lindgren 1976. Intraspecific differences in growth of beans at stress levels of phosphorus. Journal of the American Society of Horticultural Science 101: 472-475.

Williams, J.H. 1979. The physiology of groundnuts (*Arachis hypogea* L. cv. Egret). II. Nitrogen accumulation and distribution. Rodesian J. Agr. Res. 17: 49-55.

Wong, M.T.F., A. Wild and A.U. Mokwunye 1991. Overcomming soil nutrient constraints to crop production in West Africa: importance of Fertilizers and priorities in soil fertility research. Fertilizer Research 29: 45-54.

Xu, G.; Magen, H.; Tarchitzky, J. and Kafkafi, U. 2000. Advances in chloride nutrition of plants. *Adv. Agron.* 68: 97-150.

Yamada, T. 1975. Calcium and its relationship to blossom-end rot in tomato. Comm. soil Sci. Plant Analysis. 6:273-284.

Zharare, G.E., C.J. Asher and F.P.C. Blamey, 1997. Net influx of calcium and efflux of potassium in groundnut pods grown in solution culture. In Plant Nutrition—For Sustainable Food Production and Environment (T. Ando et al. Eds.) pp 177-178. Kluwer Academic Publishers, Printed in Japan.

Zeiger, E. 1983. The biology of stomatal guard cells. Annual Review of Plant Physiology 34: 441-475.

Zobel, R.W. 1991. Genetic control of root systems. In Plant roots: the hidden half (Y. Waisel et al eds.), pp. 27-38. Marcel Dekker New York, USA.

CHAPTER 16

Bambara Groundnut: Its Physiology and Introduction in India

A.L. Singh and M.S. Basu

INTRODUCTION

Bambara groundnut (*Vigna subterranea* L. Verdc.) is a legume crop grown by resource poor farmers mainly in semi-arid parts of African countries. It is an indigenous African crop that has been cultivated for centuries from Senegal to Kenya, and from the Sahara to South Africa and Madagascar, but still remains as any underutilised crop. It is a native grain legume south of the Sahara. We can call it a cousin of the groundnut as its outward appearance is same and belongs to the same family. It is a hardy crop due to its drought tolerance, resistance to pests and diseases and ability to yield on poor fertility soils. It is grown either as single crop or as intercrop in mixed cultivation with other crops of semi-arid region. It is extensively cultivated in western Africa where about 0.35 million tons of dry seed are produced on 0.40 m ha annually. Nigeria, Niger, and Ghana, are the major producers of bambara groundnuts where it is third in importance only to cowpea and groundnut (Howell 1994). In south of the Sahara, it is the most widely grown native grain legume second only to the groundnut. The IITA, Nigeria is maintaining nearly 2000 germplasm accessions having variability and yield potentials. Though it is an important legume crop of semi-arid region, its large scale cultivation is restricted in African countries only. However, attempts are being made to popularise the same in many other parts of the world wherever the climate suits for cultivation of the same. In India, the bambara groundnut was introduced only in 2002 from the University of Nottingham, UK. Since then it is being studied for its feasibility of cultivation in various parts of India (NRCG News Letter 2003a and 2003b).

There are several physiological processes occurring right from seed germination to harvest and systematic study of these is essential for the success of bambara crop in a given area. As it is a recent introduction in India some of the physiological studies and prospect of its cultivation in

National Research Centre for Groundnut, P.B.5, Junagadh 362001, India
E-mail: alsingh@nrcg.res.in and alsingh_ad1@sancharnet.in

Indian context are being highlighted here for sensitising researchers to give more emphasis on its cultivation in arid and semi-arid regions of India.

ENTRY OF BAMBARA IN INDIA

Ten landraces of bambara groundnut namely, S16-5A, S19-3, SB-4-2, Gabc, DODR-TZ, DODC-TZ, DIPC, Uniswa Red, AS-17, AHM-753 from the University of Nottingham, UK were imported by the National Bureau of Plant Genetic Resources, New Delhi based on the indent of National Research Centre for groundnut, Junagadh, Gujarat, India. After reaching in India, these genotypes were first tested through grow-out test at the Regional Center of NBPGR, Hyderabad and NRCG, Junagadh during 2002, to avoid any new diseases. After finding free from diseases and pests these were grown in pots as well as under field condition at two different locations in India i.e. Junagadh, Gujarat (Western India) and Hyderabad, Andhra Pradesh (Southern India) where these were found growing well. Later on these bambara genotypes were grown at UAS, Banglore and Bhabha Atomic Research Centre at Mumbai also.

The pod and seed characteristics of bambara groundnut genotypes being maintained at NRCG Junagadh are listed in Table 16.1. Of these, promising genotypes are being multiplied at

Table 16.1: Pod and Seed characteristic of various Bambara groundnut genotypes available at NRCG, Junagadh

S. No	*Bambara Groundnut genotypes*	*Shelling (%)*	*100 seed mass (g)*	*Seed size and testa colour*	*Other characteristics*
1	S16-5A	80	29	large round, and greyish dark spotted seed	Single seeded pod with prominent beak
2	SB-4-2	77	15	Very small, dark pink	Single seeded small pods
3	S19-3	82	25	Medium size, reddish black	Single seed, occasionally two seeded pod
4	Gabc	79	19	Small off-white seed with brown ring around the placenta	Single seeded, small greenish pods
5	DODR-TZ	76	31	Bold, red seed	Double seeded dark violet blackish colour big pods
6	DODC-TZ	78	32	Off-white bold seed	Single seeded big sized pods
7	DIPC	85	29	Bold off-white to yellow seed with black ring around the placenta	Single seed pale yellow pod
8	Uniswa Red	82	22	Red	Single seed yellow to orange-yellow pod
9	AS-17	72	31	Bold seed of off-white colour with variegation as a black stint on both side of the placenta	Single seeded pod
10	AHM 753	83	20	Small, dark red seed	Single seeded small greyish pod

NRCG, Junagadh and after obtaining sufficient seed, these will be tested at various locations mainly in the arid and semi-arid parts of India and also the seed will be made available for research as well as for cultivation. For information to the readers a project on bambara groundnut has been launched in India with the financial help of European countries at the end of 2005, to study the physiological requirements and nutritional quality of the crop for its further popularisation in India. Under this project several other promising genotypes of bambara from other countries and will be introduced in India.

CLIMATE AND SOIL REQUIREMENT

Like groundnut, the bambara groundnut also requires semi-arid to arid type of climate and it grows well in Gujarat at Junagadh, Andhra Pradesh at Hyderabad, Maharashtra at BARC Mumbai, Karnataka at Banglore during both the kharif and rabi seasons. Though it can be grown on any soil type, it prefers light textured soil as the formation of pods are of terrestrial nature and formed both below and above ground.

The FAO agricultural information series through BAMnut Model on crop biomass and pod model described prediction and potential biomass production of bambaranut for each continent (Bannayan *et al*, 2000). From this model, it is interesting to note that almost 75% of South America and 25% of Asia, two regions of the world not previously associated with bambara groundnut, are classified in the very suitable (VS), suitable (S) or moderately suitable (MS) ranges for bambara groundnut cultivation. The comparable figure for Africa, where almost all bambara groundnut is currently grown, is 65%. Within Asia, all of India, the Lao People's Democratic Republic, Philippines, Vietnam, Sri Lanka, Malaysia, Cambodia, Indonesia, Thailand, and Bangladesh fall within categories S and MS.

PHENOLOGY AND YIELD

Since bambara groundnut has been introduced in India, it is essential to study their phenology climatic and seasonal requirements and yield potential to assess its feasibility of cultivation in various parts of India. Accordingly, a systematic study in field as well as in pots started at this centre in Junagadh during 2002 to find out the seasonal requirement of the crop. Under which 10 bambara groundnut genotypes (S16-5A, SB-4-2, S19-3, Gabc, DODR-TZ, DODC-TZ, DIPC, Uniswa Red, AS-17, AHM-753) are being sown in field round the year at an interval of 15 days under natural climatic conditions and various parameters such as days of emergence, flowering, pod formation and maturity and yield are being recorded.

There are several physiological processes occurring right from seed germination to harvest, however, expansion, podding and filling are the three major phenophases in bambara groundnut which are discussed here.

Germination and Growth

In field study, a lot of variation in the germination, flowering, fruiting and maturity days were observed with respect to sowing dates of bambara groundnut (Singh *et al*, 2005). Though genotypic variations were observed, shifting the sowing dates had greater influence on the flowering and crop sown in July onwards took more time for flowering but took almost similar time for maturity as the fruiting period, in the subsequent sowing dates, was reduced which has resulted in reduced yield. Depending upon the genotypes the bambara groundnut took, 6-17 days for germination and 26-40 days for flowering and 120-135 days to maturity if sown during mid of June (Table 16.2). However, shifting the dates of sowing under field conditions had greater influence on the flowering. The crop sown in November to January took much time, sometimes more than 20 days, for their germination with poor seedling vigour, and plant stand. As a result there was unhealthy crop with delayed flowering showed poor and sometimes no yield. The time taken to flower were 40-60 days in July sown crop, 62-80 days in August sown crop, 48-64 days in September and 51-67 days in October sown crop (Singh *et al*, 2005).

Table 16.2: Time required for field emergence, flowering and maturity and respective yield of bambara groundnut at optimum sowing time in Junagadh condition

Sr. N.	Genotypes	Date of sowing	Days to emergence	Days to flowering	Days to maturity from		Pod yield g/plant
					D.O.S.	D.O.E	
1	S16 5A	10-7-03	8	44	134	127	6
2	SB 4-2	10-7-03	7	43	140	134	6
3	AHM 753	10-7-03	8	44	134	127	17
4	Uniswa Red	10-7-03	8	43	134	127	10
5	DODR TZ	10-7-03	8	43	134	127	21
6	S 19-3	10-7-03	8	43	134	127	17
7	DIPC	10-7-03	12	43	134	127	15
8	AS 17	10-7-03	17	55	134	127	18
9	DODC TZ	10-7-03	11	60	140	134	14
10	Gabc	10-7-03	13	43	140	134	3

However, Singh *et al* (2004) in a sand culture study reported that, under normal weather conditions of Junagadh with weekly mean temperature of 29.5°C (with maximum 34.2°C and minimum 26.5°C) and proper nutrient supply, the bambara groundnut genotypes, sown on 18th June the took 5-7 days for its germination (Table 16.3). It produced 10-15 leaves and 5-6 branches in 30 days, and 50-90 leaves and 20-50 branches in 60 days. Generally, the crop attains full growth, it has got indeterminate type of growth habit and the flowering and fruiting go on. The pod initiation started from 40 days and matured within 80-90 days. The genotypes DIPC and DODC-TZ showed fast germination as it took only 5 and 6 days, respectively for seedling emergence. The other genotypes took 7 days for emergence. The days of emergence depend mainly on the temperature prevailing during the sowing and vary with season. The temperature during June was congenial for bambara groundnut and early emergence was also accompanied with early vigour.

The plant showed two leaves at the time of emergence, which took 3-4 days to develop into full shape. The number of leaves increased to 3 in the second week and 4 in the third week. The branching in bambara started in the third week with 2-4 branches. Though genotypic

Table 16.3: Phenology of Bambara groundnut grown in sand culture pot experiment in Junagadh during Kharif 2003

S. N.	Genotypes	Days required for			Period of flowering (F) and podding (P) DAP		Days to maturity from		Leaves and branches $plant^{-1}$ at 70 DAP		Pods $plant^{-1}$	Pod yield (g $plant^{-1}$)
		G	F	P	F	P	DAP	DAE	L	B		
1	S 16 5A	7	37	44	37-120	44-130	130	123	80	46	21	8.0
2	DIPC	5	39	45	39-110	45-120	120	115	82	40	24	9.0
3	DODR TZ	7	37	44	37-120	44-131	131	125	62	42	36	25.0
4	Gabc	7	42	50	42-120	50-130	130	123	48	28	22	7.0
5	AHM 753	7	46	53	46-118	53-118	118	113	68	36	29	8.0
6	DODC TZ	6	56	65	56-150	65-154	154	148	96	59	18	10.0
7	Uniswa-Red	7	40	47	40-120	47-133	133	126	46	35	29	13.0
8	SB4-2	7	38	44	38-120	44-133	133	126	78	50	67	29.0
9	AS 17	7	41	47	41-110	47-117	117	110	61	38	30	16.0
10	S 19-3	7	39	45	39-110	45-117	117	110	68	37	32	22.0

Where G is germination, F is flowering, P is podding, L is leaves, B is branches, DAP is days after planting and DAE is the days after emergence

differences exist, in general, the growth was slow till 3rd week of planting, but from the 4th week onwards the growth was observed in a linear fashion and in the 4th week itself 3-5 new leaves and 2-3 branches were added. The number of leaves produced were maximum during 8-10 WAP, however, the maximum number of branches were produced during 7th WAP (Table 16.3). By the end of 10th WAP the bambara groundnut developed full canopy with 28-59 branches and 48-96 leaves. In another experimental the bambara groundnut produced 10-15 leaves and 5-6 branches in 30 days, and 50-90 leaves and 20-50 branches in 60 days (Singh *et al*, 2004).

Thus the growth and dry matter accumulation in bambara groundnut crop showed a lag phase in early growth, exponential increases in weight from vegetative to flowering stage, a linear and maximum growth rate during late vegetative to early pod filling, and levelling of weight during late pod filling stage. Among the various genotypes DODC-TZ showed most vigorous growth with maximum number of leaves (96) and branches (59) at 70 DAP. However, the Gabc and Uniswa red showed poor growth with only 28 and 30 branches and 48 and 46 leaves, respectively at 70 DAP. The growth and development of S16-5A, DIPC and SB-4-2 were higher side with 40-50 branches and 78-82 leaves at 70 DAP, however, the genotype S19-3, DODR-TZ AS-17 and AHM-753 showed medium growth with 36-42 branches and 60-70 leaves.

Flowering, Podding and Photoperiodism

There was a lot of variation in the days to flower of various genotypes ranging from 37-60 days (Table 16.2 and 16.3). The genotype S16-5A took only 37 days to flower, however, DODC-TZ took as many as 56-60 days to flower at Junagadh condition. Based upon the flowering, the genotypes were categorised into: early (S16-5A, SB-4-2, S19-3, DODR-TZ and DIPC flowering in less than 40 days), medium (Uniswa Red, Gabc, AS-17 and AHM-753 flowering in 40-50 days) and late (DODC-TZ flowering in more than 50 days) flowering days. The duration of crop was less in the genotypes with early types of flowering and in contrast the late flowering genotypes were of longer duration.

The podding in bambara groundnut started within 5-8 days of flowering and took about 40 days for development and maturity. However, the crop looks like an indeterminate type as the flowering and podding continue till the maturity of the plant. As long as moisture is there the plant add new pods and, hence, it is difficult to determine the maturity of the crop. At Hyderabad conditions, the SB 4-2 flowered earliest (44 days) whereas DIPC took 54 days to flower. The flowering, canopy development and maturity of bambara groundnut accessions introduced in India were at par with groundnut genotypes grown under the similar conditions (Basu *et al*, 2004).

In a greenhouse experiment the bambara selections from Botswana, Zimbabwe and Mali grown under 3 temperatures (20.9, 23.4 and 26.2°C) and 4 photoperiods (10, 12.5, 13.5 and 16 h) showed that the start of flowering was influenced by temperature but not by photoperiod, while the onset of pod growth was influenced by both factors (Brink, 1997). However, shifting the dates of sowing under field conditions had greater influence on the flowering and crop sown in July onwards took more time for flowering but took similar time for maturity. The time taken to flower were 40-60 day in July sown crop, 60-70 days in August sown crop and 48-67 days in September and October sown crop (Singh *et al*, 2004).

Aspects of the photoregulation of phenological development in this crop was studied in a glasshouse experiment by Linnemann (1996) and the influence of a 14 h photoperiod (which retards podding) during a period of variable length prior to an 11 h photoperiod (which induces podding) on flowering, yield and on the position of pods on the plants determined and data indicated that delay in the induction of podding resulted in higher seed yields and a more synchronised development and, hence, maturity of pods.

Maturity Duration and Yields

Under favourable environment, which generally was observed from July to October at Junagadh, the crop attained its full growth with 50-90 leaves and 20-50 branches in 60 days, the pod initiation started from 40 days and matured within 80-90 days of initiation of fruiting. Most of the genotypes showed indeterminate type of growth habit as the flowering and fruiting go on and it was difficult to find out the maturity period. However, offset of rainy season or withholding of irrigation enhanced maturity.

The bambara groundnut took 127-134 days for its maturity from emergence and 134-150 days from sowing in field experiments (Table 16.2) and 110-148 days for its maturity from emergence and 117-154 days from sowing in pot experiments (Table 16.3). Under proper nutrient supply in sand culture pot experiments, the genotypes AS-17, AHM-753 and S19-3 matured in less than 120 days, DIPC required 120 days. However, the genotypes S 19-3, DODR-TZ, SB 4-2, Gabc and, Uniswa red, took more or less 130 days duration. The maximum maturity duration of 154 days was observed in the DODC-TZ.

Among the various dates of sowing, from July to December, the crop sown on 10th July took least time for flowering followed by 27th September sown crop (Singh *et al*, 2005). Thus, for the kharif season crop the optimum sowing time was first week of July. There was also an indication of rabi season crop as the September sown crop showed all these parameters next to July sown crop. All the bambara groundnut genotypes S16-5A, SB-4-2, S19-3, Gabc, DODR-TZ, DODC-TZ, DIPC, Uniswa Red, AS-17, AHM-753 sown was found growing well at Junagadh condition when sown during 10th July. However, genotypes Gabc and S19-3 also did well during September sowing.

The number and yield of pods recorded in various genotypes ranged 18-67 pods plant^{-1} and 7-29 g pod plant^{-1}, respectively. Among these the high-yielding genotypes, were SB 4-2, DODR-TZ, S 19-3, and AS 17 which showed more than 15 g pod plant^{-1}. However, the genotypes S19-3, DIPC, Gabc, AHM-753, showed less than 10 g pod plant^{-1} and were poor-yielder. However, under field conditions in calcareous soil of Gujarat there was a lot of fluctuation in yield and the genotypes DODR-TZ, Uniswa red, DIPC DODC-TZ and AHM 753 were high-yielder with 10-25 g pod yield plant^{-1} and AS-17, AHM-753 and S19-3, were of short duration with maturity period 100-110 days (Singh *et al*, 2005).

In sand culture pot experiment in Junagadh conditions, the high-yielding bambara genotypes, were AS 17, S 19-3, AHM 753, SB 4-2 and, Uniswa red, if sown during June which showed 15-29 g pod/plant. Of these AS-17, AHM-753 and S19-3, were of short duration with maturity period 100-110 days. However, under field conditions in calcareous soil of Gujarat at Junagadh there was a lot of fluctuations in yield and the genotypes DODR-TZ, Uniswa red, DIPC DODC-TZ and AHM 753 were the high-yielder with 10-25 g pod yield per plant.

At Hyderabad condition, amongst the ten accessions evaluated, DODR-TZ from Tanzania had the highest number of matured pods per plant (38/plants) followed by AHM 753 and Uniswa Red. The lowest yield was recorded in Gabc (10 pods/plant) (Table 16.4). The weight of 100 seed mass varied from 30.6 g (DOD C) to 54.5 g (Basu *et al*, 2004).

Table 16.4: Bambara groundnut grown in grow-out test at Hyderabad, India during 2002

S No.	Genotypes	Total no of plants	Total no of pods	Pod weight (g)
1	S16-5A	9	132	52.36
2	SB-4-2	10	74	11.77
3	S19-3	7	122	35.21
4	Gabc	5	52	11.83
5	DODR-TZ	9	293	90.46
6	DODC-TZ	9	82	31.55
7	DIPC	8	126	43.79
8	Uniswa Red	10	177	47.70
9	AS-17	9	156	56.71
10	AHM 753	9	170	37.06

In general, the flowering, canopy development and maturity of bambara groundnut accessions, in Junagadh, were at par with Virginia bunch groundnut genotypes grown under similar conditions. Among the various genotypes the high-yielding one were AHM 753, S19-3, AS-17, Uniswa Red, DODR-TZ, DIPC and DODC-TZ, if sown during July with 10-21 g pod/plant (Table 16.1). Of these, AS-17, AHM-753 and S19-3 were of short duration and are most fit to the area with lesser length of growing period. The FAO agricultural information series through BAMnut Model predict India as suitable for potential biomass production of bambara nut (Bannayan *et al*, 2000).

Now, two years of study and the data on germination, flowering, fruiting, maturity days and yield have shown that in Junagadh condition the best season for growing bambara is July to October when crop is sown on 10th July. Under favourable environment, which generally was observed from July to October, the crop attained its full growth with 50-90 leaves and 20-50 branches in 60 days, the pod initiation started from 40 days and matured within 80-90 days of initiation of fruiting.

Thus three genotypes DODR-TZ, Uniswa Red, and AHM 753 could be multiplied in bulk and tested in multilocation for its suitability in other parts of India. However, the short duration genotypes AS-17, AHM-753 and S19-3 should be tested in the Rajasthan and other area where the length of growing period is less.

Physiological Traits Associated with Water Use Efficiency (WUE)

To exploit the intrinsic genetic variability, a preliminary study was conducted in selected 10 cultivars of bambara groundnut (S16-5A, S19-3, Gabc, DODR-TZ, DODC-TZ, DIPC, UNISWA Red, AS-17, AHM-753). The physiological traits associated with WUE were determined at single leaf level using gas exchange approach. Significant genotypic variability was observed in gas exchange traits along with SCMR (SPAD Chlorophyll Meter Reading, an indicator of leaf N status that influences the mesophyll capacity, associated with carboxylation efficiency of RuBisCO to fix carbon) (Table 16.5). Variations in A/g_s (intrinsic WUE) and Ci/g_s (*in vivo* carboxylation efficiency, reflects the carbon fixation ability of the Chloroplast) among the bambara cultivars, suggest the possibility of exploiting this legume crop for desirable physiological characters. The correlation matrix (Table 16.6) provides additional information on the inter-relationship between several physiological traits.

Table 16.5: Variations in SCMR, SLA (Specific Leaf Area), SLN (Specific Leaf N) and gas exchange traits in selected bambara groundnut cultivars (Basu *et al*, 2004)

Genotypes	SCMR	A	g_s	T. rate	Ci	SLA	A/gs	Ci/gs	SLN	$\Delta^{13}C$
DODR-TZ	43.35	18.5	650	7.15	248	168	28.46	38.15	2.23	18.93
UNISWA-RED	44.50	16.5	660	9.25	270	157	25.00	40.91	2.04	18.87
S-16-5A	49.00	15.5	780	11.5	285	171	19.87	36.54	2.36	19.88
Gabc	41.50	9.0	600	10.3	300	175	15.00	50.00	2.33	19.33
DIPC	43.00	12.5	400	6.15	255	170	31.25	63.75	2.05	19.42
S-19-3	53.00	11.15	515	8.85	280	159	21.65	54.37	2.34	19.85
AHM-753	41.50					183			1.89	19.96
SB-4-2	43.20					156			1.96	20.71

Table 16.6: Correlation matrix of various physiological traits (Basu *et al*, 2004)

Traits	SPAD	A	g_s	T. rate	Ci	SLA	A/gs	Ci/gs	SLN	$\Delta^{13}C$
SCMR	1.00									
A	–0.06	1.00								
g_s	0.11	0.51	1.00							
T. rate	0.31	–0.17	0.74	1.00						
Ci	0.21	–0.66	0.29	0.84	1.00					
SLA	–0.43	–0.29	0.04	0.10	0.22	1.00				
A/gs	–0.19	0.54	–0.44	–0.87	–0.96	–0.24	1.00			
Ci/gs	–0.04	–0.69	–0.96	–0.53	–0.05	0.09	0.24	1.00		
SLN	0.59	–0.32	0.40	0.62	0.62	–0.05	–0.74	–0.27	1.00	
$\Delta^{13}C$	0.15	–0.48	–0.04	0.40	0.48	–0.07	–0.39	0.25	–0.26	1.00

Index: A- Assimilation/Photosynthetic rate (mmols), g_s- stomatal conductance (mmols), T. rate- Transpiration rate (mmols), Ci-Intercellular CO_2 levels (ppm), A/g_s- Intrinsic WUE, Ci/g_s- *In vivo* Carboxylation efficiency and $\Delta^{13}C$- Carbon Isotope Discrimination.

Further, as a surrogate estimate of WUE, Carbon Isotope Discrimination ($\Delta^{13}C$) was determined in these selected cultivars. Since, the quantification of stable isotope ratios using a Continuous Flow Isotope Ratio Mass Spectrometer (IRMS) is rapid and automated, it offers a high throughput technique for the screening of large number of bambara accessions or populations in near future for broad spectrum variability in WUE and associated traits (Basu *et al,* 2003). Interestingly, the strong positive correlation of $\Delta^{13}C$ between field and container grown plants further reiterates the low G × E and stability of this trait. On a whole, $\Delta^{13}C$ provides ample scope for rapid phenotyping of large germplasm accessions for the differences in WUE and choose appropriate parental lines for crop improvement programs (Basu *et al,* 2004b). Improving WUE in bambara groundnut would assume great significance to sustain or even increase productivity under rain-fed conditions.

NUTRIENT REQUIREMENT AND FERTILISERS

The essential elements are assimilated into plant through absorption by root or other plant parts as ions from the soil and like groundnut the bambara groundnut also require all the macro-nutrient (N, P, K, Ca, S and Mg) and micro-nutrients (Fe, Mn, Zn, Cu, B, Mo and Cl).

As the pod in bambara groundnut develops both under soil and above ground, the seed nutrition is both directly through pod and from root, shoot and back to the seed.

As bambara groundnut grows well in Junagadh, it is essential to know the fertiliser requirement of the same. Accordingly eight genotypes, S16-5A, SB-4-2, S19-3, Gabc, DODR-TZ, DODC-TZ, DIPC, Uniswa Red, AS-17, AHM-753 were sown in field for studying their fertiliser requirement in Indian condition. The N, P, K, Ca and S at a rate of 25, 50, 50, 100 and 30 kg ha^{-1} were applied and performance of these genotypes was adjudged. There was genotypic variation in the response to fertiliser application in calcareous soil of Gujarat. In general, the genotype DODR-TZ and DODC-TZ and Uniswa Red respond to all the fertilisers and genotypes DIPC, AS 17, and S16-5A respond to most of the nutrients, except one or two. On the other hand, the genotypes AHM-753 and S19-3 did not respond to any of the nutrient.

In another field experiment the response of N, P, K, Ca and S fertilisers applied at 50, 50, 50, 100 and 30 kg ha^{-1}, respectively, were studied in 8 bambara groundnut genotypes, namely, S16-5A, S19-3, DODR-TZ, DODC-TZ, DIPC, Uniswa Red, AS-17, AHM-753 (Table 16.7). The study reveals that there was a great response of all the primary and secondary nutrient fertilisers on bambara groundnut, though genotypic variation were observed (Singh *et al,* 2005). The genotypes DODR-TZ and DODC-TZ and Uniswa Red, which are high-yielding, respond to all the fertilisers and genotypes DIPC, AS 17, and S16-5A respond to most of the nutrients, except one or two. On the other hand, the genotypes AHM-753, S19-3 did not respond to most of the nutrients. As the pod of bambara groundnut develops both under soil and above ground, the seed nutrition is both directly through pod and from root, shoot and back to the seed. In Africa, traditionally the bambara groundnut is grown mostly without any fertiliser and manure rather it enriches the soil itself. In this study also two genotypes AHM-753 and S19-3 did not responded to most of the nutrient probably due to having better inbuilt and efficient mechanism for mining nutrients from the soil as most of the nutrients are available in the soil. However, for most of the high-yielding genotypes, the fertilisers are essential for realising high-yield.

Table 16.7: Effect of different fertiliser treatments on yield of bambara

Sr.no.	Genotypes	Pod yield (g/plant)						Days to maturity from		Shelling (%)
		Control	N_{50}	P_{50}	K_{50}	Ca_{100}	S_{30}	DOS	DOE	
1	S16 5A	8.0	9.0	11.0	10.0	15.0	8.0	130	120	78.4
2	S19-3	5.0	10.0	6.0	6.0	5.0	6.0	130	120	86.6
3	DODR TZ	6.89	24.0	20.0	18.0	22.0	18.0	135	125	80.8
4	DODC TZ	4.42	15.0	13	31.1	28.0	26.6	145	135	79.3
5	DIPC	6.0	8.0	10	9.0	15.1	9.0	130	120	83.6
6	Uniswa Red	5.0	11.0	7.0	11.0	8.0	7.0	135	125	78.7
7	As 17	5.0	6.0	9.0	9.0	9.0	22.0	135	125	77.9
8	AHM 753	6.0	5.0	6.0	7.0	6.0	7.0	130	120	83.9

Nitrogen and S fertilisation had influenced the yield of five genotypes only of the eight genotypes studied. However, P, K and Ca each showed its effect on 6 genotypes. Two genotypes did not show any response of fertiliser and were exceptional. These genotypic variations were mainly due to their efficiency of utilisation of the mineral from the soil labile pool. The genotypes DODR-TZ and DODC-TZ, both high-yielding, were highly responsive to all the nutrients studied and, hence, application of fertiliser in these genotypes is essential. Besides increasing pod yield and size of seed, the N fertiliser also influenced the time to plant maturity (faster maturity).

Phosphorus fertilisation increased shoot, root, and nodule dry weights, seed yield, shoot P and N contents. Nitrogen fertilisation, on the other hand, had no effect on yield but influenced the time to plant maturity. Under non-limiting P conditions, bambara groundnut seems to be able to meet its N requirement from N_2-fixation and soil mineral N, but at the onset of podding, when a major shift in assimilate partitioning from vegetative growth to pod filling takes place there appears to be need of N fertilisers. The flowering stage is critical for nodule formation in bambara groundnut and N fertilisation at 50% flowering decreased nodule number and nodule dry weight and shoot N content. However, N fertilisation, irrespective of time of application or growth stage was not beneficial to plant growth and reproduction. Thus, as bambara groundnut can meet its major nitrogen requirement from N_2 fixation and soil N, it is recommended to use only 25 kg N/ha as basal to boost the plant growth initially.

Due to its high requirement, there is need for supplementary application of P, Ca, K and S fertilisers as these are important for growth and yield of bambara groundnut. However, under unfertilised conditions also some genotypes grew well, of which some did not respond to fertiliser application, however some gave still better yield if fertiliser are given. The low nutrient requirement of some of the genotypes may be responsible for its ability to thrive in poor fertility soils. However, for a positive response to these fertilisers to occur, the soil moisture content must be adequate and the fertiliser should be available in the root and pod zone. Bambara groundnut has a low P requirement and a high P efficiency compared to other crops like cereals. The P uptake and the internal P use efficiency of bambara groundnut when compared to those of maize and pigeonpea, the maize had the highest internal P use efficiency of 1.89 g DM per mg P uptake, against 0.79 for bambara groundnut and 0.56 for pigeonpea.

In a sand culture experiment during kharif season, bambara genotypes were grown at sufficiency level of all the essential elements where all the genotypes grew well with maximum flowering and fruiting. The critical level of P in leaves was around about 0.15%, however, it ranged from 0.15 to 0.20% depending upon the genotypes. The shoot P concentrations at the early podding stage seem to be a suitable guide for monitoring the P status and P requirement of bambara groundnut.

Thus the fertiliser requirement of bambara groundnut is more or less similar to groundnut and fertilisation with N, P, K, S and Ca is essential for harvesting good crop of bambara in calcareous soils of Gujarat.

IRRIGATION AND DROUGHT TOLERANCE

Bambara groundnut is an underutilised legume which provides security for many African farmers as it shows considerable drought resistance. Bambara has high intrinsic stress tolerance traits. Though bambara groundnut grows without irrigation as rainfed, higher yields can be realised through life saving timely irrigation during protracted drought. Irrigation at podding stage is most critical as it also produces both underground and aboveground pods near the root. The irrigation increases the absorption and uptake of nutrients required for growth and reproduction. Owing to its drought tolerance characteristics, early maturity and reasonable yields under stress environments, bambara groundnut can be chosen as an intercrop to fit into the groundnut based cropping system. In Africa, traditionally, many farmers grow bambara groundnut in the same field with millet, maize, sorghum or yam as an intercrop.

In a study conducted at NRCG, Junagadh a strong positive correlation between the $D^{13}C$ values of the field and pot grown genotypes of bambara groundnut was observed suggesting the stability of this trait (Basu *et al*, 2003). Therefore, this technique based on the $D^{13}C$ can be adopted for the rapid phenotyping of large germplasm accessions for the differences in WUE and choose appropriate parental lines for crop improvement programme. Improving WUE and to some extent the water uptake mechanisms in bambara groundnut would assume great significance to sustain or even increase productivity under rain-fed conditions.

NUTRITIONAL QUALITY AND USES

Bambara groundnut is an underutilised African legume crop. It is grown for its underground seed. Its pod produces edible seeds, which can be eaten unripe or stored as dried pulse for later consumption through various preparations. The unripe raw seeds are also eaten after roasting and frying. Typical fruits (pods) are rounded, about 2-3 cm in diameter, and each yield 1 or 2 seeds. The seeds are round, 1-1.5 cm in diameter, and often have a variegated pattern seed coat sometimes with an eye. The bambara is a very nutritious food and many kinds of healthy and delicious food preparations are made from its seed. The leaves are high in protein and can be fed to cattle. Though in the field, this crop is not attacked by insects and diseases, but it is attacked by insects during storage. Traditionally African people store bambara seed by mixing it with ash, sand, tobacco, or peppers to protect the same from insect pest.

The proximate composition and selected functional properties of 4 cultivars of bambara groundnut were estimated by Onimawo *et al* (1999). The crude protein ranged 17.5-21.1%, crude fat 7.3-8.5%, total ash 4-5%, crude fibre 1.8-2.0%, carbohydrate and moisture content for the different cultivars were 53.0-60.8% and 7.5-12.3%, respectively. The functional property determinations indicated that the bulk density ranged 0.65-0.75 g/ml, water binding capacity 2.1-2.9 g/2 g sample, oil binding capacity 0.9-1.6 g/2 g sample, emulsifying activity 55-1-60.0% and emulsifying stability 10-12%.

The possibility of producing a vegetable milk for local use or an extracted protein with functional properties from bambara groundnut for use in food processing applications was studied by Brough *et al* (1993). The bambara groundnut was soaked overnight, homogenised and insoluble material was removed to give a milky liquid. The beany taste could be removed by dry-frying the beans after soaking but before homogenisation. Milks produced in this way were preferred by the taste panel (composed largely of Africans). A comparison of bambara groundnut milk and milks prepared from cowpea, pigeonpea and soybean made using sensory and instrumental analysis showed that all milks were acceptable with bambara groundnut ranked first in the preference trial. The lighter colour of bambara groundnut milk was more acceptable to the taste panel. A crude protein isolate (76% protein) from bambara groundnut was subjected to standard protein functionality tests but it was inferior to the standard used (egg albumen and soya isolate). It did not show any useful properties in emulsification, foam stabilisation or gelation or in a model cake system.

Thus bambara groundnut has a great potential for incorporation into various human foods, in India, where it could provide useful plant proteins.

POTENTIAL OF BAMBARA IN INDIA

In India, the oilseeds and pulses are grown in an area of about 42 million hectare, of which groundnut and pigeonpea are prominent. Though, India is self-sufficient in cereals, we are falling short in oilseed and pulses and the annual import bill of these food commodities is nearly 100 million USD. To make India self-reliant in these commodities, the Government of India, has set up a Technology Mission on Integrated Scheme on Oilseeds, Pulses, Oilpalm and Maize (ISOPOM) under Mini mission 1. India is the world's largest groundnut growing country, however, its productivity is hampered due to lack of water availability for its cultivation. More than 80%, of about 8 million ha under groundnut cultivation, is rain fed. Thus water availability is the single most important deterrent factor in realising potential yields under rain-fed conditions which is quiet low (around 1000 kg/ha) as compared to other major groundnut growing countries of the world.

Adaptive research at target environments by introducing other competitive crops either as sole crop or as intercrop relay crop with groundnut has been priority, particularly in harsh environments. Amongst several intercrop trials, groundnut + pigeonpea, and groundnut + castor are noteworthy, as insurance crop. However, the inherent limitation of these intercrops is the longer duration of growth. Bambara groundnut, though quiet similar to groundnut, is an underutilised grain legume cultivated in Africa only. As it is mainly grown as a subsistence crop, usually by resource-poor women farmers, on poor fertility soil where other crops fail to grow, it has a great potential for similar situations in Asian countries including India. In view of this introducing bambara, a legume/oilseed crop, which responds well under drought conditions and matures closely to groundnut, was thought to be beneficial. Such a crop should also have the advantage over the traditional intercrops and serve as a better option for the resource poor dryland farmers. As it has many advantages, its introduction in India may add to the basket of legume/oilseed production of the country and, hence, reducing the import bill of oilseed and pulses.

In India, the potential areas for the cultivation of bambara are Rajasthan, Haryana, Gujarat, Maharashtra, A.P., Karnataka M.P. and Jharkhand where it can be grown as rainfed crop on the poor fertility soils during rainy season. It can be grown either as sole crop or as intercrop with other legumes and oilseeds. It can also be grown in small quantities in kitchen gardens as a vegetable crop.

CONCLUSIONS AND FUTURE RESEARCH STRATEGIES

Bambara groundnut, an indigenous African crop cultivated for centuries from Senegal to Kenya, and Sahara to South Africa and Madagascar, is quiet similar to groundnut and remains as an underutilised legume. Despite its drought tolerant properties, bambara groundnut has largely been ignored by the scientific community being regarded as a poor man's crop and cultivated in Africa only. As a result it is mainly grown as a subsistence crop, on poor fertility soil where other crops fail to grow. There is a need to introduce this crop in the Asian countries where the agro-climatic situation similar to African countries exist.

The bambara groundnut grows well under in Gujarat (western India) in semi-arid climate and kharif season with early July sowing is most suitable and three high-yielding genotypes DODR-TZ, SB 4-2 and S19-3 with 117-133 days maturity period could be multiplied in bulk and tested at multi-location for its suitability in other parts of India. The short duration genotypes AS-17, and S19-3 could be tested in Rajasthan, Maharashtra and Gujarat where the length of growing period is short. The bambara groundnut genotypes, growing well under field condition in Gujarat (Western India) and Andhra Pradesh (Southern India) both under semi-arid climate, should go for multilocational testing and the promising one may be released in India.

There is an urgent need to increase the bambara groundnut yield through developing high-yielding varieties, proper package of practices, application of nutrients and improvement of the nitrogen-fixing capacity. Also there is a need for developing location specific varieties found growing well in a set of locations. The varieties can be distributed through farmers participatory programme and field days. Also bambara groundnut has great potential for incorporation into various human foods where it could provide useful plant proteins.

Further studies on crop phenology is required to maximise the yield as introduction of an improved genotypes into a new region is largely determined by temperature and phenology, which can be used to specify the most appropriate rate and time of specific developmental process.

Though some of the bambara genotypes can grow without fertiliser, high-yielding genotypes must be fertilised with 20 kg N, 50 kg P, 50 kg K and 100 kg Ca for achieving high-yield and making the crop renumerative.

The IITA, Nigeria is maintaining nearly 2000 germplasm accessions having variability and yield potentials. There is an urgent need to import the promising genotypes for their direct introduction as well as for taking improvement programme.

Some of the areas requiring immediate research are:

- Introduction of bambara groundnut, as a new food legume crop, in the arid and semi-arid regions of India.
- Identification of suitable bambara groundnut cultivars and their seasonal and agronomic requirements for exploiting their potential in existing dry land production system.
- Promotion of crop diversification in traditionally mono-cropped rain-fed areas to ensure protein availability to resource poor small and marginal farmers.
- Testing and incorporation of bambara groundnut into various foods/feed, as a source of cheap proteins.

REFERENCES

Bannayan, M., Collinson, S.T. and Azam Ali S.N. 2000. *BAMnut model user guide*. University of Nottingham. School of Biological Sciences. Division of Agriculture and Horticulture. 43 pp.

Basu M.S., Udaya Kumar M. and M.S. Sheshshyaee, 2003. Potential of bambara groundnut in stabilizing food legume production in semi-arid tropics of India. In: proceeding of the international symposium on BAMFOOD, held at Botsawana.

Basu, M.S., Basu, S., Udaya Kumar M., Bindu Madhava, H. and M.S. Sheshshyaee, 2004. Potential of Bambara groudnut in stabilizing food legume production in semi-arid tropics of India. Plant Biology.

Brink, M 1997. Rates of progress towards flowering and podding in bambara groundnut (*Vigna subterranea*) as a function of temperature and photoperiod. Annals-of-Botany. 1997, 80: 4, 505-513.

Brough, S.H. and Azam Ali, S.N. 1992. The effect of soil moisture on the proximate composition of bambara groundnut (*Vigna subterranea* L. Verdc). *Journal of the Science of Food and Agriculture*, 60: 197-203.

Brough S.H., Azam Ali S.N. and Taylor A.J. 1993. The potential of bambara groundnut (*Vigna subterranea*) in vegetable milk production and basic protein functionality systems. Food-Chemistry. 1993, 47: 3, 277-283.

Howell, J.A. 1994. Common names given to bambara groundnut (*Vigna subterranea*: Fabaceae) in central Madagascar. Econ. Bot. 48: 217–221.

Linnemann, A.R. 1996. Phenological development in bambara groundnut (*Vigna subterranea*) transferred from 14 to 11 h photoperiods. Journal-of-Agricultural-Science. 126: 2, 175-182.

NRCG News Letter 2003a. Bambara groundnut, a promising new crop for dry land agriculture in India. Vol II. No 1 March 2003.

NRCG News Letter 2003b. Bambara groundnut, a new crop for India. Vol II. No 3 September 2003.

Onimawo I.A., Momoh A.H., and Usman A. 1999. Proximate composition and functional properties of four cultivars of bambara groundnut (*Voandezeia subterranea*). Plant-Foods-for-Human-Nutrition. 53: 2, 153-158.

Singh A.L., M.S. Basu and V. Chaudhari and C.B. Patel. 2004. Phenological requirement of bambara groundnut (*Vigna subterranea* L. Verdc.) for its cultivation in semi arid region. In: Proceeding of National Seminar on Plant Physiology held from 27-29th Dec. at Department of Botany, University of Pune, India.

Singh A.L., M.S. Basu, Vidya Chaudhari and C.B. Patel 2005. Fertilizer and seasonal requirement of bambara groundnut in India. In Physiological approaches for crop stress management under different environmental conditions—issues and strategies. Proceeding of National Symposium held from 5-7 March 2005 at WTC, TNAU Coimbatore, India.

CHAPTER 17

Seed Priming and Potash Nutrition as the Management Tools for Drought Mitigation in Upland Summer Rice

B. Haloi[1], R.K. Saud[2] and P.C. Dey[3]

INTRODUCTION

The prevailing environmental constraints (biotic and abiotic) limit crop production and productivity to a considerable extent worldwide. The environmental changes occur both at macro/micro level where the crops are cultivated. Summer rice is sensitive to the vagaries of weather variables. Because, uncertainty of rainfall coupled with high atmospheric demand leads to the development of moisture stress in the crop during the vegetative stages and sometimes it may extend upto flowering stage. Almost half of the global total of 148 M ha of rice is grown under rainfed conditions where drought at some stages limits yield. Rice productivity is low under upland conditions and any environmental stresses or aberrations such as water stress significantly reduces its productivity.

In the rainfed summer rice growing area in Assam, there is wide range of water stress environments differing in both the timing and intensity of water stress. This class of rice have been noticed to experience mild to severe drought specifically during its sowing time, during which the amount of precipitation and the amount of soil storage are quite meagre (Table 17.4). Consequently, the total available water which can be utilised by the crop pose restriction in germination capability of the cultivars chosen and also in expression of full genetic potential in grain yield. Seed priming had been suggested as one of the inducers for drought endurance through induction of embryo enlargement, increasing rate of respiration and level of RNA and protein (Tang *et al.* 1991). In an extensive study under North Bank Plains Zone situations of

[1]Principal Scientist, Crop Physiology, Regional Agricultural Research Station, Titabar
[2]Sr. Scientist, Agronomy, RARS, AAU, Titabar
[3]Sr. Scientist, Crop Physiology, RARS, AAU, Titibar

Table 17.1: Influence of drought mitigation management practices on physiological traits of summer rice (2002 season)

Drought mitigation management practices	*Physiological behaviour*					
	Days taken for seed emergence	*Root vol. (ml/3 hills)*	*Relative turgidity (%)*	*Water potential (KPa)*	*P.C. good grain*	*Grain yield (tha^{-1})*
(a) Seed priming with 3% KCl + top dressing @ 20 kg K_2O ha^{-1} at tillering PI, flowering stage	22.0	13.3	71.6	2190	66.6	2.69
(b) Seed priming + Foliar feeding (2.0% K at tillering, PI & flowering stage)	22.0	17.0	70.2	2193	61.3	2.32
(c) Seed priming + Foliar feeding (1% K in addition to 2% K at water stress period only)	22.0	13.0	70.6	2200	61.8	2.51
(d) Seed priming + irrigation application during water stress period + foliar feeding with K (2%)	22.0	16.7	75.9	2140	64.2	2.28
(e) Seed priming + irrigation application alone without K	20.0	15.3	75.7	2120	57.5	2.60
(f) Absolute control (no priming, no potash)	32.0	11.0	66.7	2170	57.4	1.50
(g) Control (seed priming, no potash nutrition	23.0	13.0	70.5	2110	59.8	2.15
SE ±	–	0.71	0.42	–	1.32	0.15
C.D. (0.05)	–	2.11	1.24	–	3.62	0.44

Assam, Haloi (1994) identified some of the local cultivars having wide range of physiological traits for adaptation under very acute water stress situation and reported use of seed primers to hasten germination under such situation. Therefore, an investigation was carried out to develop management-oriented technologies for alleviating the adverse effects of water stress situation in maximising the productivity of upland summer rice crop in Assam.

MATERIALS AND METHODS

Sets of field experiments were conducted at Regional Agricultural Research Station, Assam Agricultural University, Titabar during summer season (February to June of 2001, 2002 and 2003) in a clay loam soil with nutrient status of medium nitrogen (510-528 kg ha^{-1}), low phosphorus (12-15 kg ha^{-1} and low potash content (102 - 143 kg ha^{-1}) having pH range 4.8 to 5.8. The soil moisture regime determined gravimetrically varied from 5.9% to 18.5% within 0-10 cm and 6.2% to 20.3% within 10-15 cm of depth (Table 17.4) during crop growing period.

During the first year of experimentation pre-sowing seed priming treatments were given to water stress non-tolerable cv. Luit using Potassium Chloride (2% and 4%) and four levels of potash (0 kg, 20 kg, 40 kg, 60 kg, K_2O ha^{-1}), source being Muriate of potash applied as the basal

Table 17.2: Influence of drought mitigation management practices on physiological traits of summer rice (2003 season)

Management practices	*CSI (%)*	*Assimilatory area/unit leaf PI stage (cm^2)*	*Root vol. (ml)*	*Relative turgidity (%)*	*Total 'K' content (%)*	*P.C. good grain*	*Grain yield (t/ha)*
T_1	37.7	29.3	4.8	71.6	2.50	86.4	1.82
T_2	44.0	29.1	4.3	53.3	2.01	75.3	1.55
T_3	31.6	34.7	8.6	78.2	2.76	91.0	2.27
T_4	27.7	36.1	8.2	85.3	3.75	90.7	3.08
T_5	34.8	36.9	5.1	84.6	3.51	89.8	2.35
T_6	38.7	31.1	5.8	82.4	3.27	91.2	2.37
SE±	0.83	1.08	0.42	1.38	–	2.81	0.07
C.D. (0.05)	2.41	3.23	1.14	4.22	–	6.24	0.204

T_1 = Cowdung extract 10% + 20 kg K_2O/ha as basal.
T_2 = Control : no priming, water soaked, no basal and other treatments
T_3 = Seed priming with 3% KCl+ top dressing of K @ 20 kg K_2O/ha at tillering & PI stages.
T_4 = Seed priming + foliar feeding with 2% K at tillering & PI stages
T_5 = Seed priming + foliar feeding with 2% K at flowering stage only
T_6 = Seed priming + foliar feeding with 2% K at water stress period only
N + P_2O_2 @ 40:20 kg ha^{-1} as basal in all except T_2

dose. The required seeds were soaked for 20 hours in priming chemicals, then shade dried for at least 48 hours till the seeds attained their original weight. The seeds were then directly sown @ 75 kg ha^{-1} maintaining the row spacing of 20 cm in a randomised factorial design with three replications, main plot and sub-plot being potash levels and priming treatments respectively with plot size 4 m × 4 m. In the second and third year, however, three upland summer rice cvs viz., Luit (non tolerable), IET 16936 (drought escaping type and Inlongkiri (drought tolerable type) were subjected to priming treatments with 3% KCI with potash nutrition supplemented as basal and foliar feeding as annexed under table 17.1 and 17.2.

The days required for soil seedling emergence at varied treatments were recorded during all the three years. The leaf tissue water status were recorded at seedling stage, active tillering and panicle initiation stages as relative turgidity and water potential being determined by a Water Status Consol (Model 3005, San Dimas, CA USA) in the top two unfolded leaves (Matin

Table 17.3: Upland summer rice cultivar differences in physiological traits as influenced by drought mitigation management practices

Cultivars	*Seedling vigour*	*CSI (%)*	*Assimilatory area/unit leaf (cm^2), PI stage*	*Root vol. (ml)*	*Relative turgidity (%)*	*P.C. good grain*	*Grain yield (t/ha)*
Luit	947	30.4	25.1	7.6	67.4	84.2	2.38
Inlongkiri	1213	32.1	59.0	10.3	82.3	93.0	2.23
IET 16936	866	44.7	14.5	5.7	72.3	85.1	2.10
SE ±	22.1	1.8	5.2	0.6	1.3	3.2	0.8
C.D. (0.05)	65.6	5.3	13.8	1.8	3.1	NS	NS

Table 17.4: Precipitation and soil moisture status (%) of the experimental plot

Week of the month	10 cm depth	15 cm depth	Precipitation (mm)
February, 1st week	6.1	6.4	7.6
February, 2nd week	5.9	6.2	2.0
February, 3rd week	6.4	6.8	0.0
March, 1st week	8.3	9.7	9.2
March, 2nd week	8.7	10.2	2.8
March, 3rd week	10.2	10.8	12.2
March, 4th week	12.4	14.1	3.6
April, 1st week	14.3	14.9	0.0
April, 2nd week	12.6	13.8	1.4
April, 3rd week	15.7	16.1	15.3
May, 1st week	16.6	17.4	4.1
May, 2nd week	18.5	20.3	21.3

et al, 1989). Root volume was measured by water displacement technique at tillering stage (35 days after sowing). Chlorophyll stability index was computed following Haloi (1994) during water stress period. The assimilatory area was measured by Leaf Area Meter (Model LI 3000) during active tillering (45 days after sowing) and PI stages (55 days after sowing). The leaf potassium content was determined by Flame photometer for samples of the crop raised during 2003 (Isaac and Kerber, 1971). The chaffy grains were counted on ten matured panicles. The yield and yield-contributing parameters were recorded accordingly after harvest.

RESULTS AND DISCUSSION

It is observed from Table 17.4 that the initial soil moisture profile was low during sowing period of the upland summer rice. Such a soil moisture status restrict the germination ability of the crop leading to poor crop growth and limits productivity. So, the seeding was executed with primed treatments to overcome such situation. It was revealed from Fig. 17.1A that seed priming hastened germination capability and improved seeding establishment of the summer rice crop under water limited situation. The seedlings generated from primed seeds maintained high leaf water status as reflected from high tissue water potential and relative leaf water content (Fig. 17.1B and C). Fukai and Cooper (1995) opined that under water limiting environment, genotypes which maintained the highest leaf water potential grew best and retention of green leaves appeared to be a major character for prolonged droughts. The improved leaf areas development was also noticed in potash nutrition and primed crop in the present study (Table 17.2). Seed hardening and basal potassium application improved root volume significantly (Fig. 17.1D). This has got added advantage for the crop to extract moisture through proliferation of root system around. Such kind of water stress mitigation management is essential, because the inherent rooting pattern of a cultivar is not altered by water deficit (Lilley and Fukai, 1994).

The plants from non-primed seeds took longer period for seedling soil emergence in the second and third year of experimentation also (Table 17.1 and 17.2). In other words, the practice of priming hastened germinability of the upland summer rice under water limited environments. The practice of priming under stress situation encouraged low chlorophyll

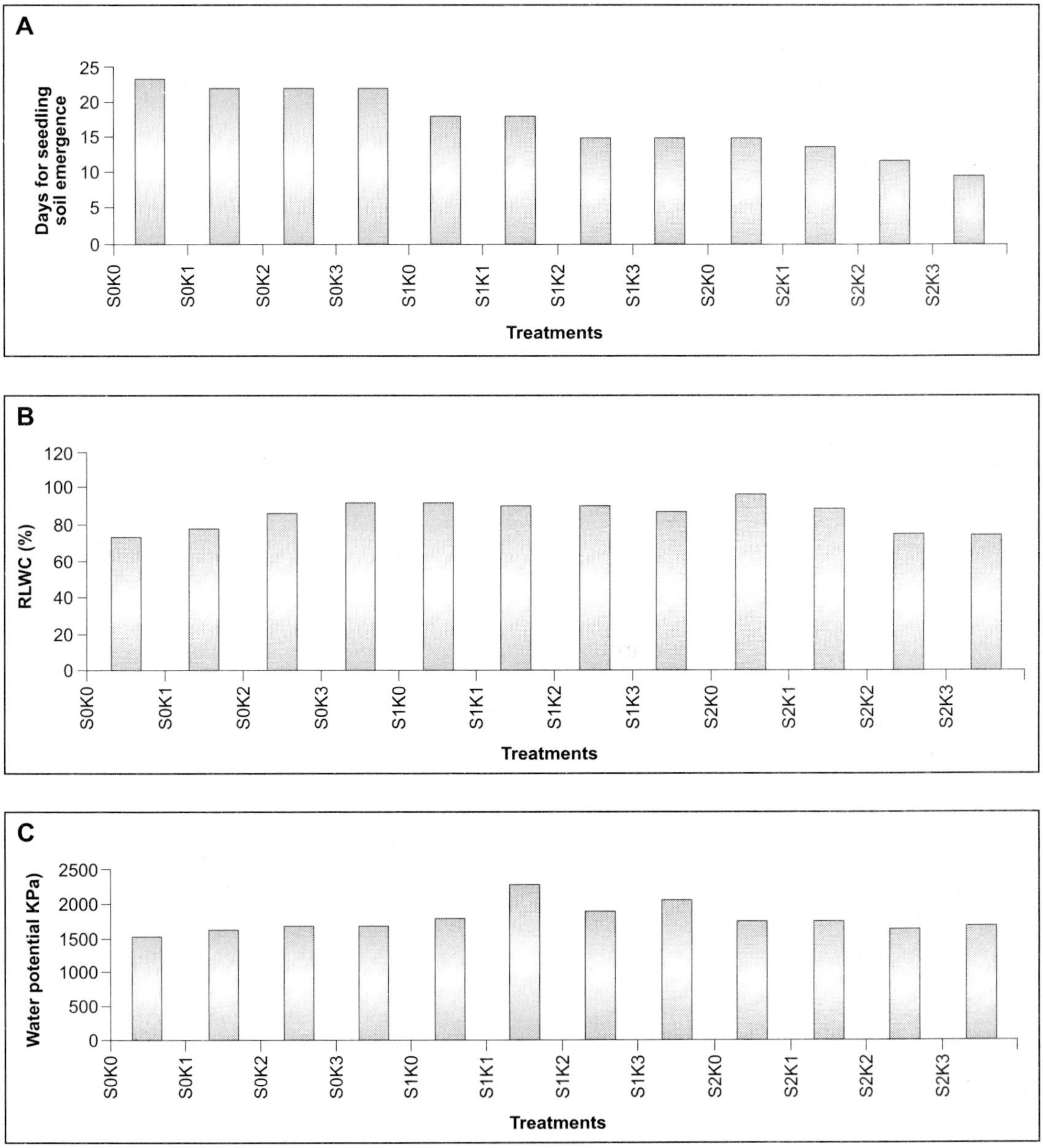

Figure Contd.

Figure Contd.

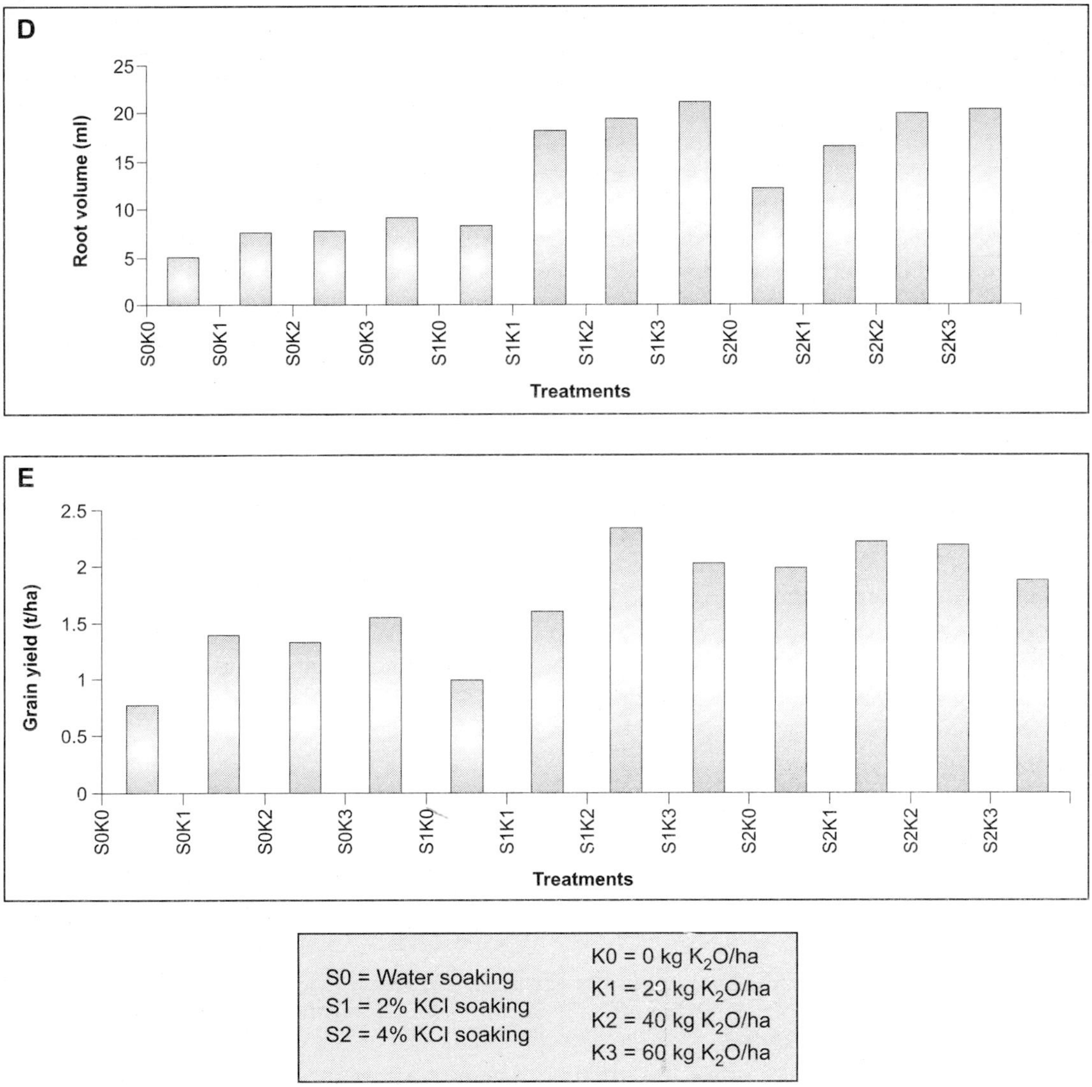

Fig. 17.1: Influence of seed priming and K nutrition on upland summer rice (2001 season).

stability index value (Table 17.2). Such low values reflect the ability of chlorophyll to withstand the adverse stress condition as also envisaged by Nandini and Radhakrishnan (1997).

Under upland water stress situation as reflected from field capacity (23.1%) and PWP range (3.7%) and Table 17.4, foliar feeding of potassium was executed during second and third year of experimentation (Table 17.1 and 17.2), Higher potash accumulation was reflected in foliar feeding plants than that of basal application and was more efficient for crop growth under adverse situation (Table 17.2). Foliar application of potassium is necessary under such situation. Because with limited K supply, there is maximum dilution of K in the plant and uptake was not restricted by other nutrients (Dobermann 1996). Pathak *et al*, (1999), however, got a good response to application of higher basal dose of K_2O in direct seeded summer rice for improving water economy of plants. Seed priming and foliar feeding by potash nutrition have been found to be a potential drought mitigation management practice for upland summer rice. The response was attained mainly though improvement in chlorophyll stability index, assimilatory areas and root volume (Table 17.1 and 17.2). The significant best treatment appeared to be T_4 with seed priming and foliar feeding with 2% K at tillering and panicle initiation stages. It was reflected from the tables that the assimilatory systems of the crops became more adaptive under water stress when they were subjected to drought mitigation management practices. This seemed to be the primary reason for considerable improvement in water status of the crops resulting in reducing the chaffy grain formation (Table 17.2), the commonly occurred events in summer rice crops. So, satisfactory grain yield was obtained following such mitigation practices (Fig 17.1E.)

Another notable observation was that the drought non-tolerable cv. like Luit was highly responsive to such mitigation treatments as reflected from its improved seedling vigour, CSI, assimilatory areas and root volume compared to that of IET 16936, the drought escaping type cultivar (Table 17.3). The drought tolerable cv. Inglongkiri could maintain the morpho-physiological make up even better.

From the ongoing study, it can be concluded that seed priming together with foliar feeding through potash nutrition could be one of the mitigation management tool against water stress environment prevalent in upland summer rice.

REFERENCES

Dobermann, A.P.C., Sta Cruz and K.G. Cassman, 1996. Fertilizer inputs., nutrient balance and soil nutrient supplying power in intensive irrigated rice system. I. Potassium uptake and potassium balance. Nutrient Recycling Agroecosyst. 46: 1-10.

Fukai, S. and M. Cooper, 1995. Development of drought resistant cultivars using physio-morphological traits in rice. Fld. Crop Res. 40: 67-86.

Haloi, B. 1994. Evaluation of rice ecotypes for drought tolerance. In: Plant Productivity under Environmental Stress (ed.) Karan Singh and S.S. Purohit., published by M/S Agrobotanical publishers, Bikaner. pp. 173-180.

Isaac, R.A. and J.D. Kerber, 1971. Atomic absorption and flame photometry: Techniques and uses in soil, plant and water analysis. Pages 17-37, in L.M. Walsh (ed.) Instrumental methods for analysis of soils and plant tissue. Soil Science Society of America, Madison W.I.

Lilley, J.M. and S. Fukai, 1994. Effect of timing and severity of water deficit on four diverse rice cultivars: Phenological development, crop growth and grain yield. Fld. Crop Res. 37: 225-234.

Matin, M.A., J.N. Brown and H. Ferguson. 1989. Leaf water potential, relative water content and diffusive resistant screening technique to drought resistant barley. Agron. J. 81: 100-105.

Nandini, K and R. Radhakrishnanm, 1997. Effect of seed hardening on growth and yield of rice under different moisture regimes. Oryza. 34 (4): 388-390.

Pathak, K., M. Saikia, and A.K. Choudhury, 1999. Influence of chemicals on water economy of upland summer rice. Oryza. 36 (3) : 296-297.

Tang, R.S., J.Y. Zhang., G.N. Wa., A.W. Wang., Y.T. Lu., W.T. Huang., D.Y. Lou., S.Y. Ye., F.T. Zhang., M.Y. Tang and Y.S. Zhou. 1991. Application of multiple effect triazol on raising stout seedling. Fld. Crop Abstr. 44: 704.

Index

A

Abiotic stress, 70
Acid soils, 202
Alkaline soils, 202
Atrataf 50 W
  stimulatory effect of, 60-65

B

Bambara groundnut,
  climate and soil requirement of, 237
  entry into India of, 236
  fertilizer requirement of, 244
  flowering, podding and photoperiodism of, 239
  germination and growth of, 237
  irrigation and drought tolerance of, 245
  maturity duration and yields of, 240
  nutrient requirement of, 243
  nutritional quality and uses of, 246
  phenology and yield of, 237, 239
  pod and seed characteristics of, 236
  Biotechnology, 136

C

Calcareous soils, 202
Calcium, 211-212
Chickpea
  drought resistance in, 70
Chilling injury, 2, 18
Climate stress, 3
Crop plants
  macronutrient stresses and, 198-226

D

Dehydration postponement, 82
Dehydration tolerance, 84
Drought escape, 81
Drought mitigation management
  influence of, 251-52
Drought resistance, 23
  physiology of, 81
  screening varieties for, 26
Drought stress, 71
Drought,
  effects of, 72

Dryland conditions,
breeding approaches, 24
features of varieties suitable for, 24-25
oil seed varieties for, 27-36

E

Export oriented farming, 137

G

Gas exposure system, 59
Green house
advantages of, 141
classification of, 141
crop yield in 148
definition of, 140
disadvantages of, 141
in India, 142
management of, 143
operations of, 146
plasticulture applications and, 144
structural design of, 144
vegetable production in, 147
Growth determination, 52
Guava (Winter crop)
absolute growth and growth rate of, 174-83
chemical composition of, 177, 181
food composition of, 176
harvesting stages of, 178
mean weight and shape index of, 179
physical characteristics of, 177, 180
physical characteristics of, 178

H

High temperature injury, 2

L

Low temperature stress, 79

M

Macronutrient stress
major causes of, 201
management of, in crop plants, 216-226
problem soils and, 201
Macronutrients,
disorder symptoms of, 205
functions of, 205
interaction of, 204-205
Magnesium, 215-16
Miscellaneous physiological disorders, 3, 19
Moisture stress
effect of, 72-74, 85
Mustard
interactive effect of ABA and cytokinin on, 163-69

N

Nagaon paper mill,
stress caused by, 95
Nitrogen, 205-207
Nutrient efficient crop genotypes,
selection of, 218-21
Nutrient requirement
and stress, 199
Nutritional and growth regulations, 15
Nutritional disorders, 5-14

O

Optimum nutrient and water use efficiency
selection of, 222-24

P

Phosphorus, 207-209
Physiological disorders,
  diagnosis and management of, 1-20
  in storage, 18
Plant modifications
  molecular aspects and, 224-25
Plant protection structure, 139
Post-harvest physiological disorders, 2, 3, 16
Potassium, 209-211
Pre-harvest physiological disorders, 1, 3
Pulses,
  improvement aspects, 36

R

Rice intensification, 152
  full potential of, 154
  physiological features of, 151, 155
  tillering potential in, 152-53

S

Salt stress
  effect of, 80
SO_2 effect on
  antioxidative enzymes activities, 55
  blackgrass, 50, 54
  chlorophyll content, 53
  growth characteristics, 53
  lipid peroxidation, 53
  rate of photosynthesis, 54
Soil and plant analysis,
  and crop response, 217
Soil solarisation, 119
SRI (System of Rice Intensification), 152, 157
  advantages of, 159
  disadvantages of, 160
  physiological features of rice, 155
  root growth with, 156
  root physiological activity in, 156
Sulphur, 213

T

Temperature stress, 78
  effect of, 78
Tissue water potential, 48

V

Vegetable crops,
  advances in, 116-137
  area under, 123
  growth regulations in, 129
  improved varieties of, 121-22
  integrated pest management and, 133
  nursery management in, 119
  pollination management in, 134
  post-harvest management of, 136
  production technology of, 123, 126
  resistance to insect pests and, 124
  river bed cultivation of, 136
  stress resistance and, 125
  water management for, 131-32
Vegetable production
  chemical weed control in, 129
  constraints of, 148
  in greenhouse, 147
  integrated plant nutrient system for, 129
  inter generic grafting in, 134
  pruning and training of, 134

status of, 118
use of mulches in, 135
Vegetable-based cropping system, 135
water management and, 133

W

Waterlogging stress, 76
effect of, 77